MINITAB Handbook

Revised Printing

Barbara F. Ryan
Minitab Inc.
State College, Pennsylvania

Brian L. Joiner
Joiner Associates Incorporated
Madison, Wisconsin

Thomas A. Ryan, Jr.
Pennsylvania State University
University Park, Pennsylvania

Duxbury Press
An Imprint of Wadsworth Publishing Company
Belmont, California

Duxbury Press
An Imprint of Wadsworth Publishing Company
A division of Wadsworth, Inc.

ISBN 0-534-91579-5
ISBN 0-534-93366-1 Contains Release 6, 7, and 8 Commands.

Cover Illustration by Joseph Ciardiello, Staten Island, NY
Cover Design by Ellen Schiff, NYC, NY

Printed in the United States of America

97 96 95 94 93 - 10 9 8 7 6 5 4 3 2

Preface

This *Handbook* is intended as a supplementary text for a first or second course in statistics (pre- or post-calculus). The *Handbook* is also useful for researchers who are starting to use Minitab.

The book is designed to be used with Minitab, a general purpose statistical system, originally developed to relieve students of the computational drudgery usually associated with statistics, so that they can focus on important concepts. This *Handbook* emphasizes aspects of statistics that are particularly appropriate to computer use, such as creative use of plots, application of standard methods to real data, in-depth exploration of data, simulation as a learning tool, screening data for errors, manipulating data, transformations, and multiple regression.

The *Handbook* includes numerous examples and exercises that show step-by-step how to use the computer to explore data. Many of these examples and exercises were obtained by the authors while consulting with researchers from a wide variety of application areas and are presented in sufficient detail to provide students with an idea of the scope and importance of the problems amenable to statistical treatment. Obviously, the computer does not replace handwork, and the *Handbook* emphasizes a mixture of hand and computer work.

The chapters and sections are designed to be as independent of each other as possible. Once the main points of Chapters 1 (Introduction to Minitab), 2 (One Variable: Displays and Summaries), and 3 (Plotting Data) have been covered, other chapters may be covered in almost any order. Chapters 2 through 5 cover descriptive statistics for a variety of statistical situations. Chapters 7 through 12 cover statistical inference for these statistical situations. For example, Chapter 5 discusses descriptive statistics for two or more sets of data, while Chapter 8 discusses the usual (normal-theory) procedures for inference on two sets of data, and Chapter 9 discusses the corresponding procedures for more than two sets. Chapter 6 provides background in distributions and simulation.

Notes on the Minitab System

Minitab was originally designed in 1972 for students in introductory statistics courses. Since that time, Minitab's use has expanded to include engineers, social and physical scientists, managers, and others who have to organize and analyze data. At the same time it continues to be used by a growing number of students. Minitab has many capabilities (such as ARIMA time series models and stepwise regression) that are not described in this *Handbook* (except in Appendix C) because they are not appropriate for the intended audience for this book.

The commands in Minitab allow the user to "speak" to the computer in simple, English-like commands. The commands generally correspond to the major steps a student might follow to solve the problem by hand. For example, to do a paired t-test, a student tells Minitab to READ the data, uses LET to compute the differences, and then calls for a single-sample TTEST on the differences.

Minitab was originally developed for an introductory pre-calculus statistics course given at Penn State. Each year over 1500 students take this course. About one hour of the first week of the semester is spent showing students how to sign on to the computer and carry out some simple Minitab commands (including, of course, HELP). Little additional class time is spent on Minitab, per se. As new statistical concepts are taught, the appropriate Minitab commands are introduced. Typically, one homework assignment per week requires Minitab. These assignments usually require the student to link together several commands in an "outline" of the analysis, to use several more commands to explore any surprises in the data, and then to use the output as part of a brief report on the data. In addition, the instructor may analyze additional data sets and hand out copies of the terminal session for discussion.

Use of the Minitab system is not limited to introductory courses. Many students use it in courses in fields of application for term papers, theses, and other projects. It is also useful in more advanced analysis courses, particularly in regression. The system is one of the most frequently used tools of the Penn State and Wisconsin statistical consulting laboratories; and it is used extensively in business, industry, and government around the world. Knowledge of Minitab also provides an excellent introduction to the use of other statistical packages, such as BMDP, SAS, SPSS, and more specialized programs.

Minitab is available on a wide variety of computers, including microcomputers, minicomputers, and mainframes. It is normally used interactively, but can be used in batch mode as well. For further information about the system, write

Minitab, Inc.
3081 Enterprise Drive
State College, PA 16801

About the Second Edition

This second edition is a substantial revision of the original *Handbook*. It reflects changes in the Minitab system, statistical practice, and introductory textbooks. Almost every chapter has been extensively revised, and material has been rearranged.

This second edition is designed to be used with Release 5.1 and later releases of Minitab. Extensive use has been made of capabilities introduced since the original *Handbook*, such as the LET command and the BY subcommand on the HISTOGRAM and DESCRIBE commands.

Release 5.1 of Minitab is upward compatible with previous releases with a few exceptions. First, the symbol for continuing a command on an additional line was changed in Release 82 to "&" (it was "*"). Second, "artificially long" columns have been abolished. This means that using DEFINE to create a column has been replaced by capabilities in SET, and that commands such as MEDIAN C, PUT IN C have been replaced by MEDIAN C, PUT IN K.

A number of commands covered in the original *Handbook* are not covered in this edition since they are being replaced by new capabilities. The old versions of these commands will continue to work for one or two releases, but you should convert to the new versions. The most important of these are listed below.

1. FORMAT subcommands of READ, SET, PRINT, and WRITE replace the formatted input and output commands.
2. USE and OMIT subcommands on COPY replace the old OMIT, CHOOSE and PICK commands.
3. LET replaces most uses of arithmetic commands such as ADD, SQRT, and MEDIAN, as well as SUBSTITUTE.
4. CODE replaces RECODE; STACK replaces JOIN; PDF and CDF replace BINOMIAL and POISSON; OW and OH replace OUTPUT WIDTH; and IW replaces SCAN.
5. Many commands now have subcommands. For example, TWOWAY has a subcommand to fit an additive model. In addition, some options have been moved from the main command line to subcommands, such as the scales for PLOT that are now specified by subcommands.

Preface to the Revised Printing

The revisions included in this printing are designed to make the *Handbook* applicable to all releases of Minitab up through Release 8. The main text, which describes Release 5, has not been altered; new material is confined to the appendixes.

Appendix C summarizes all Minitab commands through Release 7.

and Appendix D covers the changes in Releases 6 and 7 that will be most useful to the reader. (Most of the features added in Releases 6 and 7 are beyond the scope of this book.)

Anyone familiar with the commands described in the main text of the *Handbook* should be able to use Minitab on a mainframe, minicomputer, workstation, or microcomputer, since the worksheet and commands differ little from platform to platform. Minitab Release 8, available on PCs and Macintosh computers, provides additional capabilities including a full-screen data editor, windows, menus, and dialog boxes. Appendix E shows the reader how to use Release 8 in that environment, and introduces the commands that are new to that version of Minitab.

Acknowledgments

We are grateful for the many suggestions we have received from users of Minitab and the first edition of this *Handbook*. Our colleagues and students at Penn State and the University of Wisconsin deserve special mention. In writing and revising the *Handbook* we are especially grateful to Thomas P. Hettmansperger, who provided extensive suggestions (particularly in the nonparametric statistics chapter); Alice W. Pope, who checked examples and exercises and managed the production of the manuscript; Paul Velleman, who always made insightful and thought-provoking comments; and Lawrence Klimko for his careful and extensive criticisms. Reference is made throughout the book to those who kindly made their data available to us. We are grateful to The Pennsylvania State University for partial support in the writing of the first edition of the *Handbook*. We also appreciate the advice we received from reviewers of the manuscript: M. D. Butler, Robert Johnson, Elmo Keller, Jerome Klotz, George McCabe, and Donald McIsaac for the first edition, and David C. Howell, University of Vermont; Walter Read, California State University at Fresno; Alan D. Olinsky, Bryant College; Chao-Ying J. Peng, University of Wisconsin at Madison; and Paul F. Velleman, Cornell University for the second edition.

In the programming effort, special mention is due Cynthia Mable, Del Scott, Alfred Rademaker for early releases of Minitab, and to Jerry Lefkowitz, Stephen Arnold, Rueiming Jamp, Donald Kacher, Barbara Anderson, Sandra Beder-Miller, Robert Fantauzzo, James Allen, and David Balsiger for later releases. For technical suggestions, special mention is due Alison Pollack, Webb Miller, G. W. Stewart, Robert Kohm, and past Editors of the Minitab Users' Group Newsletter, Bert Gunter and Peter Piet.

Minitab began as a student-oriented adaptation of the National Bureau of Standards' "Omnitab" system, originated by Joseph Hilsenrath and developed further by David Hogben and Sally Peavy.

Contents

1 Introduction to Minitab

Minitab is an easy-to-use general purpose statistical computing system.* It is a flexible and powerful tool that has been designed especially for those who have no previous experience with computers. In this chapter we discuss how Minitab works, and, in the chapters that follow, we will show how Minitab can be used to solve various statistical problems. We begin with an example and then give an overview of what you need to know to get started.

1.1 A Simple Example: Blood Cholesterol Level Changes

A study was done to see how blood cholesterol level changed following a heart attack. Data on 12 patients are given in Table 1.1. The first measurement was made two days after the heart attack. The second measurement was made four days after the attack.

Our goals are to: (a) calculate how much the cholesterol level of each patient changed from the second to the fourth day, (b) find the average change for the 12 patients, and (c) make a histogram of the 12 changes. Minitab commands that perform these operations are shown in Exhibit 1.1. Notice that these commands are given in English, just about the same way you'd tell someone to do the calculations by hand.

*Minitab is a registered trademark of Minitab, Inc.

Table 1.1 Blood Cholesterol Levels

Cholesterol Level Two Days After	*Cholesterol Level Four Days After*
270	218
236	234
210	214
142	116
280	200
272	276
160	146
220	182
226	238
242	288
186	190
266	236

Exhibit 1.1 Minitab Commands for Cholesterol Data

```
READ THE FOLLOWING DATA INTO C1 AND C2
270  218
236  234
210  214
142  116
280  200
272  276
160  146
220  182
226  238
242  288
186  190
266  236
END
LET C3 = C2 - C1
PRINT C1 C2 C3
DESCRIBE C3
HISTOGRAM C3
```

Minitab stores the data in a worksheet that it maintains in the computer. The first command

```
READ THE FOLLOWING DATA INTO C1 AND C2
```

tells the computer to take the data from the lines that follow the READ command and put them into column C1 and column C2 of the worksheet. The command

```
END
```

signals the end of the data. Exhibit 1.2 shows the worksheet after the data are read. Just two columns are filled with data at this stage.

The next command

```
LET C3 = C2 - C1
```

says to subtract column C1 of the worksheet from column C2 and to store the differences in column C3. Exhibit 1.3 shows the worksheet at this point. Column C3 now contains the change in cholesterol level for each patient. The next three commands give printed output.

```
PRINT C1 C2 C3
```

says to print out the contents of these three columns.

Exhibit 1.2 Minitab's Worksheet After Reading the Data

C1	*C2*
270	218
236	234
210	214
142	116
280	200
272	276
160	146
220	182
226	238
242	288
186	190
266	236

Exhibit 1.3 Worksheet After the Command LET C3 = C2 − C1

C1	C2	C3
270	218	−52
236	234	−2
210	214	4
142	116	−26
280	200	−80
272	276	4
160	146	−14
220	182	−38
226	238	12
242	288	46
186	190	4
266	236	−30

```
DESCRIBE C3
```

says to calculate and print various descriptive statistics, including the average (or mean), of the numbers in C3.

```
HISTOGRAM C3
```

says to print out a histogram of the numbers in C3. Output from these three commands is shown in Exhibit 1.4.

Getting Started

In order to use Minitab, you must first learn a little bit about your computer system. Each one is different, so we cannot give you the exact details, but we can give you some general guidance.

Most people use Minitab on a computer terminal consisting of a screen and a keyboard. The screen is often called a CRT (for cathode-ray tube). Some people use typewriterlike terminals. These use paper in place of a screen. In both cases, you give the computer directions by typing them on the keyboard. Always check each line you've typed before you push the carriage return key. On most computers, you can correct errors by simply backspacing and retyping. Once the carriage return is pushed, however, it is usually more difficult to make corrections.

There is a third way to use a computer: with a keypunch machine and computer cards. This used to be the only way to use computers. Now, however, most keypunches have been replaced by terminals, which are

Exhibit 1.4 Output from the Commands PRINT, DESCRIBE, and HISTOGRAM

```
PRINT C1 C2 C3
ROW      C1      C2      C3

  1     270     218     -52
  2     236     234      -2
  3     210     214       4
  4     142     116     -26
  5     280     200     -80
  6     272     276       4
  7     160     146     -14
  8     220     182     -38
  9     226     238      12
 10     242     288      46
 11     186     190       4
 12     266     236     -30

DESCRIBE C3

                  N      MEAN    MEDIAN    TRMEAN     STDEV    SEMEAN
C3               12    -14.33     -8.00    -13.80     33.26      9.60

                MIN       MAX        Q1        Q3
C3           -80.00     46.00    -36.00      4.00

HISTOGRAM C3

Histogram of C3   N = 12

Midpoint   Count
   -80         1  *
   -60         1  *
   -40         1  *
   -20         3  ***
     0         4  ****
    20         1  *
    40         1  *
```

both friendlier and cheaper to use. If you are using Minitab with a keypunch, see your local computer experts for help.

Finally, you can use Minitab on a microcomputer. If you do, then you may have to learn about floppy disks (they look like flexible records) and disk drives. However, you will still have a keyboard and a screen.

Thus, the first step is to learn about the equipment you will use: where it is located, how to turn it on, how the keyboard works, and so forth. You may have to connect the terminal to the computer with a telephone and a device called a modem. Then you must learn the correct telephone number. These are all minor details that are fairly easy to learn, although they can cause a lot of frustration at times.

There are two basic ways to use Minitab: in interactive mode and batch mode. Most people use interactive mode, which we will discuss in this chapter. If you are using Minitab in batch mode see your local computer experts for help. Almost all Minitab commands work the same way in both modes.

The second step is to get on (often called logon or login) to your computer. You may need an account number, an identification number, and a password. You will have to learn exactly what to type, and you may have to type everything very carefully and in what seems to be a strange form. Once you have logged-on, you are in what's usually called the computer's operating system.

The third step is to ask to use Minitab. This may involve nothing more than typing the word Minitab. In any case, it should be very simple. At this point, you are in Minitab. You can now type the commands in this book.

Using Minitab Yourself

In Exhibit 1.5, we give a complete example of a Minitab session on the computer we use at Minitab, Inc. To help you see what's going on, everything we typed is shaded by a white background.

First we turned on our terminal and typed LOGIN. The computer asked for our ID and password, which we typed, and the computer responded with some information. We are now in our computer's operating system. We then asked to use Minitab.

We are now in Minitab. Minitab started by typing the prompt MTB >. This says Minitab expects a command. We typed READ C1 C2, checked our line for errors, then pushed the carriage return. Now Minitab expects data, and prompts with DATA>. We typed the first line of data, checked for errors, and pushed the return. We then typed the second and third data lines. When we were finished typing data, we typed the word END. Now Minitab knows we are finished entering data and expects another command, so prompts MTB>. *Notice:* We made an error on the second LET command and pushed the carriage return. Minitab knew something was wrong because the word LTT did not make any sense. It gave us an error message. In this case we simply retyped the command.

The last Minitab command was STOP. Type STOP when you are finished using Minitab. After typing STOP, you will be back in the computer's operating system.

The last step is to log-off the computer. Usually all you need to type is the word LOGOFF or the word LOGOUT. You may also have to turn off a switch or hang up a telephone.

If you are using a terminal with a screen, then at times you will want output on paper. Just type the word PAPER. Output from all the com-

Exhibit 1.5 Example of a Minitab Session

```
OK, LOGIN
User id? TAR
Password? XX2516

TAR (user 10) logged in Thursday, 24 May 85 16:00:24.
Welcome to PRIMOS version 19.2.6.10S
Last login Thursday 24 May 85 13:47:44.

OK, MINITAB
MINITAB RELEASE  5.1 *** COPYRIGHT - MINITAB, INC. 1985
MAY 24, 1985 *** Minitab, Inc. Headquarters
STORAGE AVAILABLE  686128
There is HELP on new features of Rel. 5.1.  Type NEWS for details.

MTB > READ C1 C2
DATA> 8 3
DATA> 10 2
DATA> 5 6
DATA> END
MTB > LET C3=C1+C2
MTB > LTT C4=C1-C2
* ERROR * NAME NOT FOUND IN DICTIONARY
        * MISSPELLED NAME, OR ERROR IN READ OR SET, OR DATA OUT OF PLACE

MTB > LET C4=C1-C2
MTB > PRINT C1 C2 C3 C4
 ROW     C1    C2    C3    C4

   1      8     3    11     5
   2     10     2    12     8
   3      5     6    11    -1

MTB > STOP

*** MINITAB RELEASE  5.1 *** MINITAB, INC. ***
STORAGE AVAILABLE  686128
```

mands you type after the PAPER command will be put into a computer file to be printed later. You can stop this process at any time by typing the word NOPAPER, and you can restart it by typing PAPER again. At the end of your session, after you type STOP, everything you put into your file will be printed on paper.

Exercises

1–1 The DESCRIBE command in Exhibit 1.4 printed ten values for each of C1, C2, and C3. These will all be discussed in Chapter 2. Now, however, see if you can guess what any of them are (i.e., try to define MEAN, MIN, etc.).

1–2 Interpret the output in Exhibit 1.4. On the average, did the cholesterol levels increase or decrease from the second to the fourth days? In how many patients did the cholesterol level decrease? In how many did it increase?

1–3 Use a computer to do the session in Exhibit 1.5 yourself.

1–4 Use a computer to READ the cholesterol data in Exhibit 1.1, then find the average cholesterol level for two days after a heart attack.

1–5 The following Minitab commands do some simple calculations.

```
READ C1 C2 C3
  15  28  30
  14  30  31
  16  30  34
  13  27  31
END
PRINT C1 C2 C3
LET C4=C1+C2+C3
READ C10 C11
  32  26
  34  27
  32  25
END
PRNT C10 C11
LET C12=C10-C11
PRINT C4 C10
```

(a) Pretend you are Minitab and carry out the commands. Keep a worksheet as you go along. What does the worksheet contain after each command? What is printed out?

(b) Type the commands on your computer and compare the results to your answers in part (a).

1.2 A Brief Overview of Minitab

Minitab consists of a worksheet where data are stored and about 150 commands. In the worksheet, you can store columns of data and single number constants. The columns are denoted by C1, C2, C3, . . . and may have names. The stored constants are denoted by K1, K2, K3, ... The total worksheet area and the number of columns and stored constants available to you depends on your particular computer. On some computers, the worksheet is enormous, on others it is quite restricted. The total area you have is printed when you use Minitab.

When you want Minitab to analyze data, you type the appropriate commands. There are commands to read, edit, and print data, to do plots and histograms, to do arithmetic and transformations, and to do various statistical analyses such as t tests, regression, and analysis of variance. A brief summary of all Minitab commands is given in Appendix C.

Some Rules

1. Every command starts with a command name, such as READ or HISTOGRAM. On most commands, this is followed by arguments. An argument is either a column number (such as C10), a column name (such as 'HEIGHT'), a constant (such as 75.34), a stored constant (such as K15), or the name of a computer file.
2. Only the first four letters of the command name and the arguments, which must be in the proper order, are used by Minitab. Other text may be added for annotation, if you wish. However, we recommend you use only letters and commas for this extra text. Never use numbers or symbols such as ; : – * & or + since these are used in special ways by Minitab.
   Using these rules, the command

   ```
   READ FOLLOWING DATA INTO COLUMNS C1 AND C2
   ```

   could be written as

   ```
   READ DATA INTO C1 AND C2
   ```

   or simply as

   ```
   READ C1 C2
   ```

   We could not have written

   ```
   READ 1ST SCORE INTO C1 AND 2ND INTO C2
   ```

   since the extra text contains a 1 in 1ST and 2 in 2ND. Minitab would try to use these as arguments.
3. A list of consecutive columns or stored constants may be abbreviated using a dash. For example, you could use

   ```
   READ C2-C5
   ```

   instead of

   ```
   READ C2, C3, C4, C5
   ```

4. Columns and stored constants may be reused any number of times. If you store new data in a column or stored constant, the previous contents are automatically erased.
5. If you type a number, do not use commas within the number. Thus, type 1041 instead of 1,041.

6. Each command must start on a new line. You need not start in the first space, however. If the entire command will not fit on one line, end the first line with the ampersand symbol (&) and continue the command onto the next line. For example,

```
PRINT C2, C4-C20, C25, C26, C30, C33 &
C35-C40, C42, C50
```

Minitab commands are described throughout this *Handbook* in a box like this one. In these descriptions, the symbol C can be replaced by any column name or number, K by any constant, and E by either a column or constant. Arguments in square brackets are optional. Any extra text used to help explain the commands is written in lower-case. Here are some examples.

PRINT the data in C

You could replace C by C18, omit the extra text, and get

```
PRINT C18
```

PRINT E, . . ., E

This says you may print any list of columns and constants. For example,

```
PRINT C1 C4 C2 K1-K4
```

SUM C [put into K]

Here storage in K is optional. If you use the form

```
SUM C10
```

then Minitab calculates and prints the sum of all the numbers in C10. If you type

```
SUM C10 store in K1
```

then Minitab stores the sum in K1, but does not print it.

Exercise

1-6 Find the errors in each of the following (there may be more than one error in each part).

(a)
```
READ C1-C3
  5   2
  6   14
  2   18
END
```
(b)
```
READ C2 C3
      983  2
    1,102  5
      992  7
END
```
(c)
```
READ C2 C5
  2   16
  4   12
  5   11
  6   12
END
PRNT C2
DESCRIBE THE 4 NUMBERS IN C2
LET C2=C2+C5
LET C10=C5+C6
```
(d)
```
READ INTO C1 AND C2
  5.6     23.0
  5.5     23.1
  5.4     23.3
END
DESCRIBE DATA IN COLUMN 1
LET C5=C1+C2   LET C6=C1-C2
```

1.3 Some Basic Minitab Capabilities

This section introduces the basic capabilities of Minitab and makes it possible for you to write simple Minitab programs of your own. More information on basic capabilities is given in Appendices B and C.

Input and Output of Data

Both the commands READ and SET let you type data into Minitab's worksheet. The difference between them is how you type the data: READ expects data one row at a time, whereas SET expects it one column at a time.

Exhibit 1.1 (p. 2) showed how to use READ to enter data into columns C1 and C2. Here is how to use SET.

```
SET into C1
  270 236 210 142 280
  272 160 220 226 242 186 266
END
SET into C2
  218 234 214 116 200
  276 146 182 238 288
  190 236
END
```

The first SET put the 12 numbers that followed it into C1. The second SET put the 12 numbers that followed it into C2. After these two SETs, the worksheet looks exactly the same as it did in Exhibit 1.2. When you enter data, use whichever command you prefer. Sometimes SET will be easier and sometimes READ will be.

As you type the data, check each line before pushing the carriage return on your terminal. If you notice an error, you can probably backspace and correct it. When you finish typing all your data, we suggest you print out a copy with the PRINT command. Again, check for errors. If you find an error, say in C2, then you could use SET to reenter all the data in C2, or you could use the special commands described on pages 15–17 to correct the data.

READ the following data into C, . . ., C

This command is followed by lines of data. Each line of data is put into one row of the worksheet.

Example

```
READ C2 C3 C5
  1   3    980
  3   0   1430
  2   4   2190
END
```

Following this, the worksheet contains

C1	*C2*	*C3*	*C4*	*C5*
	1.0	3.0		980
	3.0	0.0		1430
	2.0	4.0		2190

Numbers can be put anywhere on a line as long as they are in the correct order. Numbers must be separated by blanks or commas. (More details are on page 330.)

SET the following data into C

This command is followed by lines of data. You may put as many numbers as you want on one line and you may use as many lines as you want. All the numbers will go into the same column.

Example

```
SET C3
  1.5   2   6.3   2   1
  6.23  5.01
END
```

Following this, C3 will contain the seven numbers: 1.5, 2.0, 6.3, 2.0, 1.0, 6.23, and 5.01.

Numbers can be put anywhere on a line as long as they are in the correct order. Numbers must be separated by blanks or commas. No numbers may be put on the SET line itself. (More details are on page 330.)

END of data

Type this command after typing the data for READ or SET.

PRINT E, . . ., E

Examples

```
PRINT C1-C4 C10
PRINT K2 K4 K6
PRINT C20
```

This command may be used to print columns or constants that have been stored in the worksheet.

Stored Constants

Any operation that results in a single number answer can put that number into a stored constant. The stored constant may then be used in place of a number on any command. SUM is a command that results in a one number answer. If C1 contains the numbers 5, 3, 6, and 2, then SUM C1 calculates $5 + 3 + 6 + 2 = 16$. Since the answer is one number, we can store it in a constant. Here is an example:

```
SET C1
  5,3,6,2
END
SUM C1, PUT IN K1
LET K2 = 4
LET K3 = K1+K2-8
PRINT K1-K3
```

Doing Arithmetic with LET

The LET command makes it easy to do very complicated calculations. In most data analysis, however, you will use just simple forms of this command. Here, we give a brief introduction to LET.

LET uses the following symbols:

+ for add
- for subtract
* for multiply
/ for divide
** for raise to a power (exponentiation)

Example

```
LET K1 = 3
LET K2 = 5*13
LET K3 = K1+K2+4
SET C1
  4 6 5 2
END
LET C2 = 2*C1
LET C3 = K1*C1
LET C4 = C2+1
LET C5 = C3+C4
LET C6 = C1**2
```

After these commands, K1 = 3, K2 = 65, and K3 = 72. The following table shows what C1 – C6 contain.

C1	*C2*	*C3*	*C4*	*C5*	*C6*
4	8	12	9	21	16
6	12	18	13	31	36
5	10	15	11	26	25
2	4	6	5	11	4

Parentheses may be used for grouping. For example:

```
READ C1 C2
  8 1
  6 3
  4 4
END
LET C3 = 10*(C1+C2)
LET C4 = 10*C1+C2
LET C5 = C1/C2+1
LET C6 = C1/(C2+1)
```

At this point the worksheet contains

C1	*C2*	*C3*	*C4*	*C5*	*C6*
8	1	90	81	9	4.0
6	3	90	63	3	1.5
4	4	80	44	2	.8

Note that the expressions for C3 and C4 look similar, as do those for C5 and C6; but the results are not the same. LET follows the usual precedence rules of arithmetic; that is, operations within parentheses are always performed first, then **, then * and /, and finally + and –. If you are not sure of the sequence of a calculation, you can always use parentheses to make sure it is done right.

LET E = arithmetic expression

An arithmetic expression may be made up of columns and constants and arithmetic symbols (+ – * / **) and parentheses. Unlike other Minitab commands, no extra text may be used on a LET line (except as discussed on page 343). LET can be used to correct numbers (see below). LET can be used to evaluate very complicated expressions (see p. 41) and Section B.3).

Correcting Data in the Worksheet

There are three commands that are useful for correcting numbers you have already entered in the worksheet: LET, DELETE, and INSERT. Exhibit 1.6 shows a worksheet that contains two errors: one number is wrong and one line is omitted. The number in row 3 of C1 should be 1.3, not 3.1. To correct it, we use a special feature of the LET command:

```
LET C1(3) = 1.3
```

Exhibit 1.6 Correcting Data in the Worksheet

Worksheet Should Be		*But It Was Typed As*		*After LET*		*After INSERT*	
C1	*C2*	*C1*	*C2*	*C1*	*C2*	*C1*	*C2*
1.5	102	1.5	102	1.5	102	1.5	102
1.7	106	1.7	106	1.7	106	1.7	106
1.3	120	3.1	120	1.3	120	1.3	102
1.4	118	1.4	118	1.4	118	1.4	118
1.5	101	1.5	102	1.5	102	1.5	101
2.1	130	1.1	124	1.1	124	2.1	130
1.1	124					1.1	124

To change one number in C1, just type C1, then the row number enclosed in parentheses. Exhibit 1.6 shows the worksheet after LET was used.

The line containing 2.1, 130 was left out. It should be put between rows 5 and 6. To insert this line, we use

```
INSERT BETWEEN ROWS 5 AND 6 OF C1, C2
  2.1, 130
END
```

INSERT between rows K and K of C, . . ., C

This command inserts rows of data into the worksheet. The data are typed following the INSERT line. The row numbers K and K must be consecutive integers. (The second K is redundant but helps avoid errors.)
To put rows of data at the tops of columns, use

```
INSERT between rows 0 and 1 of C,...,C
```

To add rows of data to the end of columns, omit the two row numbers and use

```
INSERT into C,...,C
```

If you insert data into just one column, then you may string the data across the data lines, as in SET.

DELETE rows K, . . ., K of C, . . ., C

Deletes the indicated data and closes up the worksheet. You may abbreviate a list of consecutive rows by using a colon. For example,

```
DELETE 1:10, 25:30 C1
```

would delete rows 1 through 10 and 25 through 30 from C1.

How to NAME Columns

Any column may be given a name. The name serves two purposes:

1. The column may be referenced by its name. It's often easier to remember the name of a variable than the number of a column.
2. All output is labeled with the name.

Many users find that naming columns takes a little extra time but pays off in the long run by making output easier to read.

NAME C = 'name', C = 'name', . . ., C = 'name'

A name may be from one to eight characters long. Any characters may be used, with two exceptions—a name may not begin or end with a blank and a name may not contain the symbol ' (called an apostrophe or single quote). Either the column number or its name may be used in any succeeding command. When column names are used, they must be enclosed in single quotes (apostrophes). The name of a column can be changed by using another NAME command.

Example

```
NAME C1 = 'HEIGHT', C2 = 'L+W'
PRINT 'HEIGHT' C2
```

Ending a Session

STOP

If you use Minitab in interactive mode, type STOP when you are finished using Minitab. After typing STOP you will be in your computer's operating system.

If you use Minitab in batch mode, STOP should be the last Minitab command in your program.

Exercises

1–7 (a) In the following commands, SET was used to enter data. Show how to use READ to do the same job.

```
SET C1
  2  3  5  7
END
SET C2
  11 18 9 16
END
```

(b) In the following commands, READ was used to enter data. Show how to use SET to do the same job.

```
READ C1-C3
  5 6 3
  1 2 8
  5 1 1
  6 2 3
END
```

1–8 What values does C1 have after the following SET?

```
SET C1
  980    992    1,140    801
  963    1,002
END
```

1–9 Students in a small class were given three exams with the following results:

ID Number	*Exam 1*	*Exam 2*	*Exam 3*
2365	92	82	96
4879	84	84	80
4261	75	79	83
3929	98	60	72
2677	62	55	40
2417	79	72	81
3804	81	70	78

(a) Write a Minitab program that computes the average exam score for each student and prints a table containing the ID number and average score for each student.

(b) Run the program in part (a).

(c) Which student had the highest average? The lowest?

1–10 What's wrong with each of the following sets of commands?

(a)
```
SET C1-C2
  5   2
  13  6
END
```
(b)
```
LET C3=C1+C2   THIS IS TOTAL SCORE
```
(c)
```
NAME C1=SCORE1   C2=SCORE2
```
(d)
```
READ C1 C2
  5   3
  6   1
END
LET K1=C1+C2
```
(e)
```
NAME C4='EDUCATION' C2='1ST JOB'
```
(f)
```
PRINT LENGTH
```

1–11 (a) Write a Minitab program that produces a temperature table in degrees Fahrenheit (from 20° to 80° in steps of 5°) and the equivalent temperatures in degrees Celsius. To convert from Fahrenheit to Celsius, use the formula Celsius = (Fahrenheit − 32) * (5/9).

(b) Use a computer to run the program in part (a). Check your results.

1–12 In each of the following cases, there is an error in the worksheet. Show how to use LET, DELETE, and INSERT to correct the error.

(a) Data were entered as

C1	*C2*
5	83
4	65
3	50
4	58

Worksheet should be

C1	*C2*
5	83
4	85
3	50
4	58

(b) Data were entered as

C1	*C2*
10	28
8	37
8	34

Worksheet should be

C1	*C2*
10	28
8	37
9	24
8	34

(c) Data were entered as

C1	C2
13	2
15	3
11	6
12	3
15	3

Worksheet should be

C1	C2
13	2
11	6
12	3
15	3

1.4 Another Example: Men's Track

World records for men's track are given in Table 1.2. We want to see how speed, in miles per hour, changes with the length of the race.

Exhibit 1.7 shows Minitab commands to do this analysis. The first Minitab command puts the data into columns of the worksheet. Next, we give names to the data columns and to three columns we will use for our calculations.

We then convert length to miles, and time to hours. Because there are 1609.4 meters in a mile, meters divided by 1609.4 gives miles. The first LET command does this division. To convert time to hours, we divide seconds by 60 to get minutes, then divide the total minutes by 60 to get hours. The second LET does this calculation. The final LET command calculates speed. We then use PLOT to do a plot (PLOT is fully explained on p. 52) and PRINT to print out all six columns we used.

The plot shows that, overall, the speed of a race decreases as the length increases. This is not surprising since we all know we can run faster over short distances. Some other things are a bit more surprising. The

Table 1.2 World Records for Men's Track (as of September 1983)

Distance	*Time*	
in Meters	*Minutes*	*Seconds*
100		9.93
200		19.72
400		43.86
800	1	41.73
1000	2	12.18
1500	3	30.77
2000	4	51.40
3000	7	32.10
5000	13	8.42

Exhibit 1.7 Minitab Commands to Analyze the Track Data

```
READ  C1-C3
  100      0     9.93
  200      0    19.72
  400      0    43.86
  800      1    41.73
 1000      2    12.18
 1500      3    30.77
 2000      4    51.40
 3000      7    32.10
 5000     13     8.42
END
NAME C1 = 'METERS', C2 = 'MINUTES', C3 = 'SECONDS'
NAME C4 = 'MILES', C5 = 'HOURS', C6 = 'SPEED'
LET 'MILES' = 'METERS'/1609.4
LET 'HOURS' = ('SECONDS'/60 + 'MINUTES')/60
LET 'SPEED' = 'MILES'/'HOURS'
PLOT 'SPEED' VS 'METERS'

           -
           -
     22.5+      **
           -
 SPEED     -
           -
           -         *
     20.0+
           -
           -
           -
           -
     17.5+               *
           -                *
           -
           -                     *
           -                          *
     15.0+                                      *
           -
           -                                                        *
           -
        ----+---------+---------+---------+---------+---------+--METERS
            0      1000      2000      3000      4000      5000

PRINT C1-C6
ROW  METERS  MINUTES  SECONDS     MILES      HOURS      SPEED
  1     100        0     9.93   0.06213   0.002758   22.5263
  2     200        0    19.72   0.12427   0.005478   22.6862
  3     400        0    43.86   0.24854   0.012183   20.4000
  4     800        1    41.73   0.49708   0.028258   17.5905
  5    1000        2    12.18   0.62135   0.036717   16.9228
  6    1500        3    30.77   0.93202   0.058547   15.9192
  7    2000        4    51.40   1.24270   0.080944   15.3525
  8    3000        7    32.10   1.86405   0.125583   14.8431
  9    5000       13     8.42   3.10675   0.219006   14.1857
```

speeds of the 100- and 200-meter races are about the same, and quite fast. The table we printed shows those speeds are actually 22.5 miles per hour and 22.7 miles per hour. Thus, the 200-meter race is slightly faster than the 100-meter race! Perhaps this is because the 100-meter race is so short that a runner spends most of the race time building up speed. The plot also shows that speed decreases quite rapidly at first. But when we get to the longer races, there isn't that much change. For example, when we go from 100 to 1000 meters, speed decreases by about 6 miles per hour. However, when we go from 2000 to 3000 meters, speed decreases by less than one mile per hour.

Exercise

1–13 World records for women's track (as of September 1982) are given below.

Distance	*Time*	
in Meters	*Minutes*	*Seconds*
100		10.79
200		21.71
400		47.99
800	1	53.28
1500	3	52.47
3000	8	26.78
5000	15	8.26

(a) Write a Minitab program to find the speed, in miles per hour, for each race. As in Exhibit 1.7, give names to your columns, plot speed versus distance in meters, and print out the data in the worksheet.

(b) Run the program in part (a).

1.5 More Information on Minitab

Help on Minitab Commands

Information about Minitab is stored in the computer. If you forget how to use a command, you can ask Minitab for help. For example, to find out about the command SET, type

```
HELP SET
```

Minitab will respond with a brief explanation of SET.

In general, to get help on a command, type HELP followed by the command name. To find out about other help you can get from Minitab, type

```
HELP HELP
```

Saved Worksheets

Saved worksheets are a very convenient way to store data in a computer file for use in Minitab. The following example creates a file called TRACK, containing the data in Table 1.2.

```
READ C1-C3
  100     0     9.93
  200     0    19.72
  400     0    43.86
  800     1    41.73
 1000     2    12.18
 1500     3    30.77
 2000     4    51.40
 3000     7    32.10
 5000    13     8.42
END
NAME C1='METERS' C2='MINUTES' C3='SECONDS'
SAVE 'TRACK'
```

Before we typed SAVE, we gave names to three columns. These names are also stored in the file TRACK. Any time later, either in the same session or in another session, we can put these data and names back into Minitab's worksheet by typing

```
RETRIEVE 'TRACK'
```

After RETRIEVE, Minitab's worksheet is exactly the same as it was when SAVE was typed.

Many of the data sets in this *Handbook* are available as saved worksheets. To use them, all you have to do is type RETRIEVE with the appropriate file name. You do not have to type the data yourself. A list of these data sets is given on page 302. After you RETRIEVE a worksheet you can find out what is in it by typing the command INFO.

SAVE the worksheet in 'filename'

Puts the entire Minitab worksheet into a computer file. The file will contain all columns, stored constants and column names. This file can be input by Minitab's RETRIEVE command. The filename must be enclosed in single quotes (apostrophes).

One drawback with Minitab saved worksheets is that only Minitab can read them. You cannot use your computer's editor to print them or modify them. You cannot use them with other programs. However, saved worksheets are a very convenient way to store data for further Minitab analyses.

Caution: File usage and naming conventions vary enormously from computer to computer. You should check with someone who knows your local computer system if you have any difficulty using SAVE and RETRIEVE.

RETRIEVE the worksheet stored in 'filename'

Inputs data from a saved worksheet. After using RETRIEVE, the worksheet will contain the same columns, stored constants, and column names it had when SAVE was used to create the file.

Note: Anything you might have in the worksheet when you type RETRIEVE is erased before the saved worksheet is brought in.

Managing the Worksheet

Occasionally you may forget what you have in the worksheet. The INFO command can help you. See Exhibit 1.8.

INFORMATION on status of the worksheet

Prints a list of all columns used, the number of values in each, the name of each (if they have been named), and a list of all stored constants used.

Exhibit 1.8 Example of INFO Command Using TRACK Data on Page 20

```
RETRIEVE 'TRACK'
INFO
COLUMN    NAME      COUNT
C1        METERS        9
C2        MINUTES       9
C3        SECONDS       9
 CONSTANTS USED: NONE
```

You may erase columns and constants you no longer need. Sometimes you may need to do this because you run out of space. On other occasions, you will want to do it to reduce unnecessary clutter.

ERASE E, . . ., E

You may erase any combination of columns and stored constants. For example

```
ERASE C2 C5-C9 K1-K7 C20
```

Subcommands

Some Minitab commands have subcommands. These allow more control over the way the command works. For example, the HISTOGRAM command will automatically choose a scale for the display but if you want a different scale, you can specify your own by using the subcommands INCREMENT and START. Here is an example.

```
HISTOGRAM C1;
  INCREMENT 10;
  START 0.
```

The specific details of these two subcommands will be described on page 31. Here we are interested in describing the syntax. To use a subcommand, end the main command line with a semicolon. The semicolon tells Minitab there will be a subcommand on the next line. End each subcommand line with a semicolon until you are finished typing subcommands. Then end the last subcommand line with a period. Minitab waits until it sees the period to start doing the calculations.

If you accidentally forget to end the last subcommand with a period, you can put the period on the next line. For example,

```
HISTOGRAM C1;
  INCREMENT 10;
  START 0;
```

Always begin each subcommand on a new line. You may type the subcommand anywhere on the line. We often indent subcommands for clarity.

Occasionally you will discover an error after entering a subcommand in interactive mode. In this case, type ABORT as the next subcommand. This will cancel the whole command. You can then start over again with the main command.

Missing Data Code

Many data sets are missing one or more observations. When you enter these data with READ, SET, or INSERT, type the asterisk symbol (*) in place of the missing value. For example,

```
READ   C1    C2
  28    5.6
  24    5.2
  25    *
  24    51
END
```

All Minitab commands automatically take the * into account when they do an analysis.

Sometimes you enter data into the worksheet and discover that a value is wrong, and have no way to find out the correct value. You could change this value to *, using a special feature of LET. For example, if the wrong value is the fifth number in C18, use

```
LET C18(5)='*'
```

Notice: You must enclose the asterisk in single quotes (apostrophes) when using it with LET. The missing data code is explained more fully in Appendix B.

2

One Variable: Displays and Summaries

Much can be learned from data by looking at appropriate plots and tables. Sometimes such displays are all we need to answer our questions. In other, more borderline, cases they will help guide us to appropriate follow-up procedures. In fact, one great advantage of computers is their ability to make a variety of data displays quickly and easily. In this chapter we introduce some of the most useful displays and some simple summary measures, such as the mean and median.

We begin with a description of three basic types of data because the type of analysis you should use depends on the type of data you have.

2.1 Three Basic Types of Data

Not all numbers are created equal. Categorical data act merely as names; they tell us nothing about order or size. Ordinal data tell us about order but not about size. Interval data give information about size as well as order.

Categorical Data. Simple examples of categorical variables are sex, which has two values (male and female) and state in the United States, which has fifty values (Alabama, Alaska, . . ., Wyoming). When such data are stored in the computer, they often are converted to numbers. This is usually just for the convenience of the computer. Sex might be coded 1 = male and 2 = female, or as 1 = female and 2 = male, or even as 1410 = male and 2063 =

female. State might be coded in alphabetical order, going from 1 = Alabama to 50 = Wyoming.

One problem with computers is that they will do what you ask—even if it's nonsense. For example, a computer will gladly average categorical data, even though that average probably does not have any meaning. Suppose, for example, you have a data set of 30 men (coded 1) and 70 women (coded 2). A computer will calculate the average sex as 1.7. With computers, as with other tools, it is up to you to see that they are used properly. One of the goals of this book is to help you do this.

Categorical data are also called nominal or classification data.

Ordinal Data. One example is army rank: private, corporal, sergeant, lieutenant, major, colonel, and general. We know that a general is one rank higher than a colonel and a corporal is one rank higher than a private. But is the distance from private to corporal the same as the distance from colonel to general? Does distance between army ranks really have any meaning? Probably not.

Perhaps the most common occurrence of ordinal data is in surveys and questionnaires. For example,

"The President is doing a good job." Check one:

Strongly Disagree	Disagree	Indifferent	Agree	Strongly Agree

When entered into the computer, ordinal data usually are converted to numbers. For example, 1 = strongly disagree, 2 = disagree, . . ., 5 = strongly agree. Here we would know that a 4 was more favorable toward the President than a 3, but we would not have any clear idea how much more favorable.

Interval Data. These usually are based on measurements such as length, weight, or time. On an interval scale, 4 is halfway between 3 and 5. For example, the difference between a 4-centimeter rod and a 3-centimeter rod is the same as the difference between a 5-centimeter rod and a 4-centimeter rod. Unlike categorical and ordinal data, interval data occur naturally as numbers.

Exercises

2–1 For each of the following, indicate whether the data are best considered as nominal, ordinal, or interval, and justify your choice.

(a) The response of a patient to treatment: none, some improvement, complete recovery.

(b) The style of a house: split-level, one-story, two-story, other.
(c) Income in dollars.
(d) Temperature of a liquid.
(e) Area of a parcel of land.
(f) Highest political office held by a candidate.
(g) Grade of meat: prime, choice, good, or utility.
(h) Political party.

2–2 Appendix A contains a collection of data sets. Look through these data sets and select at least two examples of each of the three basic data types.

2.2 Histograms

Table 2.1 lists the pulse rates for 92 people (this is the variable called PULSE1 from the Pulse experiment described on p. 318). In Exhibit 2.1, we entered the whole PULSE data set into the worksheet using RETRIEVE, then used HISTOGRAM. HISTOGRAM grouped the pulse rates into 11 intervals, each of width 5. The first interval has a midpoint of

Table 2.1 Pulse Rates for 92 People

64	70	68	61
58	96	82	64
62	62	64	94
66	78	58	60
64	82	54	72
74	100	70	58
84	68	62	88
68	96	48	66
62	78	76	84
76	88	88	62
90	62	70	66
80	80	90	80
92	62	78	78
68	60	70	68
60	72	90	72
62	62	92	82
66	76	60	76
70	68	72	87
68	54	68	90
72	74	84	78
70	74	74	68
74	68	68	86
66	72	84	76

Exhibit 2.1 Histogram of Pulse Rates

```
RETRIEVE 'PULSE'
HISTOGRAM 'PULSE1'

Histogram of PULSE1   N = 92

Midpoint    Count
      50        1  *
      55        2  **
      60       17  *****************
      65        9  *********
      70       23  ***********************
      75       10  **********
      80       11  ***********
      85        6  ******
      90        9  *********
      95        3  ***
     100        1  *
```

50, goes from 47.5 to 52.5, and contains one observation. The second interval has a midpoint of 55, goes from 52.5 to 57.5, and contains two observations.

A histogram gives a graphical summary of the data: In Exhibit 2.1, we can see that the lowest pulse rate is around 50, the highest is around 100, and the most popular interval is the one at 70—one quarter of the pulse rates are in this interval.

When we talk about the *scale* of a histogram, we mean the intervals: how many, how wide, and where they start. Minitab automatically chooses a scale. If you want a different scale, you can specify it with subcommands. For example, suppose you wanted the intervals to be 40 to 50, 50 to 60, 60 to 70, . . ., 100 to 110. In this example the width, or increment, of each interval is 10 and the starting midpoint is 45. The following instructions can be used:

```
HISTOGRAM 'PULSE1';
  INCREMENT = 10;
  START = 45.
```

Any pulse rate that falls on a boundary between two intervals is put in the higher interval. For example, a pulse of 60 would be put in the interval from 60 to 70.

The histogram is designed primarily for interval data, although it can be used with ordinal and even categorical data as well.

HISTOGRAM C, . . ., C

Prints a histogram for each column. Observations on the boundary between two intervals are put in the interval with higher values. You may specify your own scale with the subcommands INCREMENT and START.

INCREMENT = K

Specifies the distance between midpoints or, equivalently, the width of each interval.

START with midpoint at K [end with midpoint at K]

Specifies the midpoint for the first, and optionally, the last interval. Any observations beyond these intervals are omitted from the display. (More on pp. 95–99.)

Exercises

2–3 Consider the histogram in Exhibit 2.1.

(a) How many pulse rates fell in the interval with midpoint 65?

(b) Are there more pulse rates above 75 or below it? Can you tell from this histogram?

(c) Were there any outliers (that is, any pulse rates that were much smaller or much larger than the others) in this data set?

2–4 (a) Make a histogram of the following numbers:

36, 43, 82, 84, 81, 84, 45, 60, 64, 71, 81, 78, 79, 43, 79

(b) Make a histogram for numbers that are ten times the numbers in (a). (The LET command may be useful here.) Compare this display to the one in part (a). Is the overall impression about the same?

(c) Repeat (b) with numbers five times those in (a). Compare this display to those in parts (a) and (b).

(d) Make a histogram of the numbers in part (a) using the subcommand START 37.5. The intervals will be five units wide, as they are in part (a), but will have different midpoints. The first interval will contain values from 35 through 39, the second interval will contain values from 40 through 44, the next 45 through 49, and so on. Compare the overall shape of this display to the one in part (a).

2.3 Dotplots

The Lake Data (p. 315) provide measurements on 71 lakes in northern Wisconsin. Exhibit 2.2 RETRIEVES the data and produces a dotplot of the depth of these lakes. This display is very similar to a histogram with many small intervals, which has been turned on its side. In Exhibit 2.2, there are over 50 spaces or intervals. Each space represents 1.5 feet. The numbers that label the axis give the middle of each space. Thus the space labeled 30 goes from 29.25 to 30.75 feet, the next space goes from 30.75 to 32.25 feet, and so on. An observation that falls on the boundary between two spaces goes in the lower interval.* Thus, a lake 30.75 feet deep would be put in the interval labeled 30.

A histogram groups the data into just a few intervals. For example, HISTOGRAM would use nine intervals for the 71 lake depths. A dotplot, on the other hand, groups the data as little as possible. Ideally, if we had wider paper or a printer with higher resolution, we would not group the data at all. Histograms tend to be more useful with large data sets, dotplots with small data sets. Histograms show the shape (shape is discussed in Section 2.7) of a sample; dotplots do not. Dotplots are useful if you want to compare two or more sets of data. We'll do this in Chapter 5.

Exhibit 2.2 Dotplot of Lake Depth from the Lake Data

```
RETRIEVE 'LAKE'
DOTPLOT 'DEPTH'
              : :
          ... : :    .:   .:                        .
          ::: :::.:.:::...::::::..::  :   : ..  .   :           .
          -----+---------+---------+---------+---------+---------+-DEPTH
              15        30        45        60        75        90
```

*This convention is the opposite of the one we used for HISTOGRAM. However, it seems to be the more natural one for DOTPLOT.

DOTPLOT C, . . ., C

Makes a dotplot for each column. Observations on a boundary are put in the lower (smaller values) interval. WIDTH (p. 53) controls the width of DOTPLOTS. You may specify your own scale on DOTPLOT with the subcommands INCREMENT and START.

INCREMENT = K

Specifies the distance between tick marks (the + signs) on the axis. Since there are 10 spaces between tick marks, the width of each space will be K/10.

START at K [end at K]

Specifies the first and optionally the last tick mark on the axis. Any points outside are omitted from the display. (More on pp. 95–99.)

Exercise

2–5 (a) Get a DOTPLOT of the variable AREA in the LAKE data set (see p. 315). What are the most striking features of this plot?

(b) How many lakes are under 2,000 acres? How many are under 1,000 acres?

(c) Use the commands

```
DELETE 55 C1-C5
DOTPLOT 'AREA'
```

This will give a plot which does not contain the area for the 55th lake. How does this plot compare to the one in part (a)?

2.4 Stem-and-Leaf Displays

A stem-and-leaf display is similar to a histogram, but uses the actual data to create the display, whereas HISTOGRAM uses the asterisk symbol * The display is a relatively new technique that was introduced by statistician John Tukey in the late 1960s. It is designed primarily for interval data, although it can be used with any set of numbers.

Table 2.1 gave the pulse rates for 92 people. Suppose we make a stem-and-leaf display of these data. Exhibit 2.3 shows the display after we have entered the first four pulse rates: 64, 58, 62, and 66. The digits to the

Exhibit 2.3 Stem-and-Leaf Display of First Four Pulse Rates

```
 4 |
 5 | 8
 6 | 426
 7 |
 8 |
 9 |
10 |
```

left of the vertical line are called the stems. The digits to the right are called the leaves. To create the display, we split each pulse rate into two parts: the tens digit became the stem, and the ones digit became the leaf. For example, 64 was split into 6 = stem and 4 = leaf. The 58 was split into 5 = stem and 8 = leaf, and so on. The stems for the entire data set were listed to the left of the vertical line. Each leaf was put on the same line as its stem. At this point, the line with stem = 5 has just one leaf, an 8. This represents the pulse rate 58. The line with stem = 6 has three leaves, 4, 2, and 6. These represent the three pulse rates, 64, 62, and 66.

Exhibit 2.4 gives the stem-and-leaf display for all 92 people. Reading from the top of the display, we see that the pulse rates are 48, 58, 54, 58, 54, 58, 64, 62, 66, . . ., 94, 90, 100. This display contains the same information as the original list of numbers, but presents it in a more compact and usable form. The numbers are closer to being in order. We can easily see the range of the data (from a low of 48 to a high of 100) and the most popular categories (the 60s, followed by the 70s, 80's, and 90s). The general shape of the picture is nearly symmetric. There are no gaps (stems with no observa-

Exhibit 2.4 Stem-and-Leaf Display of 92 Pulse Rates

```
 4 | 8
 5 | 84848
 6 | 42648280268628220288842088140626 88
 7 | 46020408826442060802428 2686
 8 | 402802844840276
 9 | 026600240
10 | 0
```

tions) and no outliers (observations that are much smaller or much larger than the bulk of the data).

Looking more closely, we see that all of the numbers except two are even. Why? A reasonable conjecture, and a correct one, is that pulses were counted for 30 seconds and then doubled to get beats per minute. But what about the other two, the 61 and 87? These two people may have multiplied incorrectly, or counted a half beat, or written down the wrong number. For example, the 87 could be a transposition error; perhaps 78 was correct.

Exhibit 2.5 gives the display produced by Minitab's STEM-AND-LEAF command. This differs from our hand-drawn display in several ways. First, an extra column, called depths, was added to the left of the display. Further, the message LEAF UNIT = 1.0 was added. (We will discuss both of these additions under Further Details). Second, each stem is listed on two lines, with leaf digits 0, 1, 2, 3, and 4 on the first line and 5, 6, 7, 8, and 9 on the second. This gives a display that is more spread out than our hand-drawn one.

Minitab also ordered the leaves on each line, making it easier to see what values we have. For example, now it is clear that the most common pulse rate is 68 beats per minute.

Exhibit 2.5 Stem-and-Leaf Display Produced by Minitab

```
RETRIEVE 'PULSE'
STEM-AND-LEAF 'PULSE1'

Stem-and-leaf of PULSE1     N=92
Leaf Unit = 1.0

   1    4 8
   3    5 44
   6    5 888
  24    6 000012222222224444
  40    6 6666688888888888
 (17)   7 00000022222244444
  35    7 66666688888
  25    8 0002224444
  15    8 67888
  10    9 0000224
   3    9 66
   1   10 0
```

Further Details

The *depth* of a line tells how many leaves lie on that line or "beyond." For example, the 6 on the third line from the top says there are 6 leaves on that line and above it; the 10 on the third line from the bottom says there are 10 leaves on that line and below it. The line with the parentheses will contain the middle observation if the total number of observations, N, is odd. It contains the middle two observations if N is even, as it is here. The parentheses enclose a count of the number of leaves on this line. Note that if N is even and if the two middle observations fall on different lines, then no parentheses are used in the depth column

In this example, Minitab listed each stem on two lines. In some cases, Minitab will use five lines for each stem. The number of lines per stem is always 1, 2, or 5 and is determined by the range of the data and the number of values present.

In our example, all pulse rates except one contained two digits, so it was easy to split each number into a stem and a leaf. When numbers contain more than two digits, the STEM-AND-LEAF command drops digits that don't fit. For example, the number 927 might be split as stem = 9, leaf = 2, and 7 dropped.

Decimal points are not used in a STEM-AND-LEAF display. Therefore, the numbers 260, 26, 2.6, and .26 would all be split into stem = 2 and leaf = 6. The heading LEAF UNIT tells you where the decimal point belongs: for the number 260, LEAF UNIT = 10; for 26, LEAF UNIT = 1; for 2.6, LEAF UNIT = .1; and for .26, LEAF UNIT = .01.

STEM-AND-LEAF has a subcommand, INCREMENT, which allows you to control the scale of a stem-and-leaf display. For example, suppose we wanted Minitab to produce the display of Exhibit 2.4. There the first stem contains all numbers in the 40s and the second stem contains all numbers in the 50s. Therefore, the distance or increment from one stem to the next is 10. To specify this scale use

```
STEM-AND-LEAF 'PULSE1';
  INCREMENT = 10.
```

STEM-AND-LEAF OF C, . . ., C

Prints a stem-and-leaf display for each column.

INCREMENT = K

Specifies the distance from one stem to the next. The increment must be 1, 2 or 5 with perhaps some leading or trailing zeroes. Thus, examples of allowable increments are 1, 2, 5, 10, 20, 50, 100, 200, 500, .1, .2, .5, .01, .02, .05. (More on pp. 95–99.)

Exercises

2–6 Consider the stem-and-leaf display in Exhibit 2.5.

(a) What were the lowest and highest pulse rates?

(b) How many people had a pulse rate of 58?

(c) How many had a pulse rate in the eighties?

(d) How many had a pulse rate of 64 or lower?

(e) Could you answer any of the questions in parts (a)–(d) using just the histogram in Exhibit 2.1? Could you use the histogram to get approximate answers? For each question in parts (a)–(d), give the best answer or range of possibilities you can using just the histogram in Exhibit 2.1.

2–7 (a) Do a stem-and-leaf display of the following numbers by hand:

36, 43, 82, 84, 81, 84, 45, 60, 64, 71, 81, 78, 79, 43, 79

(b) Use Minitab to do a stem-and-leaf display of the numbers in part (a). Compare it to your hand-drawn display.

(c) Multiply the numbers in part (a) by 10. Then use Minitab to get a stem-and-leaf display. Explain the differences between the displays in (b) and (c).

(d) Multiply the numbers in part (a) by 2. Use Minitab to get a display of these numbers. How does this display compare to the one in part (b)?

(e) Multiply the numbers in part (a) by 5 and get a stem-and-leaf display. Compare this display to the one in part (a).

2–8 (a) Use Minitab to get a stem-and-leaf display of the variable HEIGHT from the PULSE data set (p. 318).

(b) Convert HEIGHT from inches to centimeters (to do this, multiply by 2.54) and then get a stem-and-leaf display. Comment on the "unusual" appearance of this display. Compare it to the display in part (a), when height was measured in inches.

2–9 (a) Make a stem-and-leaf display of the variable WEIGHT from the PULSE study. Do you see any special pattern in the leaf digits of the display?

(b) What increment did Minitab choose for this display? What is the next smaller increment? Use this value with the INCREMENT subcommand to make a display. Compare this display to the one in part (a). Now use the next larger increment to make a display. Compare this display to the one in part (a).

2.5 One-Number Statistics

We often want to summarize an important feature of a set of data by using just one number. For example, we might use the mean to indicate the center or typical level of the data. We could use the range, the largest value minus the smallest value, to indicate how spread out the data are.

In this section we first discuss DESCRIBE, a command that prints a table of summary numbers. Then we show how these summaries can be computed individually, first for columns of data, then across rows.

The DESCRIBE Command

The Pulse data set, described on page 318, contains the weights of 92 people. Exhibit 2.6 provides some summaries of these numbers. The stem-and-leaf display shows several interesting things. For example, most people reported their weight to the nearest five pounds, especially the heavier people, and the heaviest person weighed 215 pounds.

Exhibit 2.6 A Summary of the Weights in the Pulse Data Set

```
RETRIEVE 'PULSE'
STEM-AND-LEAF 'WEIGHT'

Stem-and-leaf of WEIGHT      N=92
Leaf Unit = 1.0

     1    9  5
     4   10  288
    13   11  002556688
    24   12  00012355555
    37   13  0000013555688
  (11)   14  00002555558
    44   15  00000000003555555555557
    22   16  000045
    16   17  000055
    10   18  0005
     6   19  00005
     1   20
     1   21  5

DESCRIBE 'WEIGHT'

                   N       MEAN     MEDIAN     TRMEAN      STDEV     SEMEAN
WEIGHT            92     145.15     145.00     144.52      23.74       2.47

                 MIN        MAX         Q1         Q3
WEIGHT         95.00     215.00     125.00     156.50
```

The DESCRIBE command printed the following statistics:

`N = 92.` This says there were 92 people in the study who reported their weights.

`MEAN = 145.15.` This is the average of all 92 weights. The mean, often written as $\bar{x}$, is the most commonly used measure of the center of a batch of numbers.

`MEDIAN = 145.00.` To find the median, first order the numbers. If N, the number of values, is odd, the median is the middle value. If N is even, the median is the average of the two middle values. Here $N = 92$, so the median is the average of the 46th and 47th values. These are both 145, so their average is 145. The median is another value used to indicate where the center of the data is.

`TRMEAN = 144.52.` This gives a 5% trimmed mean. First the data are sorted. Then the smallest 5% and the largest 5% of the values are trimmed; the remaining 90% are averaged. Here $N = 92$ and 5% of 92 is 4.6. This is rounded to 5. Thus the five smallest values (95, 102, 108, 108, 110) and the five largest values (190, 190, 190, 195, 215) are trimmed. The remaining 82 values are averaged to give the trimmed mean.

`STDEV = 23.74.` This is the standard deviation. It is the most commonly used measure of how spread out the data are. The general formula is

$$\text{STDEV} = \sqrt{\frac{\Sigma(x - \bar{x})^2}{N - 1}}$$

Here is a simple example: For the data 1, 3, 6, 4, 6, we have $N = 5$, mean = 4, and

$$\text{STDEV} = \sqrt{\frac{(1-4)^2 + (3-4)^2 + (6-4)^2 + (4-4)^2 + (6-4)^2}{5 - 1}} = \sqrt{4.5} = 2.12$$

`SEMEAN = 2.47.` This is the standard error of the mean. It is a more advanced concept and is discussed on p. 168. The formula is STDEV/$\sqrt{N}$. For the Pulse data,

$$\text{SEMEAN} = \text{STDEV}/\sqrt{N} = 23.7/\sqrt{92} = 2.5$$

`MIN = 95.00.` MIN is the minimum or smallest value.

`MAX = 215.00.` This is the maximum or largest value.

`Q1 = 125.00.` This is the first or lower quartile.

`Q3 = 156.50.` Q3 is the third or upper quartile.

The median is the second quartile, Q2. The three numbers Q1, Q2, and Q3 split the data into four, essentially equal, parts. The concept of a quartile is very simple. However, when we try to give a formal definition, we must handle details like how to divide ten observations into four equal parts. There are several ways to do this, and you will find that different books define quartiles differently. All, however, give answers that are very close.

Here is the definition Minitab uses: First, order the observations from smallest to largest. Then, Q1 is at position $(N + 1)/4$ and Q3 is at position $3(N + 1)/4$. If the position is not an integer, interpolation is used. For example, suppose $N = 10$. Then $(10 + 1)/4 = 2.75$ and Q1 is between the second and third observations (call them x_2 and x_3) and it is three-fourths of the way up. Thus $Q1 = x_2 + .75(x_3 - x_2)$. For Q3, $3(10 + 1)/4 = 8.25$. Thus, $Q3 = x_8 + .25(x_9 - x_8)$, where x_8 and x_9 are the eighth and ninth observations. In the weight example, $N = 92$, so $(N+1)/4 = 23.25$ and Q1 is between the 23rd and 24th observations. These are both 125, so Q1 = 125. Q3 is between the 69th and 70th observation and is equal to 156.5.

`NMISS.` This is the number of values recorded as "missing." Here no weights were missing so the NMISS line was not printed by DESCRIBE. (Missing data are described on p. 26.)

DESCRIBE C, . . ., C

Prints the following statistics for each column.

`N`	Number of nonmissing values in the column
`NMISS`	Number of missing values. This is omitted if there are no missing values.
`MEAN`	
`MEDIAN`	
`TRMEAN`	5% trimmed mean
`STDEV`	Standard deviation
`SEMEAN`	Standard error of the mean
`MAX`	Maximum value
`MIN`	Minimum value
`Q3`	Third quartile
`Q1`	First quartile

One-Number Statistics for Columns

DESCRIBE prints a collection of summary statistics. Minitab also has commands that calculate and store each statistic separately.

N C [put in K]
NMISS C [put in K]
MEAN C [put in K]
MEDIAN C [put in K]
STDEV C [put in K]
MAX C [put in K]
MIN C [put in K]
SUM C [put in K]
SSQ C [put in K]
COUNT C [put in K]

In each command, the answer is printed if it is not stored.

Three of these statistics are not printed by DESCRIBE: SUM, SSQ, and COUNT. The command SUM just adds all the values in the column. SSQ is the sum of the squares of the values. For example, for the data 1, 3, 5, and 4, SSQ = $1^2 + 3^2 + 5^2 + 4^2 = 51$. The command COUNT gives the total number of entries in a column. Thus, COUNT = N + NMISS.

Column Statistics with LET. All of the statistics for columns can also be used in a LET statement. Here are some examples:

```
LET K1 = MEAN(C1)
LET C2 = C1 - MEAN(C1)
LET K2 = SUM(C1)/N(C1)
LET K3 = MEDIAN('HEIGHT')
```

Notice: You must enclose the column in parentheses when you use MEAN, SUM, MEDIAN, or some other column statistic in a LET command.

By using LET, you easily can calculate many statistics that are not built into Minitab. For example, another measure of the center of a set of numbers is the midrange. It is the average of the smallest and largest values. It can be calculated by

```
LET K1 = (MAX(C1) + MIN(C1))/2
```

Another measure of the spread of a set of numbers is the median of the absolute deviations from the median. This can be calculated by

```
LET K1 = MEDIAN(ABSO(C1 - MEDIAN(C1)))
```

One-Number Statistics for Rows

The statistics we have just described for columns are also available for rows. The command names are the same except an R (for row) has been added.

RN C, . . ., C, put in C
RNMISS C, . . ., C, put in C
RMEAN C, . . ., C, put in C
RMEDIAN C, . . ., C, put in C
RSTDEV C, . . ., C, put in C
RMAX C, . . ., C, put in C
RMIN C, . . ., C, put in C
RSUM C, . . ., C, put in C
RSSQ C, . . ., C, put in C
RCOUNT C, . . ., C, put in C

These commands compute summaries across rows rather than down columns. The answers are always stored in a column.

Example:

```
RSUM C1-C3 PUT IN C4
RMAX C1-C3 PUT INTO C5
```

C1	*C2*	*C3*		*C4*	*C5*
1	7	3		11	7
4	2	3		9	4
1	3	2		6	3
3	5	5		13	5

Exercises

2–10 Consider the output in Exhibit 2.6.

(a) How many people are exactly 150 pounds? Over 150 pounds?

(b) The mean weight was 145.15. How many people weigh less than the mean?

(c) The range of a set of numbers is defined as (maximum value) − (minimum value). Find the range for the weight data.

(d) There is a connection between the range and the standard deviation of a data set. In many data sets the range is approximately four times the standard deviation. Is this true for the weight data?

(e) Calculate the two values (MEAN minus STDEV) and (MEAN plus STDEV). In many data sets, aproximately two-thirds of the observations fall between these two values. Is this true for the weight data?

2–11 Consider the following eleven numbers:

5, 3, 3, 8, 9, 6, 9, 9, 10, 5, 10

(a) By hand, calculate each of the ten statistics printed by DESCRIBE.

(b) Use DESCRIBE to check your answers.

2–12 Suppose Minitab did not have the command STDEV. Show how to use LET and the formula on page 39 to calculate the standard deviation of the 11 observations in Exercise 2–11.

2–13 The median is said to be "resistant" to the effects of a few outlying points in a data set. That is, the median will not be very different even if there are a few unusually large values or abnormally small values in the data set. The mean, however, is not resistant to outliers.

Enter the 11 observations in Exercise 2–11 into C1. Use DESCRIBE to find the mean and median. Now use the command

```
LET C1(1) = 25
```

This changes the 5 in row 1 to a 25. Again use DESCRIBE to find the mean and median. How have they changed? Now use the command

```
LET C1(1) = 100
```

to change the first observation to a 100. Find the mean and median. How have they changed?

2–14 (a) Use Minitab to compute the standard deviation of the numbers

6, 8, 4, 10, 12, 3, 4, 10

(b) Add 29 to each number. Now compute the standard deviation. How does your answer compare with that in (a)?

(c) Multiply the data in (a) by 16 and compute the standard deviation. How does your answer compare with that in (a)? If you're not sure, divide the standard deviation in (c) by the one in (a).

2.6 Summarizing Categorical and Ordinal Data

The preceding sections described techniques that are designed primarily for interval data. This section introduces TALLY, a command that is designed especially for categorical and ordinal data. It gives four statistics: Count, cumulative count, percent, and cumulative percent.

The Furnace data set described on page 312 contains several categorical variables. These include the type of furnace in a house, the design of the house, and the shape of the chimney. Exhibit 2.7 is a TALLY of furnace type and chimney shape. The first table shows the count: 76 houses had a type 1 furnace (forced air), 7 had a type 2 furnace (a gravity system), and 7 had a type 3 furnace (a forced water system), for a total of 90 observations. The second table shows there were 39 round chimneys (coded 1), 32 square (coded 2), 18 rectangular (coded 3), and one missing value (a house that did not report its chimney shape), for a total of 89 nonmissing values.

Exhibit 2.8 shows a table that has all four statistics calculated by TALLY for the variable TYPE.

CUMCNT gives the cumulative counts. Thus, there were 76 furnaces of Type 1, 76 + 7 = 83 of Types 1 and 2, and 76 + 7 + 7 = 90 of Types 1, 2, and 3. PERCENT is (COUNT/N) × 100. Thus, (76/90) × 100 or 84.44% of the houses had forced air heat. Similarly, (7/90) × 100 or 7.78% had a gravity system. CUMPCT gives the cumulative percent. This is (CUMCNT/N) × 100. You can use subcommands to ask for each of these statistics individually or use the subcommand ALL to get all four, as we did in Exhibit 2.8. Note, in the Furnace example, that CUMCNT and CUMPCT are not very useful since there is no intrinsic order to the variable TYPE. These two statisics are most often used with ordinal data.

Exhibit 2.7 TALLY of Furnace Type and Chimney Shape

```
RETRIEVE 'FURNACE'
TALLY 'TYPE' 'CH.SHAPE'

    TYPE  COUNT    CH.SHAPE  COUNT
      1     76            1     39
      2      7            2     32
      3      7            3     18
      N=    90           N=     89
                         *=      1
```

Exhibit 2.8 Example of the TALLY Command

```
TALLY 'TYPE';
  ALL.

    TYPE   COUNT   CUMCNT   PERCENT   CUMPCT
      1       76       76     84.44    84.44
      2        7       83      7.78    92.22
      3        7       90      7.78   100.00
     N=       90
```

TALLY C, . . ., C

Prints a separate frequency table for each column. The columns must contain integers from −10000 to +10000. Any one or more of the following subcommands may be used. They specify what to print in the table. If no subcommands are given, just COUNTS are printed.

COUNTS
PERCENTS
CUMCNTS Cumulative counts
CUMPCTS Cumulative percents
ALL Same as using all four preceding subcommands

Exercises

2–15 The following questions refer to the output in Exhibits 2.7 and 2.8.
- (a) What was the most popular chimney type? The least popular?
- (b) What percent of the houses had a forced water furnace?
- (c) Why does chimney shape have $N = 89$ and chimney type have $N = 90$?
- (d) What percent of the houses had a round chimney?

2–16
- (a) Four of the variables in the Pulse data (p. 318) are most suited for TALLY. What are they? Suppose you were to use TALLY on the other variables. What would happen in each case?
- (b) Make a tally of each of the four appropriate variables.
- (c) Also make a histogram of these four variables. What are the advantages and disadvantages of TALLY and HISTOGRAM in this case?

2.7 Transformations to Symmetry

When analyzing data, it is not always clear what units should be used for a given variable. For example:

Should gasoline consumption be measured in miles per gallon or in gallons per mile?

Should a price increase be measured in dollars or in percent change?

Should the acidity of rain be measured in terms of the proportion of hydrogen ions or in terms of pH?

Choosing the appropriate units for a variable can often make its analysis much easier.

Symmetry

A histogram is symmetric if the distribution of the high values is the mirror image of the distribution of the low values. For example, histogram (c) of Exhibit 2.9 is fairly symmetric. Histogram (a), however, is "skewed" toward high values—that is, the high values are stretched out compared to the low values. Histogram (d) is skewed toward low values. Since it is usually easier to analyze symmetric data, we often transform skewed data to make it more nearly symmetric. When we transform (or reexpress) a variable, we are simply changing the unit of measurement. Transformations are used widely by experienced data analysts, especially in engineering and the natural sciences.

Of all the transformations made on data in practice, the three most popular are the square root, logarithm, and negative reciprocal.* These are listed in Table 2.2. Notice that we have used two new features of LET: the functions SQRT for square root and LOGT for log to the base 10. (These and many other functions are described in Appendix B.)

Exhibit 2.9 illustrates the effect of each transformation. Panel (a) shows the acidity of 71 lakes in Wisconsin. (This data set is described on p. 315.) Here, acidity is expressed in terms of the proportion of hydrogen ions in the water. This distribution clearly is skewed toward high values. Panel (b) shows what happened when we took the square root of each observation and then made a histogram. It is not as skewed as the original data, but still has a slight tendency toward a long tail in the high values. In Panel (c), the logarithmic transformation produced a fairly symmetric distribution.

*We use the negative reciprocal, $-1/x$, rather than just the reciprocal, $1/x$. This preserves the order of the observations. For example, if 42 is the smallest observation in our data set, then $-1/42$ will be the smallest observation in the transformed data set. If we used just the reciprocal, then 1/42 would be the largest observation in the transformed data set, and everything would be turned around.

Table 2.2 Three Transformations

Transformation	*Strength*	*Formula*	*Minitab Example*
Square Root	Moderate	$\sqrt{x}$	LET C2 = SQRT(C1)
Logarithm (base 10)	Strong	$\log_{10}(x)$	LET C2 = LOGT(C1)
Negative Reciprocal	Stronger	$-1/x$	LET C2 = -1/C1

Exhibit 2.9 Acidity of 71 Wisconsin Lakes. (Acidity was measured by the concentration of hydrogen ions.)

(a) Acidity

```
RETRIEVE 'LAKE'
HISTOGRAM 'HIONS'

Histogram of HIONS   N = 71

  Midpoint  Count
 0.0000000     29  *****************************
 0.0000002     23  ***********************
 0.0000004      5  *****
 0.0000006      4  ****
 0.0000008      3  ***
 0.0000010      3  ***
 0.0000012      1  *
 0.0000014      0
 0.0000016      2  **
 0.0000018      0
 0.0000020      1  *
```

(b) Square Root of Acidity

```
LET C20 = SQRT('HIONS')
HISTOGRAM C20

Histogram of C20   N = 71

Midpoint   Count
  0.0000       5  *****
  0.0002      24  ************************
  0.0004      20  ********************
  0.0006       8  ********
  0.0008       7  *******
  0.0010       3  ***
  0.0012       3  ***
  0.0014       1  *
```

(continued)

Exhibit 2.9 (Continued)

(c) Log of Acidity

```
LET C21 = LOGT('HIONS')
HISTOGRAM C21

Histogram of C21   N = 71

Midpoint    Count
    -8.8        4   ****
    -8.4        1   *
    -8.0        2   **
    -7.6        9   *********
    -7.2       13   *************
    -6.8       20   ********************
    -6.4        8   ********
    -6.0       11   ***********
    -5.6        3   ***
```

(d) Negative Reciprocal of Acidity

```
LET C22 = -1/'HIONS'
HISTOGRAM C22

Histogram of C22   N = 71
Each * represents 2 obs.

 Midpoint  Count
-600000000     1  *
-550000000     0
-500000000     0
-450000000     0
-400000000     3  **
-350000000     0
-300000000     0
-250000000     0
-200000000     0
-150000000     1  *
-100000000     0
 -50000000    11  ******
         0    55  ****************************
```

The reciprocal transformation in Panel (d) went much too far. Now the lower values are stretched out and the high values are clumped together. In fact, there are so many observations in the last interval that the HISTOGRAM command used each star to represent two observations in order to fit this bar on the page.

These three transformations all have the same general function: They compress the upper end of the distribution of values relative to the lower end. Thus they reduce asymmetry when the data are skewed toward high values. These transformations are also ordered, with square root being the weakest. That is, it will make symmetric a histogram that is just slightly skewed. This is followed, in strength, by log, and then negative reciprocal. The middle transformation, $\log(x)$, made the acidity data fairly symmetric. It is interesting to note that this transformation corresponds, except for a minus sign, to the usual units for measuring acidity, the pH of the liquid. By definition, pH $= -\log$(proportion of hydrogen ions).

Here we have shown how three of the most important transformations can be used to make distributions more symmetric. In Section 3.4 we will show how these same transformations can be used in other ways to make data analysis simpler and more powerful.

Exercise

2–17 If a data set is symmetric, then the mean and median are approximately equal. Use the Lake data on page 315 to compute the mean and median for each of the following: HIONS, square root of HIONS, log of HIONS, and negative reciprocal of HIONS. In each case, divide the median by the mean. If the mean and median are relatively close, this ratio will be near one. How do these ratios compare with the appearance of symmetry in Exhibit 2.9?

3

Plotting Data

The displays we have used thus far have involved only one variable at a time. Often we are interested in the relationships between two or more variables, such as the relationship between height and weight, between smoking and lung cancer, or between temperature and the yield of a chemical process. Plots allow us to investigate several variables at a time.

3.1 Scatterplots

If both variables are interval or ordinal, the most useful display is the familiar scatterplot. As an example, consider the data in Table 3.1. Exhibit 3.1 shows how we produced a scatterplot of the amount of gas used versus the outside temperature from these data.

Notice that the amount of natural gas consumed by this house decreased more or less linearly as the average temperature varied from 15°F to 45°F. The "2" on the plot means that two points fell there. That is, there were two weeks when the average temperature was about 30°F and the gas consumption was about 3.3. Reviewing the data, we see these were weeks 7 and 8.

Another Example

The Cartoon data (p. 303) were collected in an experiment to assess the effectiveness of different sorts of visual presentations for teaching. In one part of the experiment, participants were given a lecture in which cartoons were used to illustrate the material. Immediately after the lecture, a short quiz was given. In addition, each participant was given an OTIS test, a test that attempts to measure general intellectual ability. Exhibit 3.2 gives the

Exhibit 3.1 Plot of Furnace Data of Table 3.1

```
NAME  C1='BTU/HR'  C2='TEMP'
SET C1
2.10 2.55 2.77 2.40 3.31 2.97 3.32 3.29 3.62 3.67 4.07
END
SET C2
45 39 42 37 33 34 30 30 20 15 19
END
PLOT 'BTU/HR'  'TEMP'

        -           *
    4.00+
        -
BTU/HR  -
        -    *
        -            *
    3.50+
        -
        -                             2    *
        -
        -
    3.00+                                    *
        -
        -                                                 *
        -
        -                                            *
    2.50+
        -                                         *
        -
        -
        -                                                      *
          --------+---------+---------+---------+---------+--------TEMP
               18.0      24.0      30.0      36.0      42.0
```

commands we used to plot each participant's cartoon score versus his/her OTIS score.

Several interesting facts can be observed. People with higher OTIS scores tend to get higher cartoon scores, as you might expect. Note also that the highest cartoon score possible is 9, so that a person with very high ability cannot really demonstrate the full extent of his/her knowledge. The best a person can get is a 9. In this plot, the scores of people who had OTIS scores below 100 show no evidence that they were held down to the maximum score. But the scores of people with OTIS scores over 100 do show the effect of this limitation. This effect is sometimes called truncation and it results in some loss of information. If the experiment were to be done over again, it might be better to develop a slightly longer or harder test so that everyone would have a chance to demonstrate his/her full ability.

Table 3.1 Furnace Data
This table contains data from a study of the effectiveness of an energy-saving furnace modification. The data are for one house over an 11-week period. The second column gives the average amount of natural gas used per hour (per square feet of house area). The third column gives the average daily temperature during the week. The fourth column, which was coded 1 for "In" and 2 for "Out", will be explained in Section 3.2.

Week	*Gas Used BTU/Hour*	*Average Temperature (°F)*	*Furnace Modification*
1	2.10	45	Out
2	2.55	39	In
3	2.77	42	Out
4	2.40	37	In
5	3.31	33	Out
6	2.97	34	In
7	3.32	30	Out
8	3.29	30	In
9	3.62	20	Out
10	3.67	15	In
11	4.07	19	Out

PLOT C versus C

Gives a scatterplot of the data. The first column is put on the vertical (or y) axis and the second column on the horizontal (or x) axis. Ordinarily each point is plotted with the symbol *. When more than one point falls on the same plotting position, a count is given. When the count is over 9, the symbol + is used. Minitab will automatically choose "nice" scales, or you may specify your own scales with the following subcommands:

XINCREMENT = K
XSTART at K [end at K]
YINCREMENT = K
YSTART at K [end at K]

XINCREMENT is the distance between tick marks (the + symbols) on the x axis. XSTART specifies the first, and optionally, the last point plotted on the x axis. Any points outside are omitted from the plot. YINCREMENT and YSTART are for the y axis. The commands WIDTH and HEIGHT control the size of plots.

Exhibit 3.2 Plot of Immediate Cartoon Scores Against OTIS Scores

```
RETRIEVE 'CARTOON'
PLOT 'CARTOON1' VS 'OTIS'

        -
CARTOON1-                  *    * * *2*2 22   *3*2**43  2222*
        -
        -                *   **   2  242*32323*3*** 2** *2  * *
     7.5+
        -      *   *  * *   2  *2*****   *3**2  *  *22     *
        -
        -             * *    *3* *22**    *    2
        -
     5.0+            *  *2  *    *  2   ** *  *
        -
        -     *   2  ** *  2   **  ** *   *
        -
        -        *  * *        **     *            *
     2.5+
        -             *   *     * *  2
        -
        -        * * 3*  * *         *   *
        -
     0.0+       2          *
        -
         +---------+---------+---------+---------+---------+------OTIS
        72        84        96       108       120       132
```

HEIGHT of plots = K lines
WIDTH of plots = K spaces

On almost all computers, PLOT prints a plot that is 19 lines tall (plus two lines for labels) and 57 spaces wide (plus 18 spaces for labels). The commands HEIGHT and WIDTH allow you to change this size. If your output is on wide paper, you might want to make some plots very large to get better resolution. If you have many small data sets, you might want small plots.

HEIGHT and WIDTH apply to all PLOT, LPLOT, MPLOT, and DOTPLOT commands that follow them. For example, suppose you type WIDTH = 30. If, later in your session, you type PLOT C2 VS C1, this plot will be 30 spaces wide (plus 18 spaces for labels, giving a total of 48 spaces). If later still you type LPLOT C2 C1 C3, this plot, too, will be 30 spaces wide. If you want to return to 57 spaces wide, use another WIDTH command.

The command HEIGHT also applies to TSPLOT but WIDTH does not.

Exercises

3–1 The following questions refer to the plot in Exhibit 3.2.

(a) How many people got a perfect score on the cartoon test?

(b) How many people with an OTIS score of 90 or below got a perfect score on the cartoon test?

(c) Approximately what is the lowest OTIS score? The highest?

3–2 (a) Exhibit 3.1 contains a plot of gas consumption versus temperature. Suppose we want some idea how much gas would be consumed at a temperature of 25°F. Use this plot to make an estimate. Explain how you arrived at your estimate.

(b) Repeat part (a) for 10°F. How much faith do you have in your estimate?

3–3 Table 3.1 contains data on gas consumption over an 11-week period. Plot the weekly "average temperature" versus week. How did temperature change during this period?

3–4 In this problem we will examine the Trees data (p. 328) with an eye to developing a way to predict volume from measurements that are easier to make.

(a) Plot volume versus diameter.

(b) Plot volume versus height. Compare this plot to the one in part (a). Which seems to be a better predictor of volume: height or diameter? Which do you think would be easier to measure in a forest?

3–5 Let's consider the Trees data again (p. 328).

(a) Plot volume versus diameter.

(b) Use the HEIGHT commands to make the vertical axis about 50% smaller. Now plot volume versus diameter again.

(c) Plot volume versus diameter again, but this time use the WIDTH command to make the horizontal axis about 50% smaller than in (a).

(d) Compare the visual impression of the plots in (a), (b), and (c). Discuss any differences.

3.2 Plots with Symbols

A scatterplot shows the relationship between two variables. We can add information about a third variable by using different symbols for different points. For an example, let's take a closer look at the furnace modification data in Table 3.1 (p. 52). The last column indicates whether or not the energy-saving modification was installed in the furnace that week. This column is an example of a categorical variable. For Minitab, we code it using numbers.

Exhibit 3.3 continues the session begun in Exhibit 3.1. First, we added a third variable to specify the modification, using 1 for "In" and 2 for "Out." Next, we used Minitab's LPLOT command to plot these data with labels. Each observation with a 1 in C3 was plotted with the first letter of the alphabet, an A. Similarly, each observation with a 2 in C3 was plotted with a B. Two points fell on one spot and were plotted with a 2. To determine their symbols, we checked the data listing in Table 3.1. These points are for weeks 7 and 8, and thus one is an A and one is a B. We wrote these by hand on the plot.

Looking closely at this plot, we see some indication that the furnace modification might produce a slight savings in gas consumption. Overall, for a given temperature, the As seem to be just a bit lower than the Bs.

Exhibit 3.3 Labeled Plot of the Furnace Data

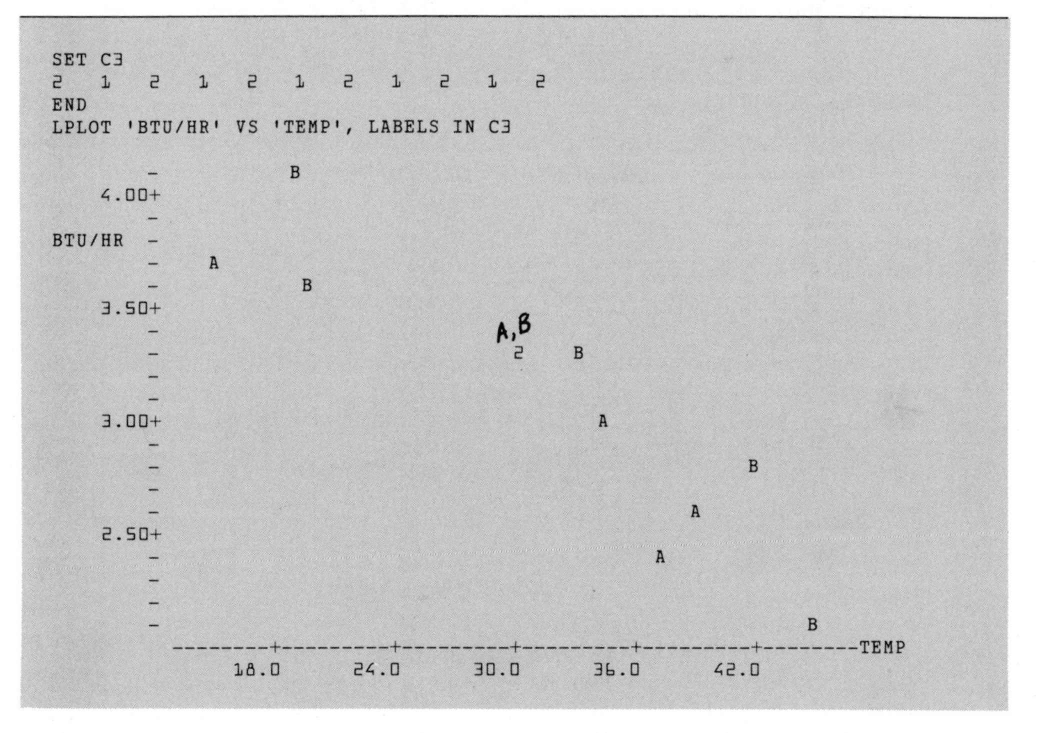

LPLOT C vs C, using labels as coded in C

XINCREMENT = K
XSTART at K [go to K]
YINCREMENT = K
YSTART at K [go to K]

LPLOT (the L is for labels or letters) plots data with labels given by

. . . −2 −1 0 1 2 3 . . . 24 25 26 27 28 . . .
. . . X Y Z A B C . . . X Y Z A B . . .

The subcommands specify scales as they do in PLOT (see p. 52). WIDTH and HEIGHT (see p. 53) control the size of LPLOTS.

Another Plot With Symbols, MPLOT

Table 3.2 gives the winning times for three Olympic races. A quick look tells us that the winners have gotten faster over the years. Suppose we want to compare the "progress" made in the different races. We could plot these winning times, but since the races are of different lengths, probably the

Table 3.2 Winning Times in Men's Olympic Track Races

Year	*100-Meter Race*	*200-Meter Race*	*400-Meter Race*
1900	10.8	22.2	49.4
1904	11.0	21.6	49.2
1908	10.8	22.4	50.0
1912	10.8	21.7	48.2
1920	10.8	22.0	49.6
1924	10.6	21.6	47.6
1928	10.8	21.8	47.8
1932	10.3	21.2	46.2
1936	10.3	20.7	46.5
1948	10.3	21.1	46.2
1952	10.4	20.7	45.9
1956	10.5	20.6	46.7
1960	10.2	20.5	44.9
1964	10.0	20.3	45.1
1968	9.9	19.8	43.8
1972	10.1	20.0	44.7
1976	10.1	20.2	44.3
1980	10.3	20.2	44.6

average speeds are easier to compare. So let's plot speed versus year for each of the three races. Recall speed = distance/time. We'll put all three plots on the same set of axes, to make comparisons easier.

Once the data have been entered into C1–C4, we can use the following commands:

```
LET C12 = 100/C2
LET C13 = 200/C3
LET C14 = 400/C4
MPLOT C12 C1, C13 C1, C14 C1
```

The resulting plot is in Exhibit 3.4. The letter A identifies the winning speeds for the 100-meter races, B identifies those for the 200, and C for the 400. Occasionally, two speeds fall on the same position and the number 2 is plotted. By looking at this plot we can get a very good idea of what happened. On the average, all three races had increased speeds over the years. These data are similar to those in Section 1.4 and, not surprisingly,

Exhibit 3.4 Winning Speeds in Men's Olympic Track Races for Various Years (A = 100-meter race, B = 200-meter race, C = 400-meter race)

```
MPLOT  C12  C1,  C13  C1,  C14  C1

   10.20+
        -                                                 2
C12     -                                              A     2 2  B
C13     -                                            A B
C14     -                         A  2       A B  B  B            A
    9.60+                                      A
        -                    A    B          B    A
        -
        -    A B  A  2    A  B A
        -      A          B    B                          C
    9.00+    B                                                C C  C
        -         B                                  C C
        -                                      C
        -                         C  C       C
        -                                         C
    8.40+                    C C
        -            C
        -      C
        -    C    C       C
        -
    7.80+
          ------+---------+---------+---------+---------+---------+C1
             1905      1920      1935      1950      1965      1980
```

reflect some of the same patterns. For example, the 100-meter and 200-meter races are run at about the same speed. Some years the 100-meter was faster and some years the 200-meter was faster. The 400-meter race, however, has always been run at a much slower pace. We also can spot a number of very "good" years, where progress was made in all three races—in 1924, 1932, 1960, and 1968. We also can see that essentially no progress was made in the three Olympics following World War II. Creative plotting can reveal a lot that is not apparent in a simple table of numbers.

MPLOT C vs C, C vs C, . . . , C vs C

XINCREMENT = K
XSTART at [K go to K]
YINCREMENT = K
YSTART at K [go to K]

MPLOT (the M is for multiple plot) puts several plots all on the same axes. The first pair of columns are plotted with the symbol A, the second pair with the symbol B, and so on. If several points fall on the same spot, a count is given.

The subcommands specify scales as they do in PLOT (p. 52). WIDTH and HEIGHT (p. 53) control the size of MPLOTS.

Difference Between LPLOT and MPLOT

The commands LPLOT and MPLOT both put data for several groups or variables on the same plot. Essentially the only difference between them is how the data are organized. Any plot that can be done by LPLOT also can be done by MPLOT and vice versa, if you reorganize the data appropriately. For example, the plot in Exhibit 3.3 can be done with MPLOT as follows:

```
NAME C1='GAS.OUT' C2='TEMP.OUT' C3='GAS.IN' &
     C4='TEMP.IN'
READ 'GAS.OUT' 'TEMP.OUT'
  2.10  45
  2.77  42
  3.31  33
  3.32  30
  3.62  20
  4.07  19
END
```

```
READ 'GAS.IN' 'TEMP.IN'
  2.55  39
  2.40  37
  2.97  34
  3.29  30
  3.67  15
END
MPLOT 'GAS.OUT' 'TEMP.OUT', 'GAS.IN' &
      'TEMP.IN'
```

In general, LPLOT requires all data for the horizontal axis to be stacked in one column, all data for the vertical axis to be stacked in a second column, and codes for group membership to be stacked in a third column. MPLOT, on the other hand, requires a separate pair of columns for each group.

Using LPLOT with Four Variables: An Advanced Example

LPLOT allows you to display three variables at a time. By being a little clever, you sometimes can put four or even more variables in one display. Some data from the U.S. Forest Products Research Laboratory provide an interesting example.

Plywood is made by cutting a thin, continuous layer of wood off logs as they are spun on their axis. Several of these thin layers then are glued together to make plywood sheets. Considerable force is required to turn a log hard enough so a sharp blade can cut off a layer. Chucks are inserted into the centers of both ends of the log to apply the torque necessary to turn the log.

The data in Table 3.3 were gathered in a study of how various factors affect the amount of torque that can be applied. The factors listed there are:

Temperature: The temperature of the logs at the time they were tested, in degrees F.
Penetration: The distance the chucks were inserted into the logs, in inches.
Diameter: The diameter of the test logs, in inches.
Torque: The average torque that could be applied before the chuck spun out, averaged over ten trial logs for each condition.

Various plots can be made of these data. For example, we might PLOT torque versus each of diameter, penetration, and temperature. We could

Table 3.3 Effect of Three Factors on Amount of Torque That Can Be Applied to Logs Being Made into Plywood

Diameter (Inches)	*Penetration (Inches)*	*Temperature (° F)*	*Torque*
4.5	1.00	60	17.30
4.5	1.50	60	18.05
4.5	2.25	60	17.40
4.5	3.25	60	17.40
4.5	1.00	120	16.70
4.5	1.50	120	17.95
4.5	2.25	120	18.60
4.5	3.25	120	18.55
4.5	1.00	150	15.75
4.5	1.50	150	16.65
4.5	2.25	150	15.25
4.5	3.25	150	15.85
7.5	1.00	60	29.55
7.5	1.50	60	31.50
7.5	2.25	60	36.75
7.5	3.25	60	41.20
7.5	1.00	120	23.20
7.5	1.50	120	25.90
7.5	2.25	120	35.65
7.5	3.25	120	37.60
7.5	1.00	150	22.55
7.5	1.50	150	22.90
7.5	2.25	150	28.90
7.5	3.25	150	35.20

Data are in the saved worksheet called PLYWOOD.

use LPLOT to display three of the variables at a time. We might, for example,

```
LPLOT 'TORQUE' versus 'PENETRTN' &
        using 'DIAMETER' for symbols
LPLOT 'TORQUE' versus 'TEMP' &
        using 'PENETRTN' for symbols
```

Suppose we try to create a display that simultaneously shows all four torque-influencing factors. Notice that diameter has just two values, 4.5 and 7.5. We easily can get a separate plot for each level. It helps if we put these two plots on the same scale. If we have paper output, we might even cut them out and paste them side by side.

We used the following commands to produce the display in Exhibit 3.5:

```
NAME C1='TEMP' C2='PENETRTN' C3='TORQUE.L' &
     C4='TORQUE.H'
READ 'TEMP' 'PENETRTN' 'TORQUE.L' 'TORQUE.H'
  1  1.00  17.30  29.55
  1  1.50  18.05  31.50
  1  2.25  17.40  36.75
  1  3.25  17.40  41.20
  2  1.00  16.70  23.20
  2  1.50  17.95  25.90
  2  2.25  18.60  35.65
  2  3.25  18.55  37.60
  3  1.00  15.75  22.55
  3  1.50  16.65  22.90
  3  2.25  15.25  28.90
  3  3.25  15.85  35.20
END
WIDTH 25
HEIGHT 16
LPLOT 'TORQUE.L'  'PENETRTN'  'TEMP';
  YSTART 15 45.
LPLOT 'TORQUE.H'  'PENETRTN'  'TEMP';
  YSTART 15 45.
```

Exhibit 3.5 Use of LPLOT to Display the Plywood Data (A = 60 B = 120 C = 150)

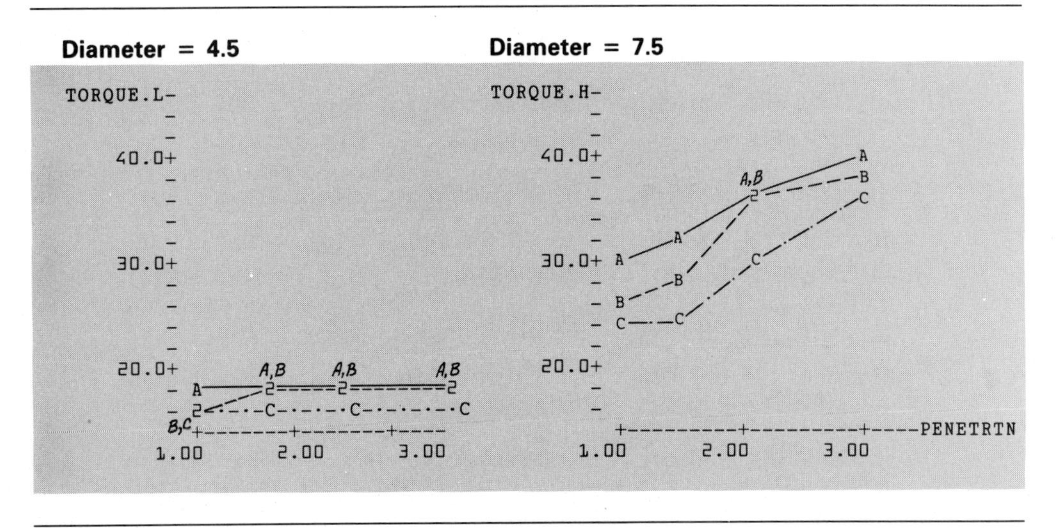

Notice we did not use the saved worksheet PLYWOOD. Instead, we reorganized the data so they would match the Minitab commands we planned to use and then typed them in ourselves.* We also coded TEMP, using 1 for 60°, 2 for 120°, and 3 for 150°. This way, LPLOT will mark the observations at 60° with an A, those at 120° with a B, and those with 150° with a C. We used WIDTH and HEIGHT to make the two plots small enough so we could tape them side by side on one piece of paper. We used the YSTART subcommand so that both plots would have the same scale for TORQUE. We labeled this plot and connected the points for each temperature by hand.

Producing a display like this requires a lot of trial and error. We have to choose which variable to put on the y axis, which to put on the x axis, and which to use for labels. We have to determine an appropriate plot size so that our display looks right. We have to determine which scales to specify ourselves and how to specify them. The display in Exhibit 3.5 was by no means our first attempt. We tried many displays before we settled on this one. (Another way to produce this same display is outlined in Exercise 3–12.)

Exhibit 3.5 provides an excellent summary of our data. The diameter of the logs is very important. At the smaller diameter (4.5 inches), the torque is much lower—it is only about 15 to 18 pounds—and neither temperature nor penetration seems to make much difference. For the larger diameter logs (7.5 inches), both penetration and temperature seem to play an important role; that is, more torque can be applied at the deeper penetrations and at the cooler temperatures. Somewhat offsetting these latter two conclusions are the practical facts that (1) deeper penetration damages more wood, and (2) cooler wood is tougher to cut.

Exercises

3–6 The following questions refer to the Track Data in Table 3.2 and Exhibit 3.4.

(a) Were there any records set in these three races in the 1980 Olympics? That is, were any of these three races run in the least time ever?

(b) In what year was the record set for the 100-meter race? For the 200-meter race? For the 400-meter race?

(c) The best time for the 100-meter race was 9.9 seconds. How fast is this in miles per hour? (*Note*: There are 1,609 meters in a mile.)

(d) Consider the 100-meter race. In which years was a new Olympic record set?

*If we wanted to use the saved worksheet and not retype the data we could have done the reorganization using the Minitab commands COPY and CODE described in Appendix B.

3–7 The following questions refer to the 7.5-inch diameter logs in Table 3.3 and Exhibit 3.5.

(a) About how much torque can be applied with a penetration of 3.25 inches at a temperature of 150°F?

(b) About how much more torque can be applied with a 3.25-inch penetration than with a 1-inch penetration? Does the amount of torque seem to depend on which temperature you use? If so, how?

(c) About how much increase in torque is there in going from 150°F to 60°F? Does the increase seem to depend on the depth of penetration? If so, how?

3–8 Exhibit 3.3 used A and B for plotting symbols. A better choice would be I for "In" and O for "Out." The number 9 is LPLOT's code for I and 15 is the code for O. Redo the LPLOT using these symbols. To change the codes, you could just retype the data in C3 or you could use Minitab's CODE command (see p. 339). Notice that this minor change in plotting symbols makes the plot easier to interpret and easier to explain to others.

3–9 (a) Use the Pulse Data (p. 318) to plot weight versus height.

(b) Use LPLOT to make the same plot but use one symbol for men and another for women. Interpret the plot.

3–10 Listed below are the monthly average temperatures for five U.S. cities.

Month	*Atlanta*	*Bismark*	*New York*	*San Diego*	*Phoenix*
January	42	8	32	56	51
February	45	14	33	60	58
March	51	25	41	58	57
April	61	43	52	62	67
May	69	54	62	63	81
June	76	64	72	68	88
July	78	71	77	69	94
August	78	69	75	71	93
September	72	58	68	69	85
October	62	47	58	67	74
November	51	29	47	61	61
December	44	16	35	58	55

They are stored in the saved worksheet called CITIES, with month coded 1, 2, . . . , 12. Get an MPLOT of temperature versus month for the five cities. (If your output is on paper, label the cities by hand and connect the points for each city.) Write a paragraph or two comparing the temperature patterns in these five cities.

3–11 During the winter months, you have probably noticed that on days when the wind is blowing it seems particularly cold (at least in colder climates). The explanation lies in the fact that wind creates a slight lowering of atmospheric pressure on exposed flesh thereby enhancing evaporation, which has a cooling effect. Thus the effective temperature (how cold it seems to you) depends on two factors: thermometer temperature and wind speed. The following table was prepared by the U.S. Army. It gives the effective temperature corresponding to a given thermometer temperature and wind speed.

Wind Chill Index

Wind Speed	*Thermometer Temperature (°F)*			
(mph)	*20*	*10*	*0*	*−10*
10	4	−9	−21	−33
15	−5	−18	−36	−45
20	−10	−25	−39	−53

(a) There is an error in this table. Can you find it? Plotting the data will help. Use MPLOT to plot the Wind Chill Index versus Wind Speed for the four temperatures. Do you see any points that do not seem to follow the overall pattern in the plot?

(b) Try to correct the point in error. What value would you substitute? Give reasons for your choice. Use both the data table and plot to help.

3–12 In Exhibit 3.5 we got two LPLOTS for the plywood data. The commands below use a trick to display these two separate plots as one plot. Basically, what we do is make the spacing on one variable, DIAMETER, wide enough so we can slip a full range of a second variable, PENETRTN, in between the levels of the first variable.

```
RETRIEVE 'PLYWOOD'
NAME C5='C.DIAM' C6='DIAM.PEN' C7='C.TEMP'
CODE (60)1 (120)2 (150)3 'TEMP' &
     put in 'C.TEMP'
CODE (4.5)0 (7.5)6 'DIAMETER' &
     put in 'C.DIAM'
LET 'DIAM.PEN' = 'C.DIAM' + 'PENETRTN'
LPLOT 'TORQUE' 'DIAM.PEN' 'C.TEMP'
```

The CODE command is described on page 339. The first CODE in the program changes all the 60s in TEMP to 1, all 120s to 2, and all 150s to 3. The second CODE command is used to space the levels of DIAMETER further apart. This allows us to slip PENETRTN in between the levels of DIAMETER. LET then combines DIAMETER and PENETRTN into one variable, which we called DIAM.PEN. Run this program and compare the output to Exhibit 3.5.

3–13 In Exercise 3–12, we plotted TORQUE versus a combined variable DIAM.PEN with letters for TEMP. Devise a similar program to plot TORQUE versus a combined variable DIAM.TEM using letters for PENETRTN. Interpret this plot and compare it to the one in Exhibit 3.5.

3–14 How do automobile accident rates vary with the age and sex of the driver? The table below gives the numbers of accidents per 100 million miles of exposure for drivers of private vehicles.

Age of Driver	*Accidents That Involved a Casualty*		*Accidents That Did Not Involve a Casualty*	
	Male	*Female*	*Male*	*Female*
Under 20	1436	718	2794	1510
20–24	782	374	1939	851
25–29	327	219	949	641
30–39	232	150	721	422
40–49	181	228	574	602
50–59	157	225	422	618
60 and over	158	225	436	443

(a) Read the four columns of accident rates into C1–C4 of the Minitab worksheet. To enter the "Age of Driver" into the worksheet, we will choose one representative age for each category in the table. For example, 18 for "Under 20", 22 for "20–24" and so on. Now use SET to put the following age categories into C5: 18, 22, 27, 35, 45, 55, and 65.

(b) Plot the casualty accident rate for males versus age and the casualty accident rate for females versus age, using MPLOT. Describe in a sentence or two how these accident rates depend on the age and sex of the driver.

(c) Repeat (b) for the noncasualty accidents. How well do the conclusions in (b) and (c) agree?

(d) Compute overall accident rates for male and female drivers by adding up the casualty and noncasualty rates separately for each sex. For example, use LET C11 = C1 + C3 and LET C12 = C2 + C4. Repeat (b) for these overall rates.

(e) For each age and sex group, compute the proportion of accidents that involve a casualty. Repeat (b) for these proportions.

(f) Suppose we want to look at the total casualty accident rate (i.e., number of casualty accidents per 100 million miles of exposure) for each age category. Would (C1 + C2) provide us with appropriate data? Can we get the appropriate figures from the data we have? Explain.

3.3 Time Series Plots

A sequence of observations taken over time is called a *time series*. Examples include the milk output of a cow recorded each week, the average temperature in Chicago recorded each day, the consumer price index recorded each month, and the number of homicides in the United States recorded each year. Minitab has a command, TSPLOT, which is designed to plot time series data.

Table 3.4 contains three time series. Each gives the number of people employed in one industry in the state of Wisconsin from January 1970 to December 1974. These three series were put in a saved worksheet called

Table 3.4 Number of Employees in Wisconsin (in Units of 1000 Employees)

Wholesale and Retail Trade

	Jan	*Feb*	*Mar*	*Apr*	*May*	*Jun*	*Jul*	*Aug*	*Sep*	*Oct*	*Nov*	*Dec*
1970	322	317	319	323	327	328	325	326	330	334	337	341
1971	322	318	320	326	332	334	335	336	335	338	342	348
1972	330	326	329	337	345	350	351	354	355	357	362	368
1973	348	345	349	355	362	367	366	370	371	375	380	385
1974	361	354	357	367	376	381	381	383	384	387	392	396

Food and Kindred Products

	Jan	*Feb*	*Mar*	*Apr*	*May*	*Jun*	*Jul*	*Aug*	*Sep*	*Oct*	*Nov*	*Dec*
1970	53.5	53.0	53.2	52.5	53.4	56.5	65.3	70.7	66.9	58.2	55.3	53.4
1971	52.1	51.5	51.5	52.4	53.3	55.5	64.2	69.6	69.3	58.5	55.3	53.6
1972	52.3	51.5	51.7	51.5	52.2	57.1	63.6	68.8	68.9	60.1	55.6	53.9
1973	53.3	53.1	53.5	53.5	53.9	57.1	64.7	69.4	70.3	62.6	57.9	55.8
1974	54.8	54.2	54.6	54.3	54.8	58.1	68.1	73.3	75.5	66.4	60.5	57.7

Fabricated Metals

	Jan	*Feb*	*Mar*	*Apr*	*May*	*Jun*	*Jul*	*Aug*	*Sep*	*Oct*	*Nov*	*Dec*
1970	44.2	44.3	44.4	43.4	42.8	44.3	44.4	44.8	44.4	43.1	42.6	42.4
1971	42.2	41.8	40.1	42.0	42.4	43.1	42.4	43.1	43.2	42.8	43.0	42.8
1972	42.5	42.6	42.3	42.9	43.6	44.7	44.5	45.0	44.8	44.9	45.2	45.2
1973	45.0	45.5	46.2	46.8	47.5	48.3	48.3	49.1	48.9	49.4	50.0	50.0
1974	49.6	49.9	49.6	50.7	50.7	50.9	50.5	51.2	50.7	50.3	49.2	48.1

These three series are all in the saved worksheet called EMPLOY.

EMPLOY, using variable names TRADE, FOOD, and METAL. Exhibit 3.6 contains a plot of the data for the food industry. TSPLOT plotted the first observation with the number 1, the second observation with the number 2, . . ., the ninth observation with the number 9, and the tenth observation with the number 0. Then it started over again. The eleventh observation was plotted with a 1, the twelfth with a 2, and so on.

There are two clear patterns in the plot: a very strong repeating cycle and a slight upward trend in the second half of the series. The cycle comes from the seasonal employment in the food industry—employment is naturally quite high in the summer and low in the winter. The cycle is repeated every 12 months. Many time series contain a repeating pattern. For example, if you record temperature every hour for ten days, you probably will see a repeating cycle every 24 hours. Temperatures generally tend to be high in the daytime and low in the night.

TSPLOT will plot data using special symbols to indicate a cycle. All you need to do is specify the length of the cycle, often called the period of the cycle. Exhibit 3.7 contains an example using the food data. In this plot, January is plotted with a 1, February with a 2, . . ., September with a 9, and

Exhibit 3.6 Plot of the Food Employment Data Using TSPLOT

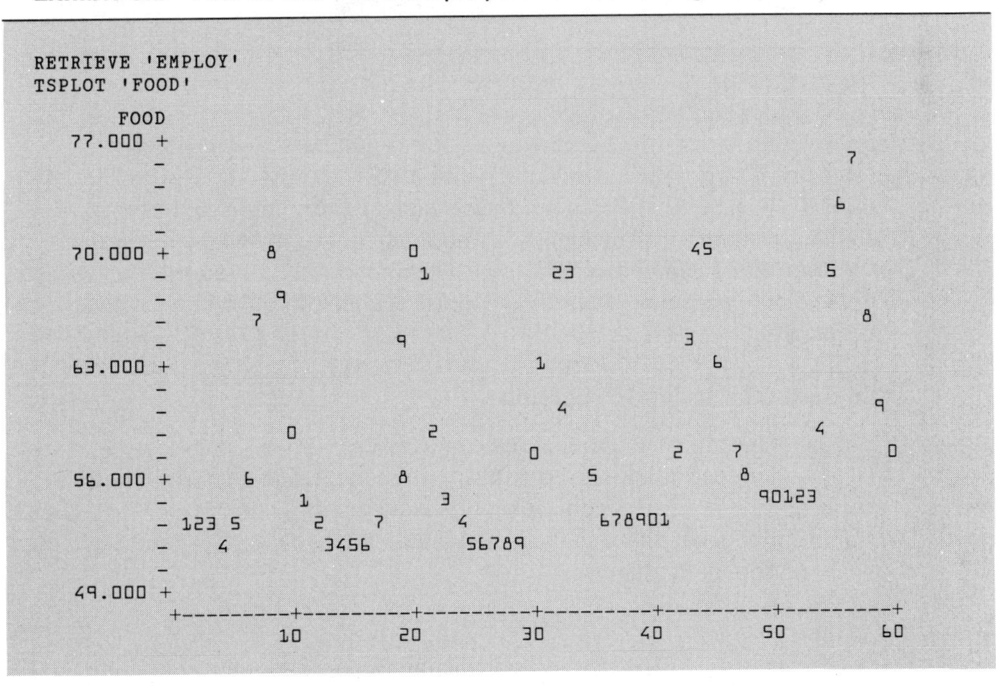

Exhibit 3.7 TSPLOT of the Food Data, with Symbols for the Period

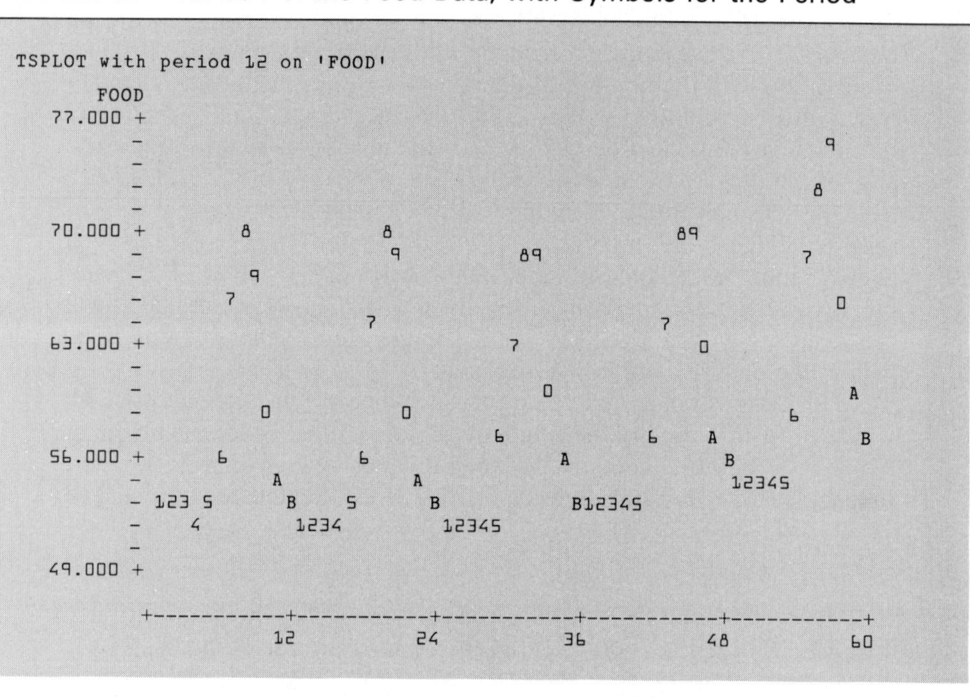

October with a 0 (for 10). Since we ran out of numbers at that point, we started using letters. Thus, November is plotted with an A and December with a B. In Exhibit 3.7 we can see the seasonal pattern more clearly than we could in Exhibit 3.6. The highest symbols each year are 8 and 9. These are for August and September. The lowest symbols are 1, 2, 3, and 4. These are for January, February, March, and April, the months after harvest and before planting.

Using TSPLOTS to Find Problems

All data must be collected either across time or across space. For example, chemical and industrial measurements are often made one after the other; agricultural data are often collected on various areas scattered across a farm. Plotting the data in ways that reflect these connections often reveals unusual features or problems.

An Example. Track etch devices were under consideration as an inexpensive way to measure low-level radiation in homes. An experiment was done to study the usefulness of one such device called an open cup. (Table 5.3, p. 107, gives a brief description of this device.) A total of 25 open cup devices were hung along a string that ran the length of a chamber. The chamber was designed to have a uniform level of radiation. The level of radiation measured by the various cups was as follows:

6.6 4.8 5.0 6.3 5.9 6.5 6.1 5.8 6.4 5.3 5.2 5.2 4.8
4.9 5.5 5.1 5.0 5.7 4.8 5.3 4.2 5.3 4.5 4.5 4.2

Exhibit 3.8 shows a TSPLOT of these data. This plot shows a systema-

Exhibit 3.8 Time Series Plot of Track Etch Data

```
SET C1
6.6 4.8 5.0 6.3 5.9 6.5 6.1 5.8 6.4 5.3 5.2
5.2 4.8 4.9 5.5 5.1 5.0 5.7 4.8 5.3 4.2 5.3
4.5 4.5 4.2
END
TSPLOT C1

          C1
   6.600 + 1
         -        6
         -      4    9
         -
         -         7
   6.000 +
         -       5
         -          8         8
         -
         -                 5
   5.400 +
         -            0         0 2
         -             12   6
         -    3              7
         -                4
   4.800 +  2           3     9
         -
         -                       34
         -
         -
   4.200 +                     1  5
           +---------+---------+---------+
                    10        20        30
```

tic pattern—the first observations at the left of the plot are larger than those at the right.

Why this pattern? These 25 observations are plotted in the order in which the track etch devices had been hung along the string. One conjecture is that the radiation level in the chamber was not constant but changed gradually from one end of the chamber to the other.

Further checking led to two other possibilities. The devices were put in the chamber in the same order as their serial numbers. Thus, a change in sensitivity of the devices during the manufacturing process could account for the observed trend. Another possible source of the trend concerned reading the devices. A track etch measurement is based on a count of the number of tracks put on the device by exposure to radiation. Thus there might have been a systematic error in the counting of the tracks. If the devices were counted in the same order as their serial numbers, a systematic drift in the counting process could account for the pattern in the TSPLOT.

As of this writing, the source of the problem is not known, and investigation is continuing. It is important to note that if the devices had not been placed in the chamber in the same order as their serial numbers, but instead were placed in a random order, we might have been able to pinpoint the source of the problem.

TSPLOT C

Plots the data in C versus the integers 1, 2, 3, etc. This type of plot is often used for time series data. You may specify the scale for the data axis by the following subcommands:

INCREMENT = K
START at K [go to K]

These subcommands have the same meaning as they do in PLOT (p. 52).

Time series data often have an associated period. For example, they may be collected monthly (period = 12) or hourly (period = 24). Plotting symbols that reflect this period are used, if you specify a period:

```
TSPLOT period = K [starting at K] C
```

You should specify the starting point if the first observation in C does not correspond to the first period. For example, if C1 contains monthly data starting in March, use

```
TSPLOT 12 3 C1
```

If the time series is too long to fit across the page, the plot is automatically broken into several pieces. The width of a page is controlled by the command OW (p. 344). The command HEIGHT (p. 53) controls the height of a TSPLOT. (The command WIDTH does not apply to TSPLOT.)

Exercises

3–15 In this section, we used TSPLOT to display the employment data for the food industry. Make similar displays of the data for wholesale and retail trade. Interpret the plot. Did employment tend to increase over this period? Is there a seasonal pattern to employment in wholesale and retail trade? Can you give any reasons for the patterns in the data?

3–16 Use TSPLOT to display the data for fabricated metals. Describe the pattern of employment in this industry. Did employment tend to increase over this period? Is there a seasonal pattern?

3–17 Just by looking at the TSPLOT in Exhibit 3.8, can you figure out the following:

(a) Which open cup device gave the highest reading? Approximately what value did the highest reading have?

(b) Two cups seem to be tied for lowest. Which are they? Approximately what value did they have?

(c) How many cups were there in all?

3–18 The data listed below are measurements from a diffusion porometer, an instrument for measuring changes in resistance across a leaf's surface. Any treatment which closes the stomates of a leaf causes the resistance to go up; when the stomates are open, resistance goes down. Four treatments were used, with six observations of each treatment:

	Treatment											
	1						*2*					
Resistance	7.6	8.9	8.4	14.5	15.8	12.8	8.2	10.5	6.8	14.3	13.4	13.2
	3						*4*					
Resistance	6.5	9.9	8.9	12.6	13.5	14.6	11.4	8.6	12.2	10.9	9.9	15.3

The observations were taken in the order they are listed. Use TSPLOT with period six to plot these data in this order. (If your plot is on paper, connect the 6 points within each treatment.) Write a concise summary of these data, including an opinion as to whether the treatments differ very much. Also discuss any problems you see with the data.

3.4 Transformations

In Section 2.7 we showed how transformations could be used to make distributions more symmetric. An even more important use of transformations is to simplify the relationship between variables. Compare, for example, the following descriptions:

When x increases, y increases in a curved fashion.

When x increases, the square root of y increases in a straight line.

The first description mentions a curve but does not say what kind of curve. The second description is more specific and, in fact, gives us the functional form of the relationship $\sqrt{y} = a + bx$. Of course, we still do not know the values of a and b. In Chapter 10 we will see how to find these.

The use of transformations is new in some areas of research, but engineers and physical scientists have long used transformations to simplify relationships. In fact, they have a saying: "Anything is a straight line after using a logarithmic transformation." Of course, this statement is an exaggeration, but it does indicate the usefulness of transformations.

Recall the three transformations discussed in Section 2.7: square root, log, and negative reciprocal. Suppose we have two variables, say x and y, and we want a simple way to describe the relationship between them. We might start by plotting y versus x. If the plot were more or less a straight line, we would have our simple description. But suppose the plot were a curve. Three curves frequently encountered in data analysis are shown in Exhibit 3.9. Plots (a) and (b) both have an upward trend—that is, as x increases, y also increases. But there is a difference; plot (a) curves down, whereas (b) curves up. Plot (c) has a downward trend—as x increases, y

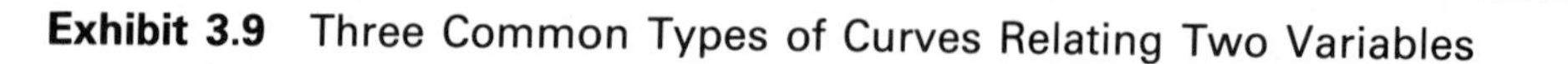

Exhibit 3.9 Three Common Types of Curves Relating Two Variables

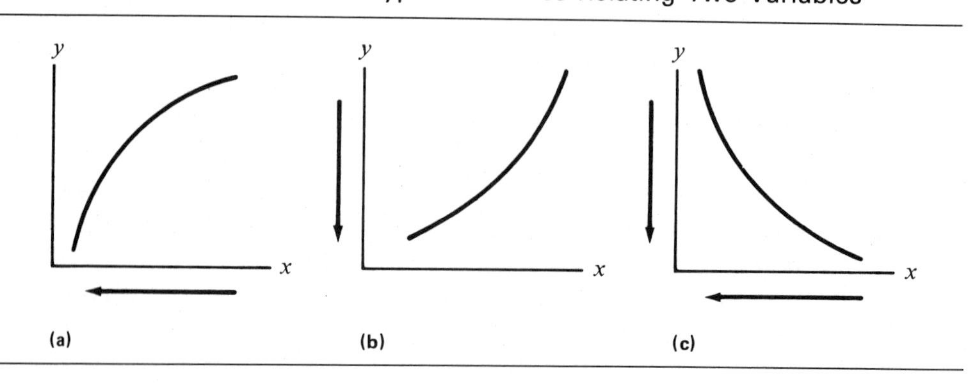

decreases—and it curves up. Our objective here is to transform x, or y, or both in order to get a straight line.

The arrows in Exhibit 3.9 indicate which transformations to try. For example, in Panel (a), the arrow points down on the x axis. This says that if we pull the upper end of the curve in the direction of the arrow, we will tend to straighten the curve. Put another way, we need to compress the high values of x to straighten the curve. We often start with $\sqrt{x}$. If this is not strong enough, we usually try $\log(x)$, and then $-1/x$, until we find a transformation that works. In Panel (b) we need to compress the high values of y. So we start with $\sqrt{y}$ and, if necessary, try $\log(y)$, and then $-1/y$. In Panel (c), there are downward arrows on both axes. This says to try transforming either x, or y, or both. For example, we might try y versus $\sqrt{x}$, or $\log(y)$ versus $\sqrt{x}$. Our goal in all cases is to get a straight line. Of course, this doesn't always happen. But it does work surprisingly often.

Example

The National Bureau of Standards conducted a series of tests to see how stopping distances are related to automobile speed. Here are some of their data:

Speed (miles per hour)	10	20	30	40	50
Stopping Distance (feet)	10	40	94	182	308

Exhibit 3.10 shows some plots of the data and the results after transformations. A plot of the original data, DISTANCE versus SPEED, is shown first. The most obvious pattern in this plot is a strong upward trend—as speed increases, stopping distance increases, which is certainly what we would expect. Overall, the plot is reasonably close to a straight line. For many purposes, all we need is a very simple description of the main features of the data. In those cases, we might report that as speed increases, stopping distance increases approximately in a straight line. For other purposes, however, this may be too much of a simplification since it ignores a secondary pattern—a slight upward curve. If we want a more detailed description of the data, we must take this curve into account.

The type of curvature in the plot of DISTANCE versus SPEED is similar to that in Panel (b) of Exhibit 3.9. Therefore, to straighten the curve, we should try a transformation on DISTANCE. The next plot shows what happened when we tried square root. This plot is very close to a straight line, so square root is probably the transformation we want. Suppose we try the stronger log transformation. Now, the curve goes the other way. This says log was too strong. Therefore, a more detailed description of the relationship between DISTANCE and SPEED would say that as speed increases, the square root of stopping distance increases in a linear fashion.

Exhibit 3.10 Plots of Distance Versus Speed, with Two Transformations

```
READ C1 C2
 10   10
 20   40
 30   94
 40  182
 50  308
END
NAME C1='SPEED'  C2='DISTANCE'
PLOT 'DISTANCE' VS 'SPEED'

      320+
         -                                                       *
DISTANCE-
         -
         -
      240+
         -
         -
         -
         -                                           *
      160+
         -
         -
         -
         -                                *
       80+
         -
         -                    *
         -
         -         *
        0+
          +---------+---------+---------+---------+---------+------SPEED
        8.0      16.0      24.0      32.0      40.0      48.0

NAME C3='SQRT(D)'
LET 'SQRT(D)' = SQRT('DISTANCE')
PLOT 'SQRT(D)' VS 'SPEED'

SQRT(D) -
        -                                                        *
        -
    16.0+
        -
        -
        -                                            *
        -
    12.0+
        -
        -
        -                                 *
        -
     8.0+
        -
        -                     *
        -
        -
     4.0+
        -          *
        -
          +---------+---------+---------+---------+---------+------SPEED
        8.0      16.0      24.0      32.0      40.0      48.0
```

Exhibit 3.10 (Continued)

```
NAME C4='LOG(D)'
LET 'LOG(D)' = LOGT('DISTANCE')
PLOT 'LOG(D)' VS 'SPEED'

             -
             -                                                       *
        2.40+
             -
  LOG(D)     -                                          *
             -
             -
        2.00+                                  *
             -
             -
             -
             -
        1.60+                      *
             -
             -
             -
             -
        1.20+
             -
             -
             -    *
              +---------+---------+---------+---------+---------+------SPEED
             8.0      16.0      24.0      32.0      40.0      48.0
```

Exercises

3–19 When a magnifying glass is used to view an object, there is a simple relationship between the apparent size of the object as seen through the lens (image size) and the distance the object is from the lens (object distance). Here are some data that were obtained in an experiment.

Object Distance (Centimeters)	12	13	14	15	16	17	18	19	20	21	22
Image Size (Centimeters)	12.0	9.4	7.2	6.2	5.2	4.5	4.0	3.6	3.2	3.0	2.7

(a) Plot image size versus object distance. How does image size vary with object distance?

(b) Suppose you were to place an object at a distance of 12.5 centimeters from the lens. What would you estimate the image size to be? Suppose you place it at 23 centimeters. Approximately what image size would you expect?

(c) Now let's investigate the relationship between these two variables. It's obviously some sort of curve, but what curve? Compare the plot in part (a) with the sketches in Exhibit 3.9. Which does it match?

(d) See if you can find a way to transform one of the variables so the resulting plot is approximately a straight line.

3–20 Experiments were run in several different scientific laboratories to determine the vapor pressure of cadmium as a function of temperature. Here are the results from two of the laboratories. These data are stored in the saved worksheet called CADMIUM.

(a) Plot vapor pressure versus temperature for Laboratory 1. Does this relationship seem to be a straight line? If not, see if you can transform one or both variables to get a straight line.

(b) Repeat (a) for Laboratory 2. Do the same transformations work for both laboratories? Can you find a transformation that works for both laboratories?

Laboratory 1		*Laboratory 2*	
Temperature (Kelvins)	*Pressure (Millionths of an Atmosphere)*	*Temperature (Kelvins)*	*Pressure (Millionths of an Atmosphere)*
525	10.1000	493	1.513
501	2.8300	523	6.421
475	0.6370	551	22.370
452	0.1590	579	68.030
413	0.0086	507	3.447
551	31.2000	537	13.550
503	2.9800	566	42.240
488	1.3900	593	109.200
569	89.7000	606	155.300
432	.0489	614	163.200
		630	294.700
		644	394.700

4

Tables

Minitab's TABLE command provides two general types of capabilities. The first type, considered in Section 4.1, makes tables of counts and percents. It is a generalization of the TALLY command (see Section 2.6). The second type, considered in Section 4.2, provides summary statistics, such as means and standard deviations, for related variables.

Statistical tests of association also can be done with the TABLE command, but the explanation of these tests is deferred to Chapter 11.

An Example: The Wisconsin Restaurant Survey

Because tables arise naturally in survey work, we will illustrate the TABLE command using data from an actual survey, the 1980 Wisconsin Restaurant Survey, conducted by the University of Wisconsin Small Business Development Center. This survey was done primarily to "allow educators, researchers, and public policy makers to evaluate the status of Wisconsin's restaurant sector and to identify particular problems that it is encountering." A second purpose was to develop data that would "be useful to small business counselors in advising managers as to how to effectively plan and operate their small restaurants."

Nineteen of Wisconsin's counties were selected for study. Lists of restaurants were drawn up from telephone directories and these were sampled in proportion to the population of the county. A sample of 1000 restuarants yielded 279 usable responses.

In a full analysis of these data, we would be concerned about possible biases from the sampling method and from the 28% response rate. Here, however, we are interested only in providing summaries of the returned questionnaires—not in making inferences to all restaurants.

The data base thus consists of 279 cases, one for each restaurant with usable data. In this book, we will use only a few of the many variables in the survey. The data set is listed on p. 321 and is saved in the worksheet called RESTRNT. The first 12 variables were taken from the questionnaires, while the other four were calculated from the 12 basic variables. Each of the 279 respondents failed to answer at least one question. Thus, there are a number of missing values, which are coded by * (see p. 26 and p. 342 about missing data).

4.1 Tables of Counts and Percents

We begin by describing the basic TABLE command. Then, as needed, we will introduce other features.

As a first example, we will classify the restaurants by type of ownership and size. The variable OWNER takes on three values: 1 = sole proprietorship; 2 = partnership; and 3 = corporation. SIZE also takes on three values: 1 = under 10 employees; 2 = from 10 to 20 employees; and 3 = over 20 employees. Exhibit 4.1 uses the TABLE command to print the appropriate table.

The first column of this table gives counts for restaurants with SIZE = 1. There were 83 restaurants with OWNER = 1 and SIZE = 1, that is, there were 83 sole proprietorships with under ten employees. Similarly, there were 16 partnerships and 40 corporations with under ten employees. Altogether, there were 139 restaurants with under ten employees.

The grand total, 261, is the total number of restaurants in this table. This is less than the 279 restaurants that are in the data base. Therefore,

Exhibit 4.1 Number of Restaurants Classified by Type of Ownership and Size

```
RETRIEVE 'RESTRNT'
TABLE 'OWNER' 'SIZE'

ROWS: OWNER        COLUMNS: SIZE

             1          2          3        ALL

 1          83         18          2        103
 2          16          6          4         26
 3          40         42         50        132
ALL        139         66         56        261

  CELL CONTENTS --
                  COUNT
```

only 261 restaurants had usable data for both OWNER and SIZE. The other 18 had missing data on either OWNER or SIZE or both.

The last row and last column are called the margins of the table, and the counts there are called marginal statistics. The other nine counts form the main body of the table.

TABLE the data classified by C, . . . , C

Prints one-way, two-way, and multi-way tables of counts. The classification variables must contain integer values between −10000 and 10000.

Percents

It often is easier to interpret a table if we convert counts to percents. TABLE has three subcommands to do this: ROWPERCENT calculates row percents, COLPERCENT calculates column percents, and TOTPERCENT calculates total percents.

Exhibit 4.2 contains an example using total percents. To calculate the total percent in a cell, Minitab divided the count in that cell by the grand total, 261, then multiplied by 100. For example, $(83/261) \times 100 = 31.80\%$ of the restaurants were sole proprietorships with fewer than ten employees. Thus, almost one third of the restaurants in the study were small: one owner and under ten employees. Over half the restaurants, 53.26%,

Exhibit 4.2 Percents for Restaurants

```
TABLE  'OWNER' 'SIZE';
  TOTPERCENT.

ROWS: OWNER       COLUMNS: SIZE

            1          2          3       ALL

 1      31.80       6.90       0.77     39.46
 2       6.13       2.30       1.53      9.96
 3      15.33      16.09      19.16     50.57
ALL     53.26      25.29      21.46    100.00

 CELL CONTENTS --
                  % OF TBL
```

have under ten employees, and just over half, 50.57%, are owned by corporations.

Exhibit 4.3 contains an example using both row and column percents. To calculate row percents, Minitab divided the count in each cell by the row total, then multiplied by 100. For example, there are 103 restaurants in row one; therefore, $(83/103) \times 100 = 80.58\%$ of the sole proprietorships had under ten employees. Only 1.94% of the sole proprietorships had more than 20 employees. The remaining 17.48% had between ten and 20 employees. On the other hand, the corporations are split fairly evenly among the three sizes: 30.30% have under ten employees; 31.82% have from ten to 20 employees; and 37.88% have over 20 employees.

To calculate column percents, Minitab divided the count of each cell by the column total, then multiplied by 100. Notice that among restaurants with under ten employees, the majority, 59.71%, are sole proprietorships, whereas among restaurants with over 20 employees, the overwhelming majority, 89.29%, are corporations.

Exhibit 4.3 Row and Column Percents for Restaurants

```
TABLE 'OWNER' 'SIZE';
  ROWPERCENT;
  COLPERCENT.

ROWS: OWNER        COLUMNS: SIZE

            1         2         3        ALL

 1      80.58     17.48      1.94     100.00
        59.71     27.27      3.57      39.46

 2      61.54     23.08     15.38     100.00
        11.51      9.09      7.14       9.96

 3      30.30     31.82     37.88     100.00
        28.78     63.64     89.29      50.57

            1         2         3        ALL

ALL     53.26     25.29     21.46     100.00
       100.00    100.00    100.00     100.00

 CELL CONTENTS --
                      % OF ROW
                      % OF COL
```

TABLE C, . . . ,C

Subcommands for counts and percents

COUNTS

Prints a count of the number of observations in each cell. If no subcommands for statistics are given, then counts are printed by default.

ROWPERCENTS

Prints row percents in each cell. The row percent for a cell is

$$\frac{\text{(number of observations in the cell)}}{\text{(number of observations in the row)}} \times 100$$

COLPERCENTS

Prints column percent as in each cell. The column percent for a cell is

$$\frac{\text{(number of observations in the cell)}}{\text{(number of observations in the column)}} \times 100$$

TOTPERCENTS

Prints total percents in each cell. The total percent for a cell is

$$\frac{\text{(number of observations in the cell)}}{\text{(number of observations in the table)}} \times 100$$

Exercises

4–1 The following refer to Exhibit 4.1.

(a) How many restaurants had corporate ownership? How many were partnerships?

(b) How many were partnerships with ten to 20 employees? How many were partnerships with over 20 employees?

(c) How many restaurants had ten or more employees? Twenty or fewer employees?

(d) How many had ten or more employees and were owned by a corporation? How many had ten or more and were not owned by a corporation?

4–2 Refer to Exhibits 4.2 and 4.3. In each part, say what percent of restaurants had the specified characteristics.

(a) What percent had fewer than ten employees? More than 20 employees? Ten or more employees?

(b) What percent were partnerships with fewer than ten employees? Corporations with ten to 20 employees?

(c) Of the partnerships, what percent had ten to 20 employees? More than 20 employees?

(d) Of the restaurants having fewer than ten employees, what percent were partnerships?

(e) Of the partnerships, what percent had ten or more employees?

4–3 (a) Make a table of the Pulse data (p. 318) in which the two classification variables are SEX and RAN. Calculate row and column percentages.

(b) What percent of the females ran in place? Males? Do either or both of these results seem surprising to you in light of the data description on page 318? Discuss.

4–4 Data on all persons committed voluntarily to the acute psychiatric unit of a Health Care Center in Wisconsin during the first six months of 1981 are given in the table. There are three variables: (1) reason for discharge with 1 = normal and 2 = other (against medical advice, court ordered, absent without leave, etc); (2) month of admission with 1 = January, 2 = February, . . . , 6 = June; and (3) length of stay in number of days.

Reason	1	1	1	1	1	1	1	1	1	1	1	1	1	1	1	1	1	1	1	1	1	1	1	1	1	1	1	1	1
Month	1	1	1	1	1	1	1	1	1	2	2	2	2	2	2	2	2	3	3	3	3	3	3	3	3	4	4	4	4
Length	1	2	5	8	8	9	10	13	25	0	1	7	7	1	8	11	11	0	1	2	4	5	13	1	25	2	4	5	12

Reason	1	1	1	1	1	1	1	1	1	1	1	1	1	2	2	2	2	2	2	2	2	2	2	2	2	2	2	2	2
Month	4	4	4	5	5	5	5	6	6	6	6	6	6	1	2	3	4	4	3	4	6	6	6	6	5	3	4	4	4
Length	25	35	45	1	11	18	19	1	1	3	15	35	75	0	9	1	2	3	1	5	2	5	6	14	1	4	6	19	25

The data are in the saved worksheet HCC.

Make a table of the data in which the two classification variables are REASON and MONTH.

(a) What month had highest number of discharges? The lowest?

(b) What percent of all discharges are normal?

(c) How many discharges are there in the winter months, January through March? In the spring months, April through June?

(d) Compare the length of stay, month by month, for the two types of discharge. Is one always longer than the other?

4.2 Tables for Related Variables

So far, we have discussed frequency tables—tables whose cells contain counts and percents. We also can create tables whose cells contain summary statistics on related variables.

Exhibit 4.4 contains a table of means. This table has the same two classification variables (also called factors), OWNER and SIZE, that we used in Exhibit 4.1. Now, however, each cell contains the mean sales of all restaurants in the cell. For example, sole proprietorships with under ten employees had average sales of $115,550; all restaurants with under ten employees had average sales of $146,300; all restaurants in the study (with usable data on OWNER, SIZE, and SALES) had average sales of $339,000.

Any number of summary statistics, on any number of variables can be put in a table. Exhibit 4.5 has three statistics in each cell: MIN gives the minimum sales in each cell, MEDIAN gives the median, and MAX gives the maximum. (Notice: Medians do not appear in the margins of the table. Minitab did this to save computing time. Median is the only statistic that does not appear in the margins.)

Exhibit 4.5 merits careful study. First, look at the cell with OWNER = 1, SIZE = 1. The minimum sales is 0. Thus, one restaurant reported no sales for 1979. Suppose we go back to the data listing on page 321. This sales figure is on line 203. This restaurant also has market value listed as 0. All other responses, however, seem reasonable. Perhaps the

Exhibit 4.4 Mean Sales for Restaurants (in thousands of dollars)

```
TABLE 'OWNER' 'SIZE';
  MEAN  'SALES'.

ROWS: OWNER       COLUMNS: SIZE

           1          2          3         ALL

 1    115.55     237.94     579.00     148.49
 2     88.12     321.25     695.00     228.12
 3    231.50     313.84     887.96     507.78
ALL   146.30     291.57     861.74     339.00

 CELL CONTENTS --
           SALES:MEAN
```

Exhibit 4.5 Sales for Restaurants (in thousands of dollars)

```
TABLE 'OWNER' 'SIZE';
  MIN 'SALES';
  MEDIAN 'SALES';
  MAX 'SALES'.

ROWS: OWNER        COLUMNS: SIZE

            1          2          3         ALL

 1       0.00      39.00     425.00        0.00
        98.00     220.00     579.00          --
       435.00     507.00     733.00      733.00

 2       2.00     200.00     480.00        2.00
        76.00     267.50     750.00          --
       250.00     550.00     800.00      800.00

 3       2.00     100.00     225.00        2.00
       133.50     274.00     565.00          --
      3450.00     720.00    8064.00     8064.00

ALL      0.00      39.00     225.00        0.00
           --         --         --          --
      3450.00     720.00    8064.00     8064.00

  CELL CONTENTS --
            SALES:MINIMUM
                  MEDIAN
                  MAXIMUM
```

restaurant owner forgot to answer these two questions, or gave unreadable answers and the person who entered the data into a computer file mistakenly typed a zero instead of an *. Or, perhaps the owner answered 0 when he meant to answer, "I don't know." Unfortunately, the original questionnaire had been destroyed by the time these problems were discovered, so the best we can do now is replace the two 0s by *, the missing data code. The appropriate commands are:

```
LET 'SALES'(203) = '*'
LET 'VALUE'(203) = '*'
```

The problem of a 0 for SALES was very easy to find. If we carefully look at the patterns in the table, we can find another unusual value. Within each OWNER category, minimum sales increase as the number of employees increases. The medians follow the same pattern. Now look at the maximums. In the first two OWNER categories, maximum sales increase as we go across the row. In the last OWNER category, however, this pattern does not hold: Corporations with under ten employees have a much larger maximum than corporations with ten to 20 employees. In fact, the sales figure of $3,450,000 is suspiciously large.

Again, we go back to the data listing on p. 321. This sales figure is on line 111. Look at the other information for this restaurant. The owner said he was very pessimistic about the future, had made only $10,000 in capital improvements, and valued his business at just $100,000. These responses do not make much sense for a restaurant that had gross sales of $3,450,000. This sales figure is almost certainly in error. Again, we will replace it by an *.

The simple table in Exhibit 4.5 brought two errors to our attention. There probably are many more still in the data set. Most data sets initially contain many errors, and one of the first steps in any analysis is to try to find them. The simple displays and summaries of Chapters 2 and 3, and the TABLE command, can help. Use the TALLY command on each categorical variable to discover values out of range. Use DESCRIBE and pehaps STEM-AND-LEAF on all interval variables to find values out of range and unusual patterns. Both DESCRIBE and STEM-AND-LEAF would have uncovered the zero on SALES. Neither, however, would have revealed the restaurant with $3,450,000 on SALES. In itself, $3,450,000 is not unusual, but in the context of OWNER and SIZE it is. This value is called a multivariate outlier; it is unusual in the context of several variables. The plotting commands of Chapter 3 and the TABLE command display several variables at once and can help find multivariate outliers.

TABLE C, . . . , C

Subcommands for summary statistics on related variables

MEANS for C, . . ., C
MEDIANS for C, . . ., C
STDEVS for C, . . ., C
MINIMUMS for C, . . ., C
MAXIMUMS for C, . . ., C
N for C, . . ., C
NMISS for C, . . ., C
STATS for C, . . ., C
DATA for C, . . ., C

The first seven subcommands calculate one summary statistic for each related variable, for each cell of the table.

N gives the number of observations in a cell that have a value for the related variable. NMISS gives the number of observations that have an * (missing value) for the related variable. (Thus, COUNT = N + NMISS.)

STATS is the same as typing N, MEAN, and STDEV.

DATA prints all values of the related variable, in each cell.

Exercises

4–5 Refer to Exhibits 4.4 and 4.5. If the desired results are not available from these exhibits, please say so.

(a) What was the median sales volume of partnerships with under ten employees? Of all partnerships? Of all restaurants?

(b) Which type of restaurant had the highest mean sales volume? The lowest?

4–6 Exhibit 4.4 contains mean SALES classified by OWNER and SIZE. Use the TABLE command to determine the total number of restaurants in the table in Exhibit 4.4. Compare this to the total number of restaurants in Exhibit 4.1. Why is there a difference?

4–7 The following are based on the Pulse Data (p. 318).

(a) Find the mean of PULSE1 classified by SEX and ACTIVITY. Discuss the output. How do men and women compare on pulse rate?

(b) Find the mean of PULSE2 classified by SEX and RAN. Discuss the output.

4–8 Use the Restaurant data (p. 321) to answer the following:

(a) Find the mean and median number of seats for each TYPEFOOD. Discuss the results.

(b) Find the mean and median amount of sales per seat for each different TYPEFOOD. Discuss the results.

(c) Find the mean and median amount of sales per employee for each of the three sizes of restaurants. Discuss the results.

4–9 Refer to the HCC data in Exercise 4–4.

(a) Obtain a table of the average length of stay in the acute psychiatric unit by reason and month.

(b) Does the average length of stay in April 1981 seem to be the same for both types of discharges?

(c) Repeat parts (a) and (b) for median length of stay.

(d) Does average length of stay seem to be constant over time for normal discharges? Other discharges? If any pattern is apparent, is it similar for both types of discharges?

4.3 Advanced Features of TABLE

In Sections 4.1 and 4.2, we introduced some of the basic features of TABLE. These are all you will need to read the rest of the book. The TABLE command, however, can produce more sophisticated displays—tables with three or more classification variables printed in a variety of layouts. In this section, we will present several examples, each illustrating a feature of TABLE.

The LAYOUT Subcommand

The output in Exhibit 4.6 uses the LAYOUT subcommand. This table is similar to the one in Exhibit 4.5; it has the same two classification variables, and it has three statistics per cell. Now, however, the two classification variables are both used on the rows and no classification variables are used on the columns. This is what the LAYOUT subcommand says to do: put the first two classification variables on the rows, with OWNER first and SIZE nested within OWNER, and put the next 0 classification variables on the columns. Since there are no classification variables on the columns, Minitab printed the three statistics across the page to save space. For some purposes, this layout is more useful than the LAYOUT 1,1, shown in Exhibit 4.5.

Three or More Classification Variables

The output in Exhibit 4.7 contains three two-way tables, one for each value of TYPEFOOD. The first table gives a breakdown of the 108 fast food restaurants (TYPEFOOD = 1 is fast food) by OWNER and SIZE; the second table gives a breakdown of the 71 supper clubs (TYPEFOOD = 2), and the third table gives a breakdown of the 73 remaining restaurants (TYPEFOOD = 3).

In general, if there are three or more classification variables, the first is used for the rows, the second is used for the columns, and the remaining are used to determine the separate tables. This arrangement can be changed with the LAYOUT subcommand. For example,

```
TABLE 'OWNER' 'SIZE' 'TYPEFOOD';
  LAYOUT 2, 1.
```

Exhibit 4.6 Sales Volume by Type of Ownership and Size (in thousands of dollars)

```
TABLE 'OWNER' 'SIZE';
  STATS 'SALES';
  LAYOUT 2,0.

ROWS: OWNER / SIZE

                SALES      SALES      SALES
                COUNT       MEAN    STD DEV
 1
       1           75     115.55      76.58
       2           18     237.94     135.76
       3            2     579.00     217.79
 2
       1           16      88.12      64.65
       2            4     321.25     157.06
       3            4     695.00     145.26
 3
       1           38     231.50     544.47
       2           38     313.84     138.90
       3           47     887.96    1261.47
ALL
      ALL         242     339.00     663.40
```

puts the first two variables on the rows and the next one variable on the columns. No variables are left over, so there is just one table. Output is in Exhibit 4.8.

In Exhibit 4.8, the last row contains marginal counts summed over both OWNER and SIZE. In general, the marginal statistics used in a table are for the rows and columns printed in the table. Therefore, marginal statistics and percents (row, column, and total) will change when you change layouts.

Sometimes the marginal statistics are not needed and just clutter up a table. This is often true when there are three or more classification variables. The subcommand NOALL can be used to suppress them. For example,

Exhibit 4.7 Example of a Table with Three Classification Variables

```
TABLE 'OWNER' 'SIZE' 'TYPEFOOD'

CONTROL: TYPEFOOD =  1
ROWS: OWNER       COLUMNS: SIZE

              1          2          3        ALL

 1           32          9          1         42
 2            6          2          0          8
 3           20         19         19         58
ALL          58         30         20        108

CONTROL: TYPEFOOD =  2
ROWS: OWNER       COLUMNS: SIZE

              1          2          3        ALL

 1           23          7          0         30
 2            2          4          2          8
 3            9         13         11         33
ALL          34         24         13         71

CONTROL: TYPEFOOD =  3
ROWS: OWNER       COLUMNS: SIZE

              1          2          3        ALL

 1           26          2          1         29
 2            5          0          2          7
 3           10          9         18         37
ALL          41         11         21         73

 CELL CONTENTS --
                  COUNT
```

```
TABLE 'OWNER' 'SIZE' 'TYPEFOOD';
  LAYOUT 2, 1;
  MEAN 'SALES';
  NOALL.
```

Exhibit 4.8 Illustration of How the LAYOUT Subcommand Changes the Appearance of a Table

```
TABLE 'OWNER' 'SIZE' 'TYPEFOOD';
  LAYOUT 2,1.

ROWS: OWNER / SIZE       COLUMNS: TYPEFOOD

                 1          2          3        ALL
 1
        1       32         23         26         81
        2        9          7          2         18
        3        1          0          1          2
 2
        1        6          2          5         13
        2        2          4          0          6
        3        0          2          2          4
 3
        1       20          9         10         39
        2       19         13          9         41
        3       19         11         18         48
ALL
      ALL      108         71         73        252

 CELL CONTENTS --
                COUNT
```

TABLE C, . . . , C

Subcommands to control the output.

LAYOUT K, K

The first K says how many classification variables to use for the rows of the table, and the second K says how many classification variables to use for the columns. For example,

```
TABLE C4 C5 C8-C12;
  LAYOUT 3, 2.
```

says to use the first three variables, C4, C5, and C8, for the rows, and the next two variables, C9 and C10, for the columns. There are two variables left over, C11 and C12. A separate table is produced for each combination of entries in C11 and C12.

If no LAYOUT subcommand is given, LAYOUT 1,0 is used for tables with one classification variable, and LAYOUT 1,1 is used for tables with two or more classification variables.

The first K of LAYOUT may be 0,1, . . . ,10. The second K may be 0, 1, or 2.

NOALL

suppresses the printing of all marginal statistics.

5 Comparing Two or More Sets of Data

Many important questions involve comparisons between two or more sets of data.

Did the region with the high advertising budget have substantially greater sales than the region with the low budget?

How much stronger are the connections made with the new glue than those made with the old glue?

Is there any evidence that women are paid less than men for the same jobs? If so, how much less?

How much difference is there in the survival rate for patients on the new drug as compared to the survival rate on the standard treatment?

In this chapter we introduce some methods by which two or more sets of data can be compared. The emphasis is on exploratory and descriptive methods at this point. We begin by discussing the important distinction between *paired* data and *independent* samples.

5.1 Paired and Independent Data

A Pennsylvania medical center collected some data on the blood cholesterol levels of heart-attack patients. A total of 28 heart-attack patients had their cholesterol levels measured two days after the attack, four days after,

Table 5.1 Blood Cholesterol After a Heart Attack

Experimental Group			
2 Days After	*4 Days After*	*14 Days After*	*Control Group*
270	218	156	196
236	234	*	232
210	214	242	200
142	116	*	242
280	200	*	206
272	276	256	178
160	146	142	184
220	182	216	198
226	238	248	160
242	288	*	182
186	190	168	182
266	236	236	198
206	244	*	182
318	258	200	238
294	240	264	198
282	294	*	188
234	220	264	166
224	200	*	204
276	220	188	182
282	186	182	178
360	352	294	212
310	202	214	164
280	218	*	230
278	248	198	186
288	278	*	162
288	248	256	182
244	270	280	218
236	242	204	170
			200
			176

(Note: * indicates a missing value)
These data are in the saved worksheet called CHOLEST.

and 14 days after. In addition, cholesterol levels were recorded for a control group of 30 people who had not had a heart attack. Part of these data were given in Section 1.1; the full set is given in Table 5.1.

This data set contains both paired and independent variables. The columns headed "2 Days After" and "4 Days After" are *paired* since the numbers on a given row are related—they are both for the same patient.

For example, the first patient had a "2 Days After" score of 270 and a "4 Days After" score of 218. Because the observations are paired, we can determine how much each patient's cholesterol changed during that period.

Now, let's look at an example of independent or unpaired data. The cholesterol levels in the last column of Table 5.1 are for a control group of 30 people. The people in this group were not linked in any known way to any of the patients in the experimental group. Thus we say that the "2 Days After" and control group data are *independent.*

The distinction between paired and independent data is relatively clear in this example, as it is in most data sets. However, it is an important concept you should keep in mind because it determines which statistical procedures are appropriate for a given set of data.

Exercises

5–1 For the cholesterol data in Table 5.1, determine which of the following variables are paired and which are not:

(a) The "2 days after" and "14 days after."

(b) The "4 days after" and "control group."

(c) The "4 days after" and "14 days after."

(d) The "14 days after" and "control group."

5–2 In each of the following, indicate whether the data are paired or independent samples. Justify your answer.

(a) A national fast food chain used a special advertising campaign at half of its stores. It then compared sales at those stores with sales at its other stores.

(b) A total of 24 sets of mice were used in a study of a nutrition supplement. Each set consisted of two mice from the same litter. In each set, one mouse was given the supplement and one was not. After three weeks, the weight gain of each mouse was recorded for analysis.

(c) A tool manufacturer took a sample of ten new wrenches. He treated five of them with a special chemical. The other five were left untreated. The manufacturer then compared the strength of the treated and untreated tools.

(d) A total of 50 people were used in a study to compare two treatments for the rash produced by poison ivy. A small patch of skin on the left arm of each person was exposed to poison ivy. Once a rash had developed, treatment A was used. The improvement was assessed and recorded. Two months later, the same 50 people were again exposed to poison ivy. This time treatment B was used.

5–3 There are a number of data sets given in Appendix A. Find three examples of paired data and three examples of unpaired data. (Do not use any of the examples that were used in this chapter.)

5–4 In the Cartoon data (p. 303), which of the following comparisons involve paired data and which involve independent samples?

(a) Difference between color and black and white, as measured by the immediate cartoon score, and by the immediate realistic score.

(b) Difference between the OTIS scores in hospital A and hospital B.

(c) Difference between the OTIS scores of pre-professionals and professionals in hospital C.

(d) Difference between immediate cartoon and realistic scores for those participants who saw color slides.

(e) Difference between immediate and delayed cartoon scores for those participants who saw color slides.

5.2 Comparing Two Independent Sets of Data

In this section we show how HISTOGRAM, DOTPLOT, STEM-AND-LEAF, and DESCRIBE can be used to compare two independent sets of data.

Such data can be organized in two ways, *stacked* and *unstacked*. If each set of data is in a separate column of the worksheet, the data are said to be *unstacked*. If the two sets of data are both in the same column, with another column used to identify each set, the data are said to be *stacked*. We will consider unstacked data first, then stacked data, and then, at the end of this section, show how you can go back-and-forth between these two types.

Unstacked Data and the SAME Subcommand

As an example of unstacked data, we will use the two-day and control readings in Table 5.1. The data for the 28 patients with two-day readings are in the first column and the data for the 30 people in the control group are in the fourth column. Thus, these data are unstacked.

Exhibit 5.1 RETRIEVES the data and does two DOTPLOTS. Suppose we try to compare them. It's difficult, since they are on different scales; that is, the distance between the tick marks (the +'s) on the 2-DAY dotplot is 40 (equivalently, each space represents 4), whereas the distance for the CONTROL dotplot is just 15 (equivalently, each space represents 1.5). In addition the two plots start and end at different values: the scale for

Exhibit 5.1 Dotplots for the 2-DAY and CONTROL Cholesterol Levels

```
RETRIEVE 'CHOLEST'
DOTPLOT '2-DAY' 'CONTROL'
                                        .                   .
  .    .          .       .. ... : :          .:.::: .    . .           .
-----+---------+---------+---------+---------+---------+-2-DAY
    160       200       240       280       320       360

                    .
                    :                      .
... .  .     . : : ...        .::   ..    .    .         .  .    . .
---+---------+---------+---------+---------+---------+---CONTROL
  165       180       195       210       225       240
```

the 2-DAY scores starts at 140 (with the first labeled tick mark at 160) and ends at 364, whereas the CONTROL scale starts at 160.5 and ends at 244.5.

One way to make the two dotplots comparable is to use the subcommand SAME. This tells Minitab to figure out a scale that works for both dotplots. Exhibit 5.2 does a DOTPLOT with SAME and also uses DESCRIBE.

From this we easily can see the important differences between the two groups. The readings for the 2-DAY group are much more spread out than the readings for the CONTROL group. The 2-DAY group goes from 142 to 360. This gives a range of 360 − 142 = 218. The CONTROL group has a range of just 242 − 160 = 82. The center for the 2-DAY group is also much higher, with a median of 268 compared to 187 for the CONTROL group. In fact, over half of the 2-DAY levels were greater than the maximum level of 242 in the CONTROL group.

SAME

This subcommand can be used with the commands STEM-AND-LEAF, HISTOGRAM, and DOTPLOT. It is useful when you want to compare several columns of data. If you use SAME with STEM-AND-LEAF or HISTOGRAM, then the same scale will be used for all columns you list on the main command. If you use SAME with DOTPLOT, then not only are all plots given the same scale, but they are also put in the same display. If you do not use SAME, then each column listed on the main command line is scaled separately.

Exhibit 5.2 Dotplots and Descriptive Summaries for the 2-DAY and CONTROL Cholesterol Levels (Note the use of the subcommand SAME.)

```
DOTPLOT '2-DAY' 'CONTROL';
  SAME.

                        .          .
 .   .      .    .. ... : :     .:.::: .   . .         .
-----+---------+---------+---------+---------+---------+-2-DAY
           :   .
           :   :
     .:...::: .:... .  : ..
-----+---------+---------+---------+---------+---------+-CONTROL
    160       200       240       280       320       360


DESCRIBE '2-DAY' 'CONTROL'

                 N      MEAN    MEDIAN    TRMEAN     STDEV    SEMEAN
2-DAY           28    253.93    268.00    254.15     47.71      9.02
CONTROL         30    193.13    187.00    192.00     22.30      4.07

               MIN       MAX        Q1        Q3
2-DAY       142.00    360.00    224.50    282.00
CONTROL     160.00    242.00    178.00    204.50
```

Stacked Data and the BY Subcommand

The Pulse data (p. 318) provide a convenient example of stacked data. The 92 students were asked to flip a coin. If the coin came up heads, they were to run in place for one minute; if it came up tails, they were to do nothing. Then everyone took his/her pulse again. These 92 pulse rates are all in one column, named PULSE2. Another column, named RAN, indicates whether or not the person ran in place. It is coded 1 = yes and 2 = no. Exhibit 5.3 compares these two groups.

The BY subcommand tells HISTOGRAM to produce a separate histogram for each value of RAN. The first histogram is for the 35 observations where RAN = 1; the second histogram is for the 57 observations where RAN = 2. Notice that these two histograms are on comparable scales: They both have an increment of ten and start at 50. When we use the BY subcommand with stacked data, Minitab assumes we want to compare the displays and automatically puts them all on the same scale. Here are some things we can see from this exhibit:

Those who ran in place tended to have higher pulse rates than those who did not, as we would expect.

Exhibit 5.3 Histograms and Descriptive Summaries for the Second Pulse Rate

```
RETRIEVE 'PULSE'
HISTOGRAM 'PULSE2';
  BY 'RAN'.

Histogram of PULSE2   RAN = 1   N = 35

Midpoint   Count
      50       0
      60       1  *
      70       3  ***
      80      13  *************
      90       3  ***
     100       6  ******
     110       3  ***
     120       4  ****
     130       1  *
     140       1  *

Histogram of PULSE2   RAN = 2   N = 57

Midpoint   Count
      50       2  **
      60       9  *********
      70      24  ************************
      80      17  *****************
      90       5  *****
     100       0
     110       0
     120       0
     130       0
     140       0

DESCRIBE 'PULSE2';
  BY 'RAN'.

            RAN        N     MEAN   MEDIAN   TRMEAN    STDEV   SEMEAN
PULSE2        1       35    92.51    88.00    91.68    18.94     3.20
              2       57    72.32    70.00    72.24     9.95     1.32

            RAN      MIN      MAX       Q1       Q3
PULSE2        1    58.00   140.00    76.00   106.00
              2    50.00    94.00    66.00    79.00
```

Some who did not run in place had higher pulse rates than some who did.

There was a greater range of pulse rates among those who did run (58 to 140 for a range of 82), than among those who did not (50 to 94 for a range of 44).

The average pulse rate is 92.51 for those who ran, and 72.32 for those who didn't. Thus, running in place increased the pulse rate by about 20 beats per minute, on the average.

BY C

This subcommand can be used with the commands STEM-AND-LEAF, HISTOGRAM, DOTPLOT, and DESCRIBE. A separate display is produced for each different value in C. All are put on the same scale. The column C must contain integers from −10000 to +10000.

Stacking and Unstacking Data

In most cases you will be able to analyze your data the way it naturally comes, that is, either stacked or unstacked. However, on some occasions you may need to change the structure of your data from one form to the other. Appendix B shows how you can do this using the commands STACK and UNSTACK. When preparing your data for computer analysis, you should try to anticipate what you will need and structure your data accordingly.

Exercises

5–5 The following refer to Exhibit 5.2.

(a) About how many of the 2-DAY readings are larger than the highest CONTROL reading? About how many are smaller than the lowest CONTROL reading?

(b) What percentage of the 2-DAY readings are outside the range of the CONTROL readings?

(c) How much smaller is the median of the 2-DAY readings than the median of the CONTROL readings?

5–6 The data listed below were obtained in a study of tool life. Ten specimens were untreated and ten were treated by a new process thought to improve wear resistance. The specimens were not paired in any way. Tool wear was measured as volume loss in millionths of a cubic inch.

Untreated tools	.56	.50	.69	.59	.47	.42	.45	.47	.50	.50
Treated tools	.13	.13	.18	.23	.18	.31	.35	.23	.31	.33

(a) Display the two sets of data using the commands DOTPLOT, HISTOGRAM, and STEM-AND-LEAF. Use the subcommand SAME. Compare the usefulness of the three types of displays.

(b) DESCRIBE the two sets.

(c) Write a brief practical summary of the results from parts (a) and (b).

5–7 Comparisons between two groups can be made for several variables in the Pulse data (p. 318).

(a) Make stem-and-leaf displays of the first pulse rates using the subcommand BY to separate the males and the females.

(b) Repeat part (a), using dotplots.

(c) Does there seem to be any overall difference between the male and female pulse rates? Explain.

(d) Repeat (a) or (b), but this time separate the smokers from the nonsmokers.

5–8 Sometimes it is useful to split data into two groups based on an interval scale variable. In this exercise we will look at the Peru data (p. 317). Suppose we create a new variable, which has two values:

1 for those with 0 to 6 years since migration,

2 for those with 7 to 100 years since migration.

This will allow us to compare recent migrants with those who have had longer to get resettled. To create this variable, first suppose the Peru data have been entered into C1–C10. Then use CODE (see p. 339 for a full description of CODE) as follows:

```
CODE (0:6) TO 1 (7:100) TO 2 IN C2, PUT IN C12
```

(a) Use DOTPLOT and DESCRIBE to compare the weights of these two groups of people.

(b) Compare the systolic blood pressure of the two groups.

(c) Compare the ages.

(d) Interpret the results from (a), (b), and (c).

5–9 A seventh-grade student (named Kristy) was given an experiment to run. She was to complete the same basic task a series of times. Half the time she was to do the task in a quiet room and half the time she was to do it with the radio and TV on loud, while chewing gum.

	Time to Complete Task (Seconds)									
Quiet (Trials 1–10)	60	50	40	21	28	17	19	14	12	12
Noisy (Trials 11–20)	36	30	35	21	20	15	18	9	10	10

(a) Use DOTPLOT and DESCRIBE to compare the two sets of times. Discuss the results. Is one environment appreciably better than the other?

(b) The 20 trials were done in the order indicated: first trial 1, then trial 2, and so on. Use the command TSPLOT to plot these data in time order. What do you see?

(c) Discuss the planning of the experiment and how it could have been improved. Would you change your conclusions of part (a) after seeing the plot in (b)? Can you offer any advice to future analysts who might be comparing data like these?

5.3 Paired Data

A shoe company wanted to compare two materials (A and B) for use on the soles of boys' shoes. We could design an experiment to compare the two materials in two ways. One way might be to recruit ten boys (or more if our budget allowed) and give five of the boys shoes with material A and give five boys shoes with material B. Then after a suitable length of time, say three months, we could measure the wear on each boy's shoes. This would lead to independent samples. Now, you would expect a certain variability among ten boys—some boys wear out shoes much faster than others. A problem arises if this variability is large. It might completely hide an important difference between the two materials.

The other way, a paired design, attempts to remove some of this variability from the analysis so we can see more clearly any differences between the materials we are studying. Again, suppose we started with the same ten boys, but this time had each boy test both materials. There are several ways we could do this. Each boy could wear material A for three months, then material B for a second three months. Or we could give each boy a special pair of shoes with the sole on one shoe made from material A and the other from material B. This latter procedure produced the data in Table 5.2.

First, we will look at each material separately. Exhibit 5.4 contains an example. After looking at this output, would you feel comfortable saying that material A is definitely better than B? The mean wear for A is 10.63, which is certainly lower than 11.04, the mean for B. But, on the other hand, four of the ten shoes using material B did better than 10.63. Overall, this analysis does not provide very much evidence to say which material is better, A or B.

Using the Pairing

Exhibit 5.5 contains a display that makes use of the pairing. The MPLOT was used to plot MATL-A versus BOY and MATL-B versus BOY, both on the same axes. This plot shows that wear varied greatly from boy to boy.

Table 5.2 Wear for Boys' Shoes, Using a Paired Design

Boy	*Material A*	*Material B*
1	13.2	14.0
2	8.2	8.8
3	10.9	11.2
4	14.3	14.2
5	10.7	11.8
6	6.6	6.4
7	9.5	9.8
8	10.8	11.3
9	8.8	9.3
10	13.3	13.6

Boy 6 had the lowest wear—both shoes were well below 8. Boy 4 had the highest wear—both shoes were well above 14. For a given boy, however, both shoes had about the same amount of wear. Using statistical terminology, we would say, "The among boy variability was high but the within boy variability was low."

Perhaps the most important feature of the plot is the order of the letters; that is, in eight out of ten cases, the As are lower than the Bs, by roughly the same amount. Material A seems to wear better than material B for most of the boys.

Exhibit 5.4 Wear for Boys' Shoes. (The data first were read into C1–C3.)

```
NAME C1='BOY' C2='MATL-A' C3='MATL-B'
DOTPLOT C2 C3;
  SAME.

             .          .   .   .       ...              ..     .
         +---------+---------+---------+---------+---------+---------+-------MATL-A
            .               .  .  .         :   .           . . .
         +---------+---------+---------+---------+---------+---------+-------MATL-B
      6.0       7.5       9.0      10.5      12.0      13.5

DESCRIBE C2 C3

                   N     MEAN   MEDIAN   TRMEAN    STDEV   SEMEAN
MATL-A            10   10.630   10.750   10.675    2.451    0.775
MATL-B            10   11.040   11.250   11.225    2.518    0.796

                 MIN      MAX       Q1       Q3
MATL-A         6.600   14.300    8.650   13.225
MATL-B         6.400   14.200    9.175   13.700
```

Exhibit 5.5 Plot to Compare Wear for Boys' Shoes (This continues the session in Exhibit 5.4.)

```
MPLOT 'MATL-A' 'BOY', 'MATL-B' 'BOY'

             -                        2
      14.0+        B
             -                                                B
MATL-A       -     A                                          A
MATL-B       -
             -
      12.0+                            B
             -
             -               B                     B
             -               A         A           A
             -
      10.0+                                  B
             -                               A
             -                                          B
             -          B                               A
             -          A
       8.0+
             -
             -
             -                              A
             -                              B
        --+---------+---------+---------+---------+---------+----BOY
         0.0       2.0       4.0       6.0       8.0      10.0
```

In addition to looking at the original data, we can calculate a new quantity for each boy—the difference between the two materials. Many paired analyses focus on this difference. Exhibit 5.6 calculates the difference (B − A), then prints, and then describes it. We see that (B − A) has two negative values, −.2 and −.1, and eight positive values. These two negative values are from the two boys who had less wear with material B. The positive values are from the boys who had less wear with A. The mean of the differences was about .4. That is, on the average, material A did .4 wear units better than B. This could be expressed in percentage terms. We calculate (average difference)/(average wear for B) = .41/11.04 = .04 = 4%. That is, on the average, material A resulted in 4% less wear than material B.

The paired data analysis does not change the fact that the difference between the two materials is small. The wear for material A is just 4% lower than it is for material B. What the analysis does reveal is the relative consistency of the differences. In eight of ten cases, material A was slightly

Exhibit 5.6 The Difference, B − A, Between the Two Materials for Boys' Shoes (This continues the session in Exhibit 5.5.)

```
LET C4 = C3-C2
NAME C4 = 'B-A'
PRINT C1-C4

 ROW    BOY   MATL-A   MATL-B        B-A

   1      1     13.2     14.0    0.80000
   2      2      8.2      8.8    0.60000
   3      3     10.9     11.2    0.30000
   4      4     14.3     14.2   -0.10000
   5      5     10.7     11.8    1.10000
   6      6      6.6      6.4   -0.20000
   7      7      9.5      9.8    0.30000
   8      8     10.8     11.3    0.50000
   9      9      8.8      9.3    0.50000
  10     10     13.3     13.6    0.30000

DESCRIBE 'B-A'

                   N      MEAN    MEDIAN    TRMEAN     STDEV    SEMEAN
B-A               10     0.410     0.400     0.400     0.387     0.122

                 MIN       MAX        Q1        Q3
B-A           -0.200     1.100     0.200     0.650
```

better than B, and in the two cases where B did better, it was not very much better.

Concise Summary

The results from our analysis might be concisely described as follows:

Two materials, A and B, for making shoe soles were tested for wear in a paired design. Ten boys wore a special pair of shoes that had material A on one sole and B on the other. The wear on each shoe was recorded. On the average, the difference between the two materials was small; the mean and median differences were both about .4 wear units in favor of A. This represents a 4% difference. The advantage of A, however, was fairly consistent. Eight out of the ten boys had less wear with material A.

Among the ten boys, there was considerable variation in wear, with one boy having as little as 6.4 wear units and another having 14.3 units, more than twice as much.

Exercises

5–10 (a) Use MPLOT to compare the 2-DAY score with the 4-DAY score of the Cholesterol data in Table 5.1. First RETRIEVE the data. Then add a new variable, PERSON. Finally, use

```
MPLOT '2-DAY' 'PERSON' '4-DAY' 'PERSON'
```

Interpret the plot.

(b) About how much change is there, on the average, between the second- and fourth-day scores?

(c) Repeat the analysis in parts (a) and (b), comparing the second- and fourteenth-day scores. (Notice: Some fourteen-day scores are missing.)

(d) What effect might the missing data have on your answer to part (c)? Do you think patients who have a missing value at 14 days after a heart attack might differ from the rest of the patients? Do the data show any evidence of this?

5–11 In this exercise we will compare the immediate cartoon scores with the immediate realistic scores in the Cartoon data (p. 303). There were 179 people in this experiment. The MPLOT of Exhibit 5.5 is a good display when there are small to moderate numbers of observations (no more than 50 or 100), but will not work well with 179 observations. The technique of Exhibit 5.6 is more useful with large data sets. Thus, compute the difference between the cartoon and realistic scores for each participant, then make a histogram of these differences. Also, DESCRIBE the differences. Discuss your results.

5.4 Several Independent Groups

The procedures in Section 5.2 for comparing two independent sets of data work equally well for several sets. We will treat stacked data first, since it is more common.

Stacked Data

The Pulse data (p. 318) again provide a convenient example. Exhibit 5.7 shows how pulse rate depends on a person's usual level of physical activity. From the output we see that one person's activity level was recorded as 0. (Does this mean no activity?) We could not find out what the correct value was, so, in future analyses, we would change the 0 to *, for missing data.

Exhibit 5.7 PULSE1 for Each ACTIVITY Level in the Pulse Data Set (1 = low activity, 2 = moderate activity, 3 = high activity)

```
RETRIEVE 'PULSE'
DOTPLOT 'PULSE1';
  BY 'ACTIVITY'.

 ACTIVITY
 3
                           .       .
                       . . : :    : . : . : .     .  ..   .
         ---+---------+---------+---------+---------+---------+---PULSE1
                                  :
                          .       : .
 ACTIVITY              . :     : : : : . . . .        .
 2                  :    : :.: : : : : : : : : : : :    : . . . :   .
         ---+---------+---------+---------+---------+---------+---PULSE1
 ACTIVITY
 1
                                                      .
                         .   .        .   .   . .     :
         ---+---------+---------+---------+---------+---------+---PULSE1
 ACTIVITY
 0
         .
         ---+---------+---------+---------+---------+---------+---PULSE1
           50        60        70        80        90       100

DESCRIBE 'PULSE1';
  BY 'ACTIVITY'.

           ACTIVITY         N      MEAN    MEDIAN    TRMEAN     STDEV    SEMEAN
PULSE1            0         1    48.000    48.000    48.000         *         *
                  1         9     79.56     82.00     79.56     10.48      3.49
                  2        61     72.74     70.00     72.35     10.98      1.41
                  3        21     71.57     70.00     71.21      9.63      2.10

           ACTIVITY       MIN       MAX        Q1        Q3
PULSE1            0    48.000    48.000         *         *
                  1     62.00     90.00     70.00     90.00
                  2     54.00    100.00     65.00     80.00
                  3     58.00     92.00     63.00     77.00
```

Now let's look at the rest of the output. The three groups are certainly unbalanced—61 people, well over half, reported moderate activity while just 21 reported high activity, and only nine reported low activity. Perhaps these are the appropriate categories for the 92 participants but it is also possible that most people just classified themselves as "average" because they were not sure what else to do. In a more careful study, we might give each participant a questionnaire asking about different aspects of physical activity. Then, based on the answers, we would make a classification ourselves.

The three groups overlap a lot. In fact, when we first look at the dotplots, they do not seem very different. The output from DESCRIBE helps sharpen our perception a little: the low activity group, with a median pulse of 82, is somewhat higher than the other two groups, which both have a median around 70. This result is consistent with medical findings; people who exercise tend to have a lower resting pulse rate than those who are sedentary.

Unstacked Data

In Section 3.3 (pp. 68–70), we looked at data from a device used to measure radiation. Using TSPLOT, we found a problem with those data. Table 5.3

Table 5.3 Radiation Exposure Rates (picocuries per liter) from Radon, Measured by Four Different Devices

Badge: The bare device (looks like a 3/4 inch piece of clear tape)

Open Cup: The badge is partially shielded by being placed on the inside bottom of a paper cup

Membrane: Same as open cup, with plastic membrane cover used to keep out all dust

Filter: Similar to membrane, but specially designed to filter out all but radon radiation

Filter	*Membrane*	*Open Cup*	*Badge*
26	45	36	21
21	33	34	23
16	26	33	27
28	46	29	29
27	25	30	24
19	32	36	26
21	33	28	27
26	22	34	26
25	30	34	24
16	26	33	25
20	37	33	31
25	34	29	28
25	44	28	30
21	43	27	30
23	35	27	31
26	38	30	29
21	38	29	29
23	45	28	29
17	39	31	30
25	39	27	28

These data are in the saved worksheet called RADON.

gives some more data from that study; no problems have been found with this second set.

In this part of the study, four different types of devices were used: filter, membrane, open cup, and badge. Twenty devices of each type (for a total of 80 devices) were put in a chamber and subjected to constant radiation. The amount of radiation measured by each device was recorded. How do the four devices compare?

These data were entered into C1–C4, which were given appropriate names. Since the data for each device are in a separate column, this data set is unstacked. Exhibit 5.8 shows dotplots and descriptive statistics. There we see that the filter measurements tended to be quite a bit lower than the others.

The means of the measurements seem to progress in roughly equal steps from filter, to badge, to open cup, to membrane. The difference

Exhibit 5.8 Dotplots and Descriptive Statistics for Radiation Measurements

```
NAME C1 = 'FILTER' C2 = 'MEMBRANE' C3 = 'OPEN CUP' C4 = 'BADGE'
DOTPLOT C1-C4;
  SAME.

                      :       :.
          :.    .. :  :     :: . .
     -----+---------+---------+---------+---------+---------+-FILTER
                       .    .:      .  . : ..   .: :      .. : .
     -----+---------+---------+---------+---------+---------+-MEMBRANE
                               . ..      . .
                               : :: :  .  : :   :
     -----+---------+---------+---------+---------+---------+-OPEN CUP
                                  : .
                     .  . : .: : :: : :
     -----+---------+---------+---------+---------+---------+-BADGE
         18        24        30        36        42        48

DESCRIBE C1-C4

                N      MEAN    MEDIAN    TRMEAN     STDEV    SEMEAN
FILTER         20    22.550    23.000    22.611     3.663     0.819
MEMBRANE       20     35.50     36.00     35.67      7.21      1.61
OPEN CUP       20    30.800    30.000    30.722     3.054     0.683
BADGE          20    27.350    28.000    27.500     2.815     0.629

              MIN       MAX        Q1        Q3
FILTER     16.000    28.000    20.250    25.750
MEMBRANE    22.00     46.00     30.50     42.00
OPEN CUP   27.000    36.000    28.000    33.750
BADGE      21.000    31.000    25.250    29.750
```

between the means of the filter and membrane measurements is quite large —the membrane mean is about 60% higher.

Now let's look at the variation within each type of device. The measurments for the filter vary from a low of 16 to a high of 28, giving a range of 12. The standard deviation is 3.66. The variation for the open cup and badge are similar to the filter: a range of about 10 and a standard deviation of about 3. The variation for the membrane, however, is about twice as much: a range of 24 and a standard deviation of 7.21.

Exercises

5–12 In the Cartoon experiment (p. 303), there are three levels of education: preprofessionals, professionals, and college students. Do these three groups have the same general ability as measured by their OTIS scores? First consider whether the data are stacked or unstacked, then use DOTPLOT and DESCRIBE to compare the three groups.

5–13 In Chapter 4, we discussed data from a restaurant survey. The restaurants were classified by the variable TYPEFOOD as fast food, supper club, and other.

(a) Do these three groups spend about the same percent of their sales on advertising? Do appropriate displays and statistics using the variable ADS.

(b) Repeat part (a), using the variable WAGES.

(c) Repeat part (a), using the variable COSTGOODS.

(d) Interpret and summarize the results of parts (a), (b), and (c).

5–14 Several years ago there was a dispute between teachers and school administrators in the Madison, Wisconsin public schools. One day many teachers called in "sick" as a protest. The following figures, giving a breakdown by school, were reported in a local newspaper, *The Capital Times*, on December 12, 1975.

Elementary School	*Number of Teachers in the School*	*Number Who Reported Sick*
Allis	35	7
Badger	23	22
Crestwood	22	12
Elvehjem	30	15
Emerson	27	21
Falk	28	16
Franklin	35	23

Elementary School	*Number of Teachers in the School*	*Number Who Reported Sick*
Glendale	31	23
Gompers	29	17
Hawthorne	30	23
Hoyt	18	8
Huegel	18	16
Kennedy	35	7
Lake View	27	3
Lapham	36	21
Leopold	50	30
Lindbergh	19	13
Longfellow	25	2
Lowell	29	7
Marquette	20	12
Mendota	31	18
Midvale	26	14
Muir	24	14
Odana	24	0
Orchard Ridge	34	15
Randall	29	16
Sandburg	18	13
Schenk	23	21
Sherman	22	10
Shorewood	29	16
Spring Harbor	15	11
Stephens	19	4
Thoreau	23	3
Van Hise	17	0

Middle School	*Number of Teachers in the School*	*Number Who Reported Sick*
Cherokee	47	31
Gompers	50	44
Jefferson	50	21
Lincoln	39	22
Marquette	52	47
Orchard Ridge	48	35
Schenk	49	39
Sennett	53	18
Sherman	45	35
Van Hise	48	33

High School	*Number of Teachers in the School*	*Number Who Reported Sick*
East	150	36
LaFollette	127	70
Memorial	120	82
West	140	83

Note: These data are in the saved worksheet called SCHOOLS. The first variable codes the type of school using 1 for elementary schools, 2 for middle schools, and 3 for high schools.

(a) Compare the number of teachers who reported sick in the three types of schools. Do these three groups seem different? Use appropriate displays and statistics.

(b) Calculate the percentage of teachers who reported sick in each school. Again, use appropriate displays and statistics to compare the three types of schools. Now do they appear different? Compare this analysis to the one in part (a).

5.5 Independent Groups, Two Classification Variables

In Section 5.4, we analyzed PULSE1 classified by ACTIVITY. Often we want to analyze data classified by two (or even more) variables. For example, let's see how the second pulse rates depended on both the usual activity level and whether or not the student ran in place.

The TABLE Command and Hand-Drawn Plots

Numerical summaries are easy with the TABLE command. Exhibit 5.9 gives an example using the Pulse data. (Note: We changed the 0 activity level for the 54th participant to *, so all analyses in this section are for the 91 participants with good data.) Exhibit 5.10 gives a plot of the mean pulse rates taken from the table. We did the plot by hand. Many times it is easier, faster, and clearer to do hand work than to use the computer.

The displays used in Section 5.2 (two groups) and in 5.4 (several groups) also work for data with two classifications. All that is required is the following little trick.

To continue with the same example, suppose we display PULSE2 classified by both ACTIVITY and RAN. These two variables divide the data set into $3 \times 2 = 6$ groups. What we want to do is create a new variable that indicates these six groups. We will use the codes given in Table 5.4.

Exhibit 5.9 Numerical Summaries with Two Classification Variables Using the Pulse Data

```
TABLE 'RAN' 'ACTIVITY';
   COUNT;
   MEAN 'PULSE2'.

ROWS: RAN      COLUMNS: ACTIVITY

              1         2         3       ALL

 1            3        25         7        35
         85.333    98.080    75.714    92.514

 2            6        36        14        56
         80.667    72.167    70.429    72.643

ALL           9        61        21        91
         82.222    82.787    72.190    80.286

 CELL CONTENTS --
                    COUNT
             PULSE2:MEAN
```

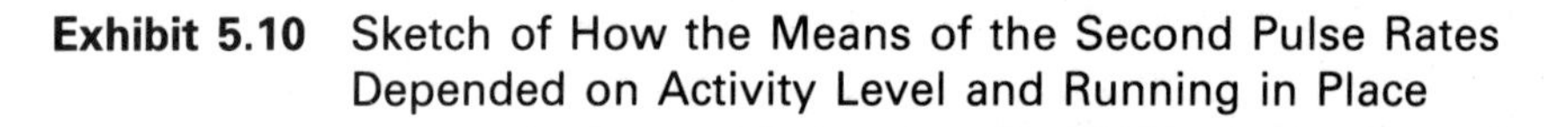

Exhibit 5.10 Sketch of How the Means of the Second Pulse Rates Depended on Activity Level and Running in Place

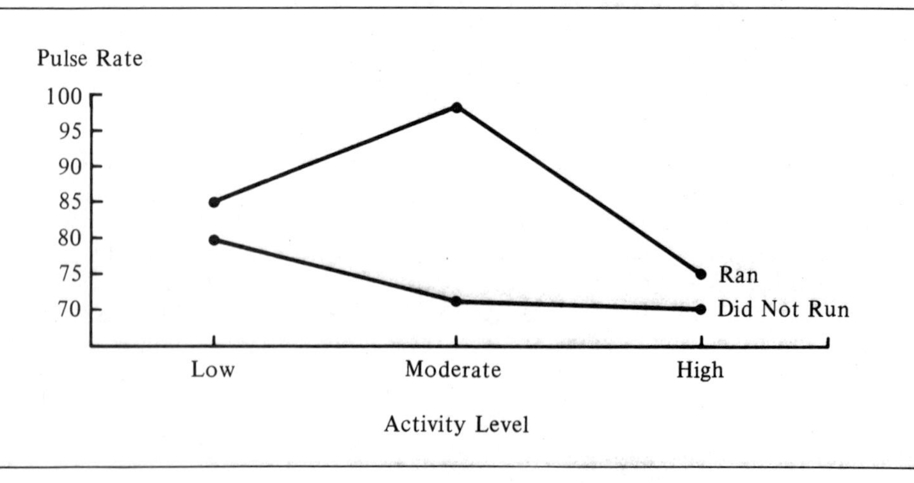

Table 5.4 Codes for the Six Groups Determined by ACTIVITY and RAN

	Activity Level		
	Low (1)	*Moderate (2)*	*High (3)*
Ran (1)	1	3	5
Did Not Run (2)	2	4	6

There are many ways to create this new variable—including just typing in the codes for all 92 participants. Here is a little trick that does it in just one LET command:

```
LET C10='RAN'+2*('ACTIVITY'-1)
```

You might try some values to see how this works. For example, "did not run" with "high activity" is calculated by $2 + 2 * (3-1) = 6$.

Once we have this new variable, we can use it in commands such as:

```
DOTPLOT 'PULSE2';
  BY C10.
DESCRIBE 'PULSE2';
  BY C10.
```

The output from this DOTPLOT is shown in Exhibit 5.11.

Let's look at this trick in general. Suppose we have a data set in which C1 has the values 1, 2, and 3, and C2 has the values 1, 2, 3, and 4. Then there are 12 possible categories altogether. To compute a new variable that classifies the data into the 12 categories we could use

```
LET C6 = C1 + 3*(C2-1)
```

or

```
LET C7 = C2 + 4*(C1-1)
```

The displays produced using C6 will be in a different order than those produced using C7. In general it is useful to do both—often one order gives a more informative picture than the other.

As another example, if C5 has the values 1, 2, 3, 4, 5, and 6, and C6 the values 1, 2, then we could use

```
LET C10 = C5 + 6*(C6-1)
```

or

```
LET C10 = C6 + 2*(C5-1)
```

Exhibit 5.11 Dotplots for PULSE2 Classified by Both RAN and ACTIVITY

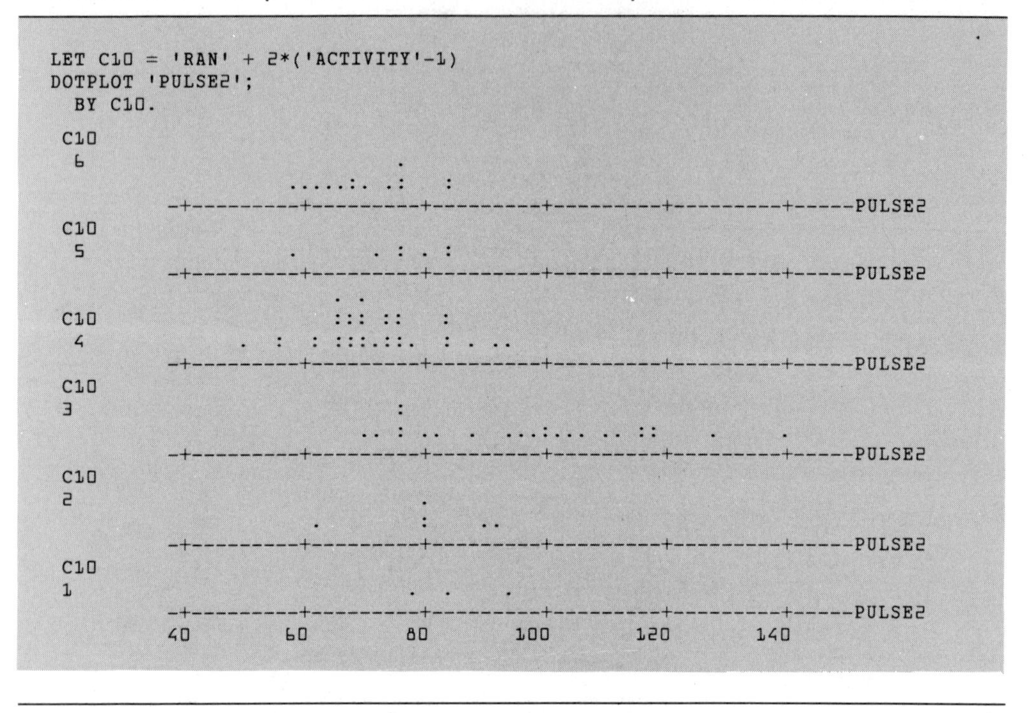

In general, if one column (say C1) has the values 1 to M and the other column (say C2) has the values 1 to N, then we can use either

```
LET C10 = C1 + M*(C2-1)
```

or

```
LET C10 = C2 + N*(C1-1)
```

Exercises

5–15 Refer to Exhibit 5.9.

(a) What is the mean pulse rate for those with the highest activity level who ran in place? How many people were there in the group?

(b) Which group had the highest mean pulse rate? The lowest? The most people? The least?

(c) How many people ran in place? What was their mean pulse rate? How much higher was it than those who did not run?

5–16 The following refer to Exhibit 5.11.

(a) Which group had the highest pulse rate? The lowest? What are these values (approximately)?

(b) One group seems to have a lot of high pulse rates. Which group? How many of the people in this group had a pulse rate higher than the maximum pulse rate in the other five groups? What percent is this of the people in this group?

5–17 Exhibits 5.9 and 5.10 summarize the variable PULSE2 when the data are classified by RAN and ACTIVITY. Do a similar summary for the variable PULSE1 classified by SMOKES and ACTIVITY. Interpret the results.

5–18 Using PULSE1 classified by SMOKES and ACTIVITY, produce a display of dotplots like the one in Exhibit 5.11.

5–19 Produce displays like those in Exhibits 5.9 and 5.10 for the variable PULSE1 classified by SEX and ACTIVITY. Interpret the results.

5.6 Comparing Several Related Groups

In Section 5.3, we analyzed a paired experiment to compare the amount of shoe wear for two different materials. Here we show how MPLOT and LPLOT can be used to compare more than two groups.

Unstacked Data and the MPLOT Command

The data in Table 5.5 give the elasticity of billiard balls made under three different conditions. As in the shoe study (pp. 101–104), the experimenters tried to remove variability by using a paired or blocked design. The term *blocked* is often used instead of paired when there are more than two groups in the experiment. The experiment was carried out as follows:

Table 5.5 Elasticity of Billiard Balls

Batch	*Control*	*Additive A*	*Additive B*
1	51	75	39
2	45	89	43
3	49	73	51
4	66	84	34
5	53	66	54
6	41	85	43
7	58	73	42
8	56	71	41
9	60	78	37
10	63	65	44

Ten batches of melted plastic were prepared. (These are the ten blocks of material.) Each batch was divided into three equal portions. One portion was chosen at random and set aside as a control. A second portion was chosen at random from the remaining two, and mixed with additive A. The third portion was mixed with additive B. In this way, the experimenters hoped to balance out any variations in the plastic from batch to batch. Elasticity was measured on a scale from 0 to 100, with the higher numbers representing greater elasticity. (Higher elasticity is considered more desirable.)

To plot these data, we entered them into C1-C4 and used

```
MPLOT C3 VS C1, C4 VS C1, C2 VS C1
```

(Minor details sometimes make a difference. Here we noticed that the above ordering of the columns on the MPLOT command gave nicer symbols: A for additive A, B for additive B, and C for Control.)

The results, in Exhibit 5.12, seem clear enough from the plot. Additive A improved elasticity, while additive B reduced it. There was no appreciable batch-to-batch variability, so there was little, if any, gain from balanc-

Exhibit 5.12 MPLOT of Elasticity of Billiard Balls

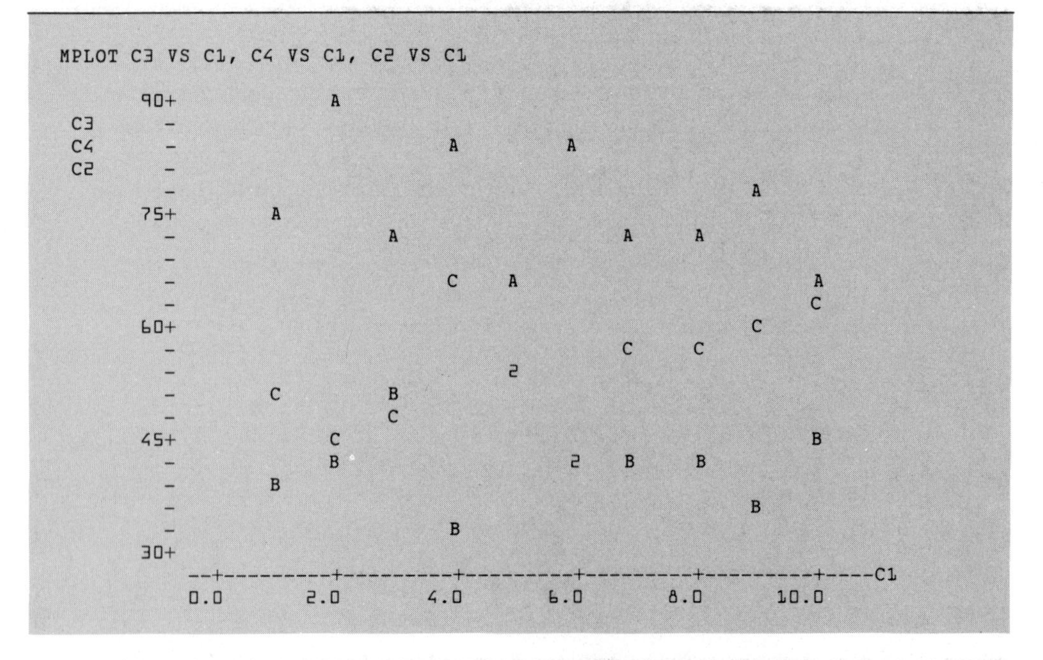

ing the experiment to make sure that each condition was tried in each batch. This contrasts with the boys' shoe experiment, in which there was considerable gain from the pairing.

Stacked Data and the LPLOT Command

An experiment was performed at the University of Wisconsin to compare the yield of six varieties of alfalfa. Four fields were used. Each field was divided into six plots, one plot for each variety. The four fields can be considered to be four blocks. They are analogous to the ten batches in the billiard ball experiment and the ten boys in the shoe experiment. The six alfalfa varieties are analogous to the three conditions for making billiard balls. The total yield from each combination is given in Table 5.6. The following commands show how we originally entered these data. We then saved the worksheet as ALFALFA. You can use this saved worksheet if you want to analyze these data.

```
NAME C1 = 'YIELD' C2 = 'VARIETY' &
     C3 = 'FIELD'
SET 'YIELD'
  3.22  3.04  3.06  2.64  3.19  2.49
  3.31  2.99  3.17  2.75  3.40  2.37
  3.26  3.27  2.93  2.59  3.11  2.38
  3.25  3.20  3.09  2.62  3.23  2.37
END
SET 'VARIETY'
  1 2 3 4 5 6 1 2 3 4 5 6 1 2 3 4 5 6
  1 2 3 4 5 6
END
SET 'FIELD'
  1 1 1 1 1 1 2 2 2 2 2 2 3 3 3 3 3 3
  4 4 4 4 4 4
END
SAVE 'ALFALFA'
```

Next we typed

```
TABLE 'VARIETY' 'FIELD';
  DATA 'YIELD'.
```

This prints a table similar to the one in Table 5.6 and allows us to see if we made any errors when we typed in the data. Then we made two plots:

```
LPLOT 'YIELD' VS 'VARIETY' symbols 'FIELD'
LPLOT 'YIELD' VS 'FIELD' symbols 'VARIETY'
```

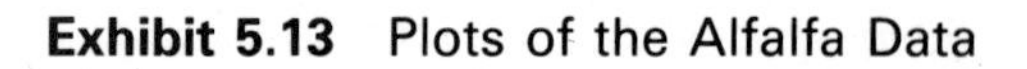

Exhibit 5.13 Plots of the Alfalfa Data

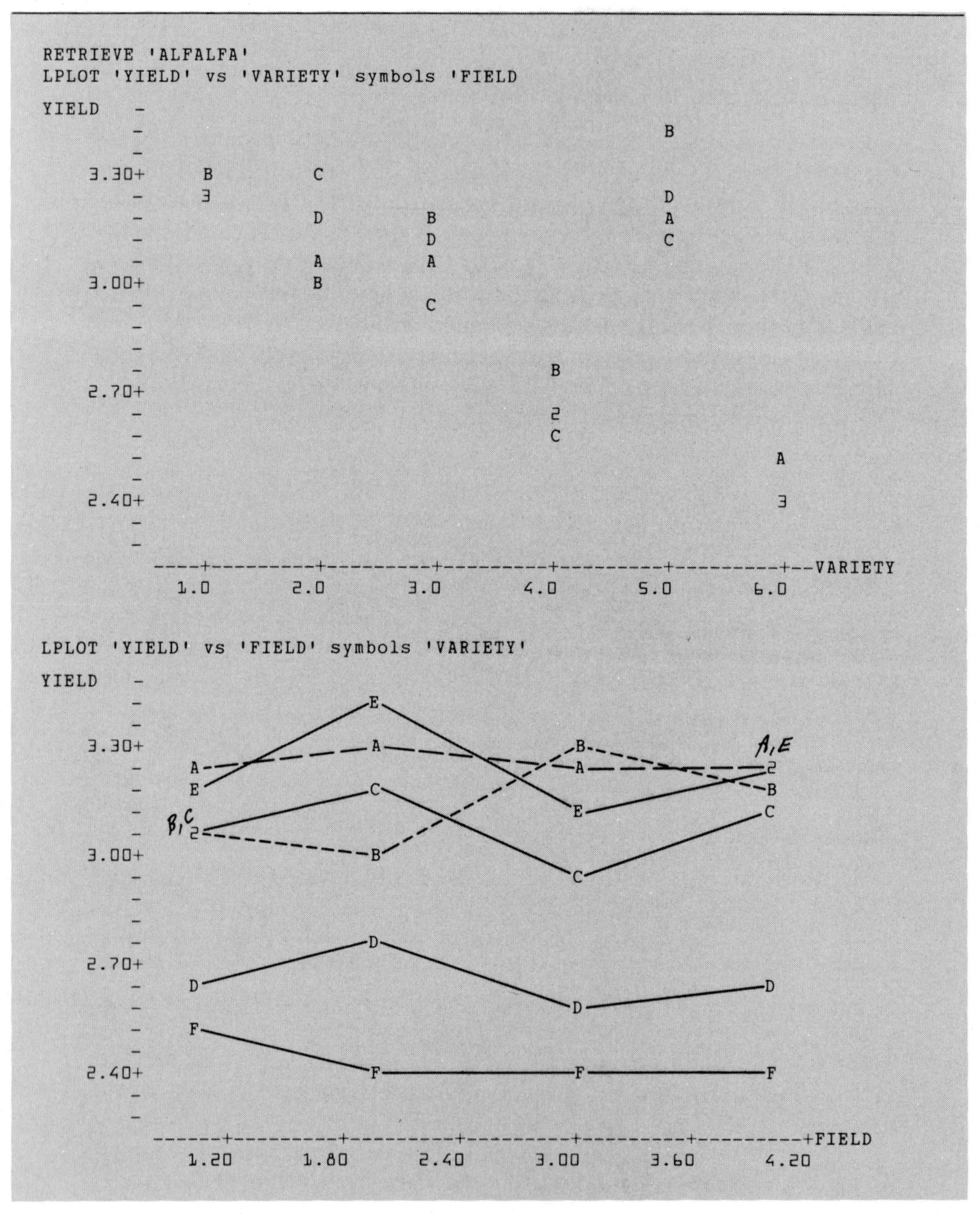

Table 5.6 Alfalfa Yield, in Tons per Acre

	Field			
Alfalfa Variety	*1*	*2*	*3*	*4*
Atlantic	3.22	3.31	3.26	3.25
Buffalo	3.04	2.99	3.27	3.20
Culver	3.06	3.17	2.93	3.09
Lohontar	2.64	2.75	2.59	2.62
Narragansett	3.19	3.40	3.11	3.23
Rambler	2.49	2.37	2.38	2.37

These data are in saved worksheet ALFALFA.

The output is shown in Exhibit 5.13. Each LPLOT gives slightly different insight into the structure of the data. The first one clearly shows that the next to last variety, Narragansett, had the best yield overall, although it was not the best in each field. No one field had the highest yield on all six varieties. For Atlantic, field B had the highest yield and A had the lowest, whereas for Rambler, field A had the highest yield and B was tied for the lowest.

The second LPLOT shows how the six varieties did on each field. Here we connected the letters by hand. This sometimes makes it easier to see patterns. We see that on every field, A and E were both better than C, which was better than D, which was better than F. Variety B does not show such a clear pattern. It is certainly better than both D and F. On Field 3, it was tied for best, but on Field 2 it was below A, C, and E.

Note: We also could have produced these two plots with the MPLOT command. To do the first plot in Exhibit 5.13, we READ the data in Table 5.6 into C1–C4, one field per column. Then SET the integers 1, 2, . . . , 6 into C5 to represent variety. We then use

```
MPLOT C1 C5, C2 C5, C3 C5, C4 C5
```

To do the second plot in Exhibit 5.13, we READ the data into C1–C6, one variety per column, and SET 1, 2, 3, 4 into C7 to represent field. Then we use

```
MPLOT C1 C7, C2 C7, C3 C7, C4 C7, C5 C7, C6 C7
```

As we mentioned in Chapter 3, the only difference between MPLOT and LPLOT is in how you organize the data in the worksheet.

6 Statistical Distributions and Simulation

6.1 Bernoulli Trials

There are many cases where an outcome can be classified into two categories: a coin falls either heads or tails; the next person to walk into a store either buys something or does not; the next child born in a hospital is either a girl or a boy; a tomato seed either germinates or does not. These two categories are often labeled, somewhat arbitrarily, as "success" (coin falls heads, person buys something, etc.) and "failure" (coin falls tails, person does not buy, etc.).

Now suppose we imagine a sequence of, say, 20 trials: We toss a coin 20 times or we observe the next 20 people who walk into a store. If the chance of a success remains the same from trial to trial, independent of the outcomes on all previous trials, we say we have Bernoulli trials. We use the letter p to represent the probability of a success on a single trial and the letter q to represent the probability of a failure.

One way to learn about Bernoulli trials is to toss a coin 20 times, or observe the next 20 people who walk into a store. Unfortunately, even studies as simple as these can be time consuming. So, instead of actually doing a study, we can use the computer to simulate the results. Minitab's RANDOM command will simulate a sequence of Bernoulli trials as well as data from other distributions.

Example

Suppose we spin a roulette wheel which has 38 different possible stopping points. If it stops on any of 18 points, we win. Otherwise, we lose. Thus,

RANDOM K observations into each of C, ..., C

Bernoulli p = K

Simulates observations from K Bernoulli trials into each column. The value *p* is the probability of a success on each trial. Each success is assigned the value 1, and each failure is assigned the value 0.

the probability of a success is 18/38 = .4737. Exhibit 6.1 uses the command RANDOM to simulate 20 spins. The results from these 20 trials were stored in C1 and then printed. Here a 1 represents a win and a 0 represents a loss. We did not do too well in this simulation—we lost on 13 spins and won on just 7.

Random Sequences

It is important to understand that randomness has to do with the process that generates observations, not the observations themselves. For example, tossing a coin is a good way to generate a random sequence; so is observing the sex of the next 20 children born in a hospital. However, even if the data come from a random process, this still does not guarantee that a particular set of outcomes will look random. It is possible to toss a coin 30 times and get a sequence of 15 heads followed by 15 tails, or a sequence that alternates head, tail, head, tail, ... Such clearly patterned sequences are possible, but not very likely if the process that generated the observations is truly random.

Exhibit 6.1 Simulation of Twenty Spins of a Roulette Wheel (1 Represents a Win, 0 Represents a Loss)

```
RANDOM 20 C1;
  BERNOULLI .4737.
PRINT C1
C1
 0    1    1    0    0    0    0    0    1    0    0    1    0    0    1
 0    0    1    0    1
```

Exhibit 6.2 gives nine sequences we generated with RANDOM. In the first three we used $p = .2$, in the second three we used $p = .5$, and in the last three we used $p = .8$. In all nine sequences, the 0s and 1s have the appearance of randomness. No clear patterns are evident. The sequences in C7–C9 have more 1s than the sequences in C1–C3. This is because the value of p was larger for C7–C9 than it was for C1–

Exhibit 6.2 Simulation of 9 Bernoulli Sequences with Different Probabilities of Success

```
RANDOM 20 C1-C3;
  BERNOULLI .2.
RANDOM 20 C4-C6;
  BERNOULLI .5.
RANDOM 20 C7-C9;
  BERNOULLI .8.
PRINT C1-C9
 ROW   C1   C2   C3   C4   C5   C6   C7   C8   C9

   1    0    1    0    0    0    0    1    0    0
   2    0    0    0    1    1    1    1    1    1
   3    0    0    0    1    0    1    1    0    0
   4    0    0    1    0    1    1    0    1    1
   5    0    0    0    1    1    1    1    1    1
   6    0    0    1    1    0    1    1    1    1
   7    0    0    0    0    1    0    1    1    1
   8    0    0    0    1    1    1    1    0    1
   9    0    0    0    1    1    0    1    1    1
  10    1    0    1    0    0    1    1    1    1
  11    1    0    0    1    0    0    1    1    1
  12    0    0    0    0    1    1    1    1    1
  13    0    1    0    1    0    1    1    1    0
  14    0    1    0    1    1    1    1    1    1
  15    0    0    0    0    0    1    1    1    0
  16    0    0    0    0    0    0    1    1    0
  17    0    0    0    0    1    1    1    1    0
  18    0    1    0    0    0    0    1    1    1
  19    0    0    0    0    1    1    1    1    1
  20    0    0    1    0    1    1    1    1    1
```

C3. The value of p determines how many 1s you are likely to get. It does not, however, have anything to do with randomness.

Now suppose we collected some real data. How could we know if we have a random sequence? In practice, we can never know for certain. We can try to understand how the data arose or how they were collected and then try to determine if it was likely they came from a random process. We also can look at the sequence itself. If we see a clear pattern we might suspect that the data did not come from a random process.

The Base for the Random Number Generator

BASE = K

To simulate random data, Minitab uses a mathematical function that creates a very long list of numbers which appear to be random. When you use RANDOM, Minitab haphazardly chooses a place to start reading from this list. Each time you use the RANDOM command, Minitab chooses a different place to start and therefore you get a different set of simulated data.

Occasionally you may want to generate the same set of "random" data several times. To do this, type the BASE command, with the same value of K each time, before using RANDOM. The value of K tells Minitab where to start reading in its list of random numbers. (Note: K determines where in the list to start. It is not the first random number Minitab gives you.) If you type two RANDOM commands in the same program, the second RANDOM will continue reading the list of random numbers where the first RANDOM stopped.

K should be a positive integer. The BASE command applies to all distributions used with RANDOM.

Exercises

6–1 In the roulette wheel example of Exhibit 6.1, how long was the longest winning streak? The longest losing streak? What would have been the best time to quit; that is, when was the gambler in his best financial condition? What were his winnings or losings at that point?

6–2 Suppose a basketball team wins a game, on the average, 55% of the time. Suppose we make the simplifying assumptions that there is a 55% chance of winning each game and that these odds are not influenced by what the team has done so far. Then we can simulate a "season" for the team using Minitab. Suppose a season contains 30 games. Simulate one season.

(a) Did the team have a winning season (win more games than it lost)?

(b) How long was the team's longest winning streak? How long was its longest losing streak?

(c) Simulate four more seasons and answer the questions in (a) and (b) for each season. Roughly how much variation should you expect from season to season in this team's performance, even when its basic winning capability is unchanged?

6–3 Suppose someone gives you a coin which is loaded so that heads comes up more often than tails. You want to know exactly how loaded the coin is. To estimate the probability of a head, p, you could toss the coin, say, 100 times.

(a) How could you simulate this, assuming that p is really .6? How could you estimate p from the data? Will your estimate be exactly equal to p?

(b) Suppose you asked five friends to estimate the probability of a head by having each toss the coin 100 times. Will their estimates all be equal to p? Will they all get the same estimate? How can you simulate the results for all five friends?

(c) Perform the simulation for you and your five friends. How much do the six estimates vary?

6–4 The baseball World Series pits the winner of the American League against the winner of the National League. These two teams play until one team wins a total of four games.

(a) Pretend that the two teams are perfectly matched and there is a 50–50 chance of each team winning any given game. Further, suppose the games are independent; that is, the chance of winning a particular game does not depend on whether other games have been won or lost. Simulate 20 World Series by using the instructions

```
RANDOM 20 series into C1-C7;
  BERNOULLI p = .5.
PRINT C1-C7
```

Now pretend a 1 means the National League won that game and 0 means the American League won. For each series, figure out which team won the series (i.e., won four games first). In how many series did the American League win? The National League? How many series lasted only four games? How many lasted five games? Six games? Seven games?

(b) Repeat (a) but now pretend the National League had a probability of $p = .6$ of winning each game. How do these results compare with those in (a)?

6.2 The Normal Distribution

Probability Density Function

The normal distribution is probably the most important distribution in statistics. Many populations in nature are well approximated by the normal. As an example, consider the data in Table 6.1. Exhibit 6.3 contains a bar chart (this is very similar to a histogram) of the chest girth proportions. The area of each bar is equal to the proportion of men with the corresponding chest girth. For example, the bar at 32 has a base of 1 and a height of .0673. This gives an area of $1 \times .0673 = .0673$, which is the proportion of men with a chest girth of 32 inches.

In Exhibit 6.4 we drew a curve that approximates the bar chart of 6.3. The technical term for this curve is the probability density function (pdf) of the normal distribution. The equation for this curve is given in most statistics textbooks and on page 152 of this *Handbook*. This formula

Table 6.1 Chest Girth for 1516 Soldiers of the 1835 Army of the Potomac

Chest Girth (in inches)	*Number of Soldiers with this Girth*	*Proportion with this Girth*	*Cumulative Proportion with this Girth*
28	2	.0013	.0013
29	4	.0026	.0040
30	17	.0112	.0152
31	55	.0363	.0515
32	102	.0673	.1187
33	180	.1187	.2375
34	242	.1596	.3971
35	310	.2045	.6016
36	251	.1656	.7672
37	181	.1194	.8865
38	103	.0679	.9545
39	42	.0277	.9822
40	19	.0125	.9947
41	6	.0040	.9987
42	2	.0013	1.0000

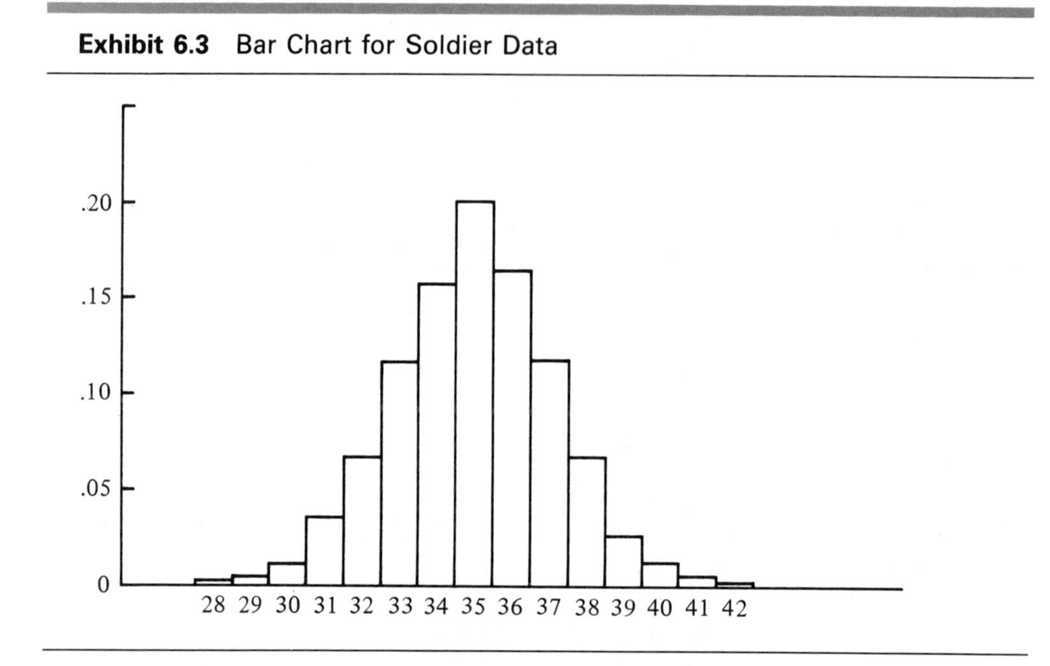

contains two parameters—the mean μ and the standard deviation σ. The center of the normal curve is at μ and the spread is specified by σ. Large values of σ give curves that are spread out; that is, short and fat. Small values of σ give curves that are tall and thin. The curve in Exhibit 6.4 is the pdf of a normal distribution with $\mu = 35$ and $\sigma = 2$. The following commands helped us sketch this curve.

```
SET C1
  28 28.5 29 29.5 30 30.5 31 31.5 32 32.5
  33 33.5 34 34.5 35 35.5 36 36.5 37 37.5
  38 38.5 39 39.5 40 40.5 41 41.5 42
END
PDF C1, put into C2;
  NORMAL 35, 2.
PRINT C1 C2
```

Since our data go from 28 to 42, we calculated the normal pdf for values from 28 to 42. The normal pdf is essentially zero for values below 28 and above 42 (you might check this out yourself using the PDF command). We needed to decide how many values between 28 and 42 to use; we needed enough to be able to draw a smooth curve. Going by

Exhibit 6.4 Bar Chart and Approximating Normal Distribution for Soldier Data

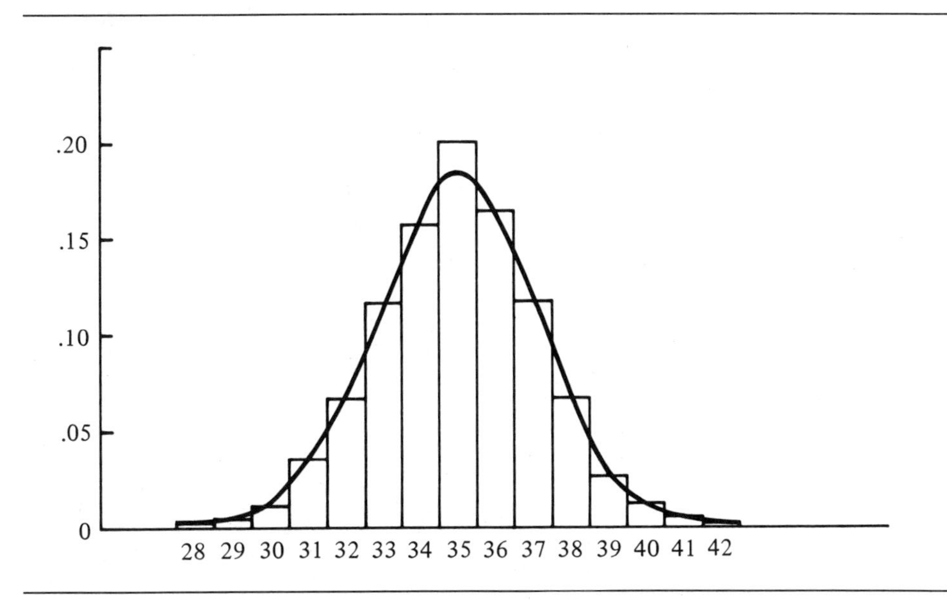

.5s seemed about right, so we used SET to put these values into C1. PDF calculated the height of the normal curve at each value and stored these heights in C2. The subcommand NORMAL specifies which parameters to use: $\mu = 35$ and $\sigma = 2$. We then printed the values. The results are in Exhibit 6.5. We plotted these points on the bar chart and drew a smooth curve through them.

PDF for values in E [put results into E]

NORMAL with mu = K, sigma = K

Calculates a probability density function. The subcommand NORMAL states you want the pdf for a normal distribution with the specified mean and standard deviation. If you do not use any subcommand, PDF calculates values for the "standard" normal distribution with $\mu = 0$ and $\sigma = 1$. If you do not store the results, they are printed. If you do store them, they will not be printed.

Exhibit 6.5 PDF for a Normal Distribution with $\mu = 35$ and $\sigma = 2$

```
SET C1
28.0  28.5  29.0  29.5  30.0  30.5  31.0  31.5  32.0  32.5  33.0
33.5  34.0  34.5  35.0  35.5  36.0  36.5  37.0  37.5  38.0  38.5
39.0  39.5  40.0  40.5  41.0  41.5  42.0
END
PDF C1, put into C2;
  NORMAL 35, 2.
PRINT C1 C2
ROW     C1          C2

  1    28.0    0.000436
  2    28.5    0.001015
  3    29.0    0.002216
  4    29.5    0.004547
  5    30.0    0.008764
  6    30.5    0.015870
  7    31.0    0.026995
  8    31.5    0.043138
  9    32.0    0.064759
 10    32.5    0.091324
 11    33.0    0.120985
 12    33.5    0.150568
 13    34.0    0.176033
 14    34.5    0.193334
 15    35.0    0.199471
 16    35.5    0.193334
 17    36.0    0.176033
 18    36.5    0.150569
 19    37.0    0.120985
 20    37.5    0.091325
 21    38.0    0.064759
 22    38.5    0.043139
 23    39.0    0.026995
 24    39.5    0.015870
 25    40.0    0.008764
 26    40.5    0.004547
 27    41.0    0.002216
 28    41.5    0.001015
 29    42.0    0.000436
```

Cumulative Distribution Functions

A cumulative distribution function (cdf) gives the cumulative probability associated with a pdf. In particular, a cdf gives the area under the pdf, up to the value you specify. The cdf is analogous to the cumulative proportions in Table 6.1.

Exhibit 6.6 gives some examples of how Minitab's CDF command can be used to compute cumulative probabilities. In panel (a) we want the area under the normal curve up to the value 33. The CDF command calculates this directly as .1587. In panel (b) we want the area above 36. The CDF command only gives areas below a value. Therefore, we used CDF to find the area below 36. This is .6915. The area under the entire normal curve is 1. Thus 1 − .6915 = .3085 is the area above 36. In panel (c) we want the area between 33 and 36. This is given by (area below 36) minus (area below 33) = .6915 − .1587 = .5328.

Exhibit 6.7 gives two examples to show how well the normal curve approximates the soldier data. In panel (a) we want the proportion of soldiers with a chest girth of 33. The observed proportion is .1187 and is the area of the bar at 33. To approximate it with the normal curve, we use the region under the curve that approximates the bar which is centered at 33. This is the region between 32.5 and 33.5, and is shaded in Exhibit 6.6. Note that we cannot just use the value of the PDF at 33. To calculate the area of this bar, we use the CDF command and subtract.

Panel (b) approximates the proportion of soldiers with chest girths from 31 to 33 inches. The observed proportion is the sum of the areas

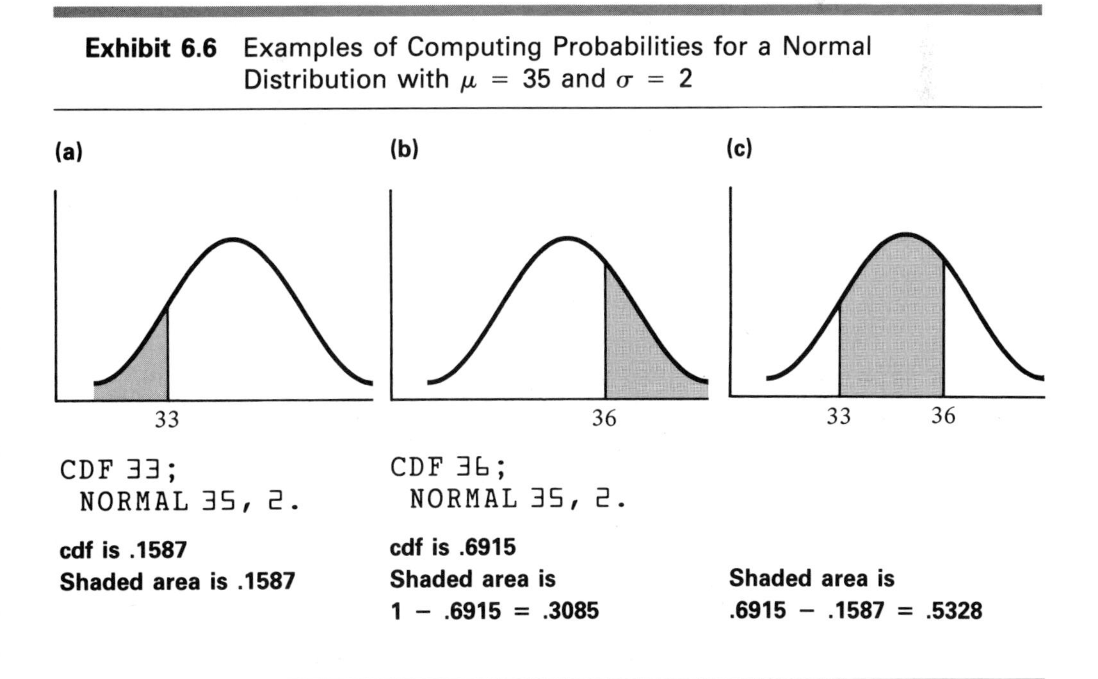

Exhibit 6.6 Examples of Computing Probabilities for a Normal Distribution with $\mu = 35$ and $\sigma = 2$

Exhibit 6.7 Soldier Data with Approximating Normal Curve

(a) Proportion with chest girth equal to 33

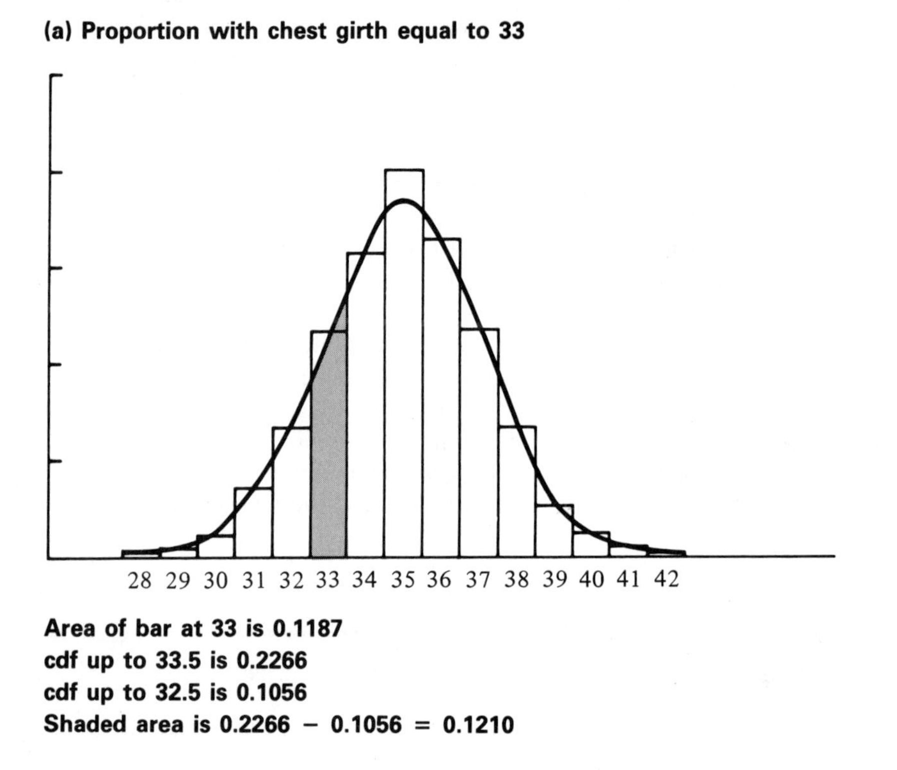

Area of bar at 33 is 0.1187
cdf up to 33.5 is 0.2266
cdf up to 32.5 is 0.1056
Shaded area is 0.2266 − 0.1056 = 0.1210

of the bars at 31, 32, and 33. This sum is .2223. The normal approximation is .2144, which is very close.

CDF for values in E [put results into E]

NORMAL with mu = K, sigma = K

Calculates the cumulative distribution function. Answers are printed if they are not stored.

The Inverse of a Cumulative Distribution Function

The command CDF calculates the area associated with a value. The command INVCDF does the opposite—it calculates the value associated with an area. Exhibit 6.8 gives an example using a normal curve with

Exhibit 6.7 (Continued)

(b) Proportion with chest girth from 31 to 33

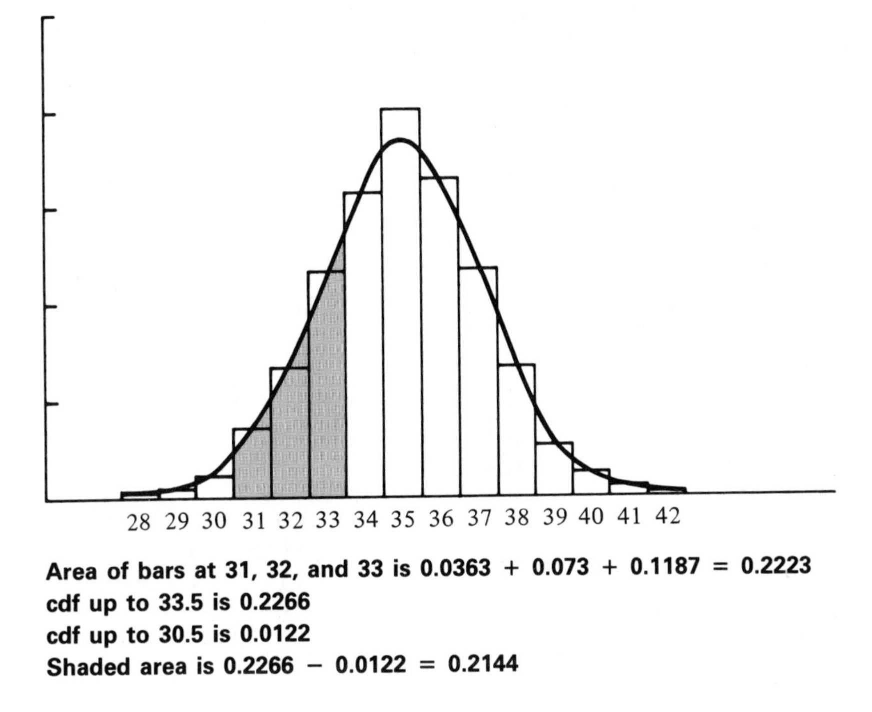

Area of bars at 31, 32, and 33 is 0.0363 + 0.073 + 0.1187 = 0.2223
cdf up to 33.5 is 0.2266
cdf up to 30.5 is 0.0122
Shaded area is 0.2266 − 0.0122 = 0.2144

$\mu = 35$ and $\sigma = 2$. We want to find the value x which has an area of .25 below it. The command

```
INVCDF .25;
  NORMAL 35, 2.
```

prints the answer 33.65. Thus, one quarter of the area under this normal curve lies below 33.65.

> **INVCDF for values in E [put results into E]**
>
> **NORMAL with mu = K, sigma = K**
>
> Calculates the inverse cumulative distribution function. Note that the values given to INVCDF are probabilities and thus must be between 0 and 1. Answers are printed if they are not stored.

Exhibit 6.8 Normal Curve with $\mu = 35$ and $\sigma = 2$

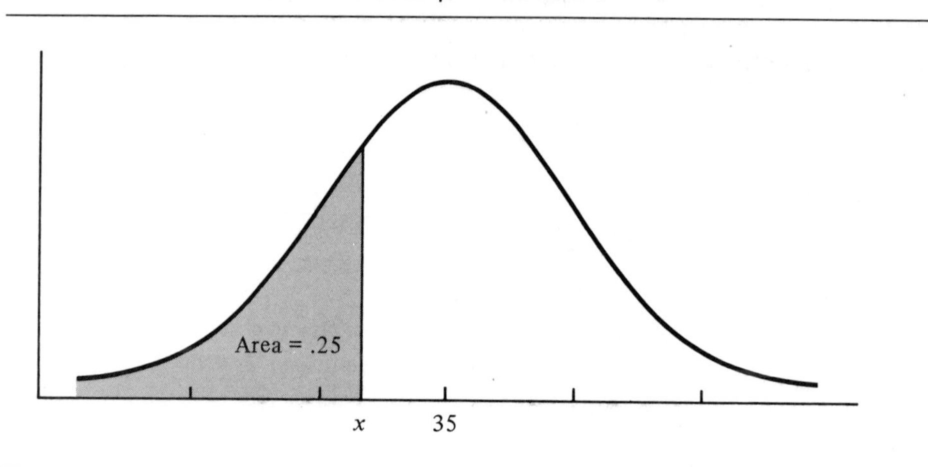

Simulating Data from a Normal Distribution

Suppose we had a large population of soldiers whose chest girths had a normal distribution with a mean of 35 and a standard deviation of 2. Then suppose we took a random sample of 20 men. Would you expect any of the 20 to have a chest girth over 37 inches? Over 40 inches? Under 30 inches? Would you expect the mean of the sample to be close to the mean of the population? How close? What do you think the histogram of the 20 observations might look like?

Exhibit 6.9 simulates 20 observations. In this sample, three men had chest girths over 37 inches, none had girths over 40 inches, and none had girths under 30 inches. The sample mean was 35.109 inches, which is very close to the population mean of 35 inches. Of course, if we were to take a second sample of 20 soldiers, we would get different results. Properties such as the mean girths of the 20 soldiers vary from sample to sample. In Chapter 7 we will study how much such properties can vary.

RANDOM K observations into each of C, ..., C

NORMAL mu = K, sigma = K

Puts a random sample of K observations into each column. If no subcommand is given, random data are simulated from a "standard" normal distribution with $\mu = 0$ and $\sigma = 1$.

The BASE command (p. 123) can be used to get the same sequence again.

A Useful Trick for Simulating Data

Suppose we wanted 50 random samples, each containing 20 observations. We could use

```
RANDOM 20 observations into C1-C50
```

or we could use

```
RANDOM 50 observations into C1-C20
```

With the first command, each sample is in a separate column. With the second command, each sample is in a separate row. Sometimes it

Exhibit 6.9 Simulation of Twenty Observations from a Normal Distribution with $\mu = 35$ and $\sigma = 2$

```
RANDOM 20 OBS INTO C1;
  NORMAL MU=35 SIGMA=2.
STEM-AND-LEAF C1

Stem-and-leaf of C1          N  =   20
Leaf Unit = 0.10

     2    31  59
     3    32  1
     3    32
     4    33  1
     5    33  8
     6    34  4
     9    34  589
   (3)    35  144
     8    35
     8    36  014
     5    36  5
     4    37  014
     1    37  7

DESCRIBE C1

                 N      MEAN    MEDIAN    TRMEAN     STDEV    SEMEAN
C1              20    35.109    35.317    35.162     1.833     0.410

               MIN       MAX        Q1        Q3
C1          31.518    37.761    34.017    36.516
```

is better to have samples in rows and sometimes it is better to have them in columns. It all depends on what we want to do with the samples.

As an example, suppose we wanted to simulate 100 samples, each of size 20, from a normal distribution with $\mu = 35$ and $\sigma = 2$ and then determine how many of these samples have at least one observation over 40. We could use the following program:

```
RANDOM 100 C1-C20;
  NORMAL 35, 2.
RMAX C1-C20 into C21
PRINT C21
STEM-AND-LEAF C21
```

In this case, putting the 100 samples into rows, rather than columns, saved us a great deal of work.

Exercises

6–5 In Table 6.1, how many soldiers had chest girths of 37 inches? What proportion of soldiers had girths of 37 inches? What proportion had girths of 37 inches or less? Of 37 inches or more? Which chest size was the most common?

6–6 For a normal distribution with $\mu = 35$ and $\sigma = 2$, use the information in Exhibit 6.6 to find:

(a) The area above 33;

(b) the area below 36;

(c) the area between 33 and 35;

(d) the area between 35 and 36.

6–7 Suppose a normal distribution has $\mu = 100$ and $\sigma = 10$. Use the CDF command to find each of the following. In each case, make a sketch like those in Exhibit 6.6.

(a) The area below 90;

(b) the area above 110;

(c) the area below 80;

(d) the area above 120;

(e) the area between 90 and 110;

(f) the area between 80 and 120;

(g) the area between 80 and 110.

6–8 In this exercise you will compute and plot the pdf and the cdf for a normal distribution with $\mu = 100$ and $\sigma = 10$. The normal pdf is essentially zero for values smaller than -3σ and for values larger than $+3\sigma$. Therefore, we will use the values 70, 72, 74, ..., 130. We can use the commands

```
SET C1
  70   72   74   76   78   80   82   84   86   88   90
  92   94   96   98  100  102  104  106  108  110  112
 114  116  118  120  122  124  126  128  130
END
PDF C1, PUT IN C2;
  NORMAL 100, 10.
CDF C1, PUT IN C3;
  NORMAL 100, 10.
PLOT C2 C1
PLOT C3 C1
```

Run this program, sketch the two curves, and comment on their shape.

6–9 (a) Make a histogram for the SAT math scores for the first set of data in the Grades data set (p. 309). Do you think these scores might have been a random sample from a normal distribution? Why or why not?

(b) Make a histogram for the other set of data in the Grades data set. Compare both histograms. How much do they vary in shape? Do they both look bell-shaped? Do they both look like they are random samples from a normal population? Do they look like they both came from the same population?

6–10 In samples from the normal distribution, about 68% of the observations should fall between $\bar{x} - s$ and $\bar{x} + s$. About 95% should fall between $\bar{x} - 2s$ and $\bar{x} + 2s$. Simulate a random sample of size 100 from a normal distribution. What percentages do you find in these two regions (you choose μ and σ)?

6–11 Repeat Exercise 6–10 using real data. For example, use the SAT math scores from sample A of the Grades data (p. 309).

6–12 In this exercise we look at the distribution of a linear combination of normal variables. Simulate 200 observations from a normal distribution with $\mu = 20$ and $\sigma = 5$ into C1. Simulate another 200 with $\mu = 7$ and $\sigma = 1$ into C2. Then compute the following:

```
LET C3=3*C1
LET C4=2*C2
LET C5=3*C1+10
LET C6=2*C2+5
LET C7=C1+C2
LET C8=3*C1+2*C2+3
```

Now use DESCRIBE C1–C8 and HISTOGRAM C1–C8. Try to give formulas indicating how the means and standard deviations of the various columns are related. Use DESCRIBE to support your formulas. What sort of distributions do the various columns seem to have?

6–13 In this exercise, we look at a mixture of two normal populations.

(a) The heights of women are approximately normal with $\mu = 64$ and $\sigma = 3$. The heights of men are approximately normal with $\mu = 69$ and $\sigma = 3$. Suppose you took a sample of 200 people, 100 men and 100 women, and drew a histogram of the 200 heights. Do you think it would look normal? Should it look normal? Use Minitab to simulate this process. Simulate 100 men's heights and put them in C1. Simulate 100 women's heights and put them in C2. Then use STACK C1 ON C2 PUT IN C3 to put the combined sample into C3. Get a histogram of the data. Does it look normal?

(b) Part (a) is an example of a mixture of two populations. Let's try a slightly more extreme example. Take a sample of 200 observations: 100 of them from a normal population with $\mu = 5$ and $\sigma = 1$, and the other 100 from a normal population with $\mu = 10$ and $\sigma = 1$. Get a histogram of the 200 observations. Does this histogram seem to indicate that there might be two populations mixed together in the sample?

6.3 The Binomial Distribution

Suppose we have a series of Bernoulli trials (see Section 6.1, p. 120) and are interested only in the total number of successes. We then can use the binomial distribution to compute probabilities.

As an example, suppose you bought four light bulbs. The manufacturers claim that 85% of their bulbs will last at least 700 hours. If the manufacturer is right, what are the chances that all four of your bulbs will last at least 700 hours? That three will last 700 hours, but one will fail before that?

Consider another situation. You've just entered a class in ancient Chinese literature. You haven't even learned the alphabet yet, but they've given you a pop quiz. You'll have to guess on every question. It's a multiple choice test, with each of the 20 questions having three possible answers. To pass, you must get at least 12 correct. What are the chances you'll pass?

Questions like these can sometimes be answered with the help of the binomial distribution. However, for the binomial distribution to hold, the trials must be independent. That is, success on one trial must not change the probability there will be a success on some other trial. Thus, there must be no special reason why successes or failures should tend to occur in streaks. (Is this condition likely to be met in the light bulb example? In the Chinese literature example?)

Calculating Binomial Probabilities with PDF

A binomial distribution is specified by two parameters: n = the number of trials and p = the probability of success on each trial. The number of successes can be 0,1,2, ..., n. The PDF command will calculate the probability for any possible number of successes. To find out the probability that three of your four light bulbs will be successes (last more than 700 hours) and one will fail, we use

```
PDF 3;
  BINOMIAL n=4, p=.85.
```

Then Minitab printed the answer .3685.

If we omit the number of successes from the main command line, PDF will print out the probabilities for all values 0,1,2, ..., n. Exhibit 6.10 contains an example.

PDF for values in E [put results into E]

BINOMIAL with n = K and p = K

Calculates probabilities for the binomial distribution with the specified parameters, n and p. If you do not store the results, they will be printed. If you use no arguments on the PDF line, a table of all probabilities is printed. For example,

```
PDF;
  BINOMIAL n=10, p=.4.
```

prints a table with probabilities for 0,1,2, ..., 10 successes. If any probabilities at the beginning or end of the list are very small (less than .00005), they are not printed.

Cumulative Distribution Function

The CDF command calculates cumulative probabilities. A cumulative probability is the probability that your result will be less than or equal to a particular value. As an example, suppose we calculate the probability you will fail the test in ancient Chinese literature. Here n = 20 and p = .3333. You will fail the test if you get less than or equal to 11

Exhibit 6.10 Probabilities for a Binomial with $n = 4$, $p = .85$

```
PDF;
  BINOMIAL N = 4, P = .85.

BINOMIAL WITH   N = 4,   P = 0.85
    K    P(X = K)
    0      0.0005
    1      0.0115
    2      0.0975
    3      0.3685
    4      0.5220
```

questions correct. (You will pass if you get 12 or more right.) The command

```
CDF 11;
  BINOMIAL 20, .3333.
```

calculates this probability to be .9870. This is the probability you will fail. The probability you will pass is therefore 1 − .9870 = .0130. You thus have a very small chance of passing.

CDF for values in E [put results into E]

BINOMIAL with n = K and p = K

Calculates the cumulative distribution function (cumulative probabilities) for the binomial. If you do not store the results, they will be printed.

If you use no arguments on the CDF line, a table of all cumulative probabilities is printed. Values that are essentially 0 (less than .00005) and values that are essentially 1 (over .99995) are omitted from the table.

The Inverse of a CDF

The command CDF calculates the cumulative probability associated with a value. The command INVCDF does the opposite: it calculates the values associated with a cumulative probability. As an example, Exhibit 6.11 contains the CDF for a binomial with $n = 5$ and $p = .40$. The cumulative probability for 3 is .913. Therefore,

Exhibit 6.11 Cumulative Probabilities for Binomial with $n = 5$, $p = .40$

```
CDF;
  BINOMIAL N = 5 P = 0.40.

BINOMIAL WITH  N = 5,   P = 0.40
   K  P(X LESS OR = K)
   0              0.0778
   1              0.3370
   2              0.6826
   3              0.9130
   4              0.9898
   5              1.0000
```

```
INVCDF .913;
  BINOMIAL 5, .4.
```

gives the value 3.

Notice that Exhibit 6.11 contains six probabilities, one corresponding to each possible number of successes. These are the only probabilities for which INVCDF can give an exact answer. Exhibit 6.12 shows what happens when there is no exact answer. The value below and the value above the requested probability are both printed. If you put storage on the CDF command, then the larger value (here it is 3) is stored.

Exhibit 6.12 Example of INVCDF for a Binomial Distribution Where the Probability Requested (.8) Cannot be Exactly Attained (Instead the Next Values Up and Down are Given)

```
INVCDF .8;
  BINOMIAL 5 .4.

BINOMIAL WITH N = 5,   P = 0.4

   K  P(X LESS OR = K)     K  P(X LESS OR = K)
   2            0.6826     3            0.9130
```

INVCDF for probabilities in E [put results into E]

BINOMIAL with n = K and p = K

Calculates the inverse cumulative distribution function. Answers are printed if they are not stored.

If there is no exact answer for a probability you specify, both the value below and value above are printed. If you request storage, the value above is stored.

Simulating Binomial Data

Binomial data may be simulated with the RANDOM command.

RANDOM K observations into each of C, ..., C

BINOMIAL n = K, p = K

Simulates random observations from a binomial distribution. The BASE command (p. 123) also applies.

Normal Approximation to the Binomial

For many combinations of n and p, the binomial distribution can be well approximated by a normal distribution that has the same mean and standard deviation; that is, by a normal distribution which has $\mu = np$ and $\sigma = \sqrt{npq}$.

Exhibit 6.13 lets us see how the normal approximates a binomial with $p = 1/2$ and $n = 16$. The approximating normal has $\mu = 16(1/2) = 8$ and $\sigma = \sqrt{(16)(1/2)(1/2)} = 2$. Recall that MPLOT plots the first pair of columns with the letter A and the second pair with the letter B. When two points overlap, it plots a 2. In this MPLOT, there are many overlaps. We converted the binomial probabilities, by hand, to a bar chart. This chart is similar to the one we used for the soldier data in Exhibit 6.4 (p. 127). The area of the bar at a given value is equal to the probability of that value. To get the normal curve, we drew a smooth curve though the values we plotted. Note the quality of the normal approximation.

If we were to use a normal approximation for a binomial with $p = 1/2$ and $n = 30$, the approximation would look even better. In general, for any fixed value of p, the larger n is, the closer the binomial distribution

Exhibit 6.13 Binomial and Approximating Normal Curve (A = binomial; B = normal)

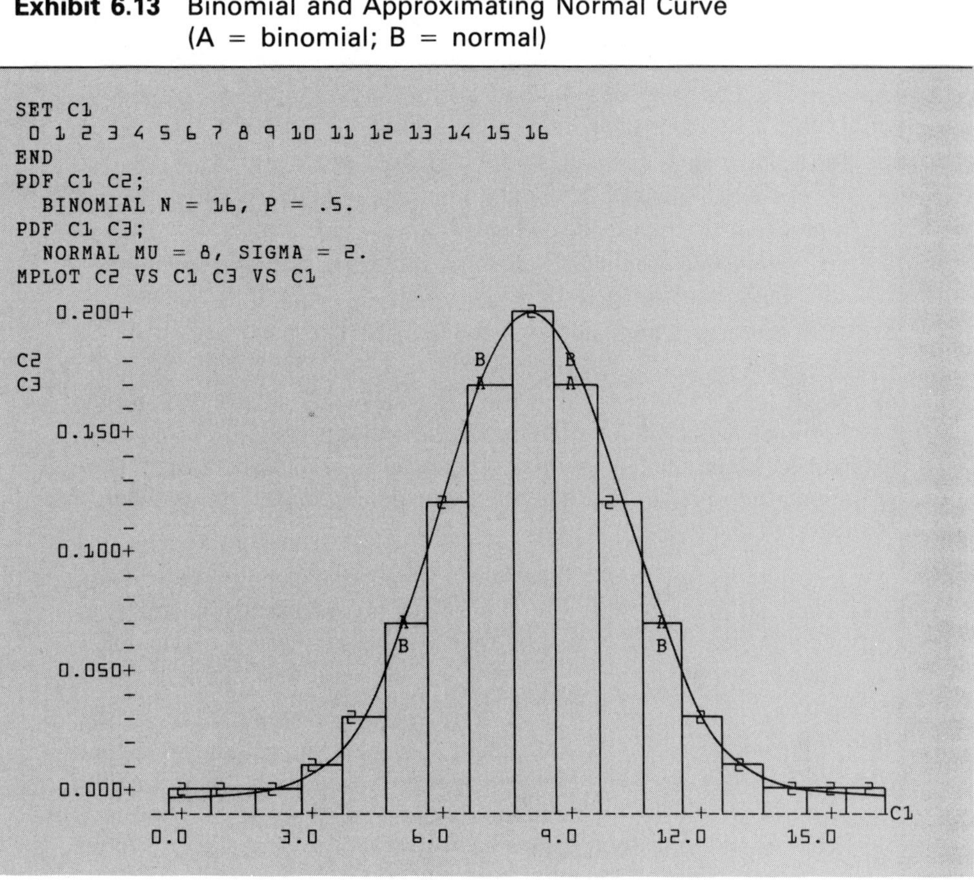

is to a normal. Here we used $p = .5$. In the exercises, we'll look at other values of p.

Exercises

6–14 A total of 16 mice are sent down a maze, one by one. From previous experience it is believed that the probability a mouse turns right is .38. Suppose their turning pattern follows a binomial distribution. Use the PDF and CDF commands to answer each of the following.

(a) What is the probability that exactly seven of these 16 mice turn right?

(b) That eight or fewer turn right?

(c) That more than eight turn right?

(d) That eight or more turn right?

6–15 This exercise has you compute and plot some binomial probabilities so you'll have a better idea what they look like.

(a) First, let's fix n at 8 and vary p. Use $p = .01, .1, .2, .5, .8$, and $.9$. For each value of p, compute the binomial probabilities and get a plot like the one in Exhibit 6.13.

(i) How does the shape of the plot change as p increases?

(ii) Compare the two plots where $p = .1$ and $p = .9$. Do you see any relationship? What about the two plots where $p = .2$ and $p = .8$?

(b) Next, let's fix p at 0.2 and vary n. Use $n = 2, 5, 10, 20$, and 40. For each value of n, compute the binomial probabilities and get a plot as in part (a). How does the plot change as n increases? How about the spread? The shape? Does the "middle" move when n increases?

6–16 Two instruments are used to measure the amount of pollution in Lake Erie. They sit side by side but they never agree exactly. Sometimes instrument A gets a higher reading than B, and sometimes it's the other way around. Suppose the instruments are identical and each one has a 50–50 chance of giving the higher reading on any given occasion. Also suppose the readings are independent.

(a) What are the chances that instrument A gives the higher reading 15 or more times out of 20?

(b) What would you think if you measured the water quality 20 times with each instrument and A was bigger 15 times out of 20? What are some of the possible causes of such an event?

6–17 Suppose X is a binomial random variable with $n = 16$ and $p = .75$.

(a) Write a Minitab program to calculate the mean of X using the formula $\mu = \Sigma\, x \times P(X = x)$. Does the answer agree with the answer you get when you use the formula $\mu = np$?

(b) Write a Minitab program to calculate the variance of X using the formula $\sigma^2 = \Sigma(x - \mu)^2 \times P(X = x)$. Use the value of μ from part (a). Does the answer agree with the formula $\sigma^2 = npq$?

(c) Simulate 1000 binomial observations with $n = 16$ and $p = .75$. Use DESCRIBE to estimate μ and σ. Do these results agree reasonably with those in (a) and (b)?

6–18 In the Pulse experiment (p. 318), students were asked to toss a coin. If the coin came up heads they were asked to run in place. Tails meant they did not run in place. Do you think all students who got a head ran in place? Compare the data with the output from the instruction

```
CDF;
  BINOMIAL n = 92, p = .50.
```

Does it seem very likely that only 35 of 92 students would get heads if they all flipped coins? Can you make a conjecture as to what might have happened?

6–19 Imagine flipping six fair coins and recording the number of heads. Simulate this process 100 times.

(a) Use the "Useful Tricks" described on page 133. Get a histogram of your results.

(b) How often did all six coins come up heads? How often would you expect this event to occur in the 100 simulations? (PDF can help you answer this.)

(c) How often did you get more heads than tails? How often would you expect this event to occur in the 100 simulations? (PDF can help you answer this.)

(d) Does your histogram look symmetric (more or less)? Would you expect it to?

(e) Convert your observed frequencies of 0, 1, 2, ..., 6 heads to relative frequencies. Compare these relative frequencies with the theoretical probabilities from the PDF command.

(f) Repeat (a) through (e), but assume the coin is biased so the probability of a head is .9.

6–20 *Acceptance Sampling*. When a company buys a large lot of materials, they usually don't check every single item to see that they all are satisfactory. Instead, some companies pick a sample of items, then check these and if they do not find many defective items, they go ahead and accept the whole lot. In this problem we'll look at the kind of risk they run when they do this. Suppose the inspection plan consists of looking at 40 items chosen at random from a large shipment, then accepting the entire shipment if there are 0 or 1 defective items, and rejecting the shipment if there are 2 or more defective items.

(a) If in the entire shipment 25% of the items are defective, what is the probability the shipment will be accepted?

(b) Compute the probability of acceptance if the shipment has .1% defective, .5% defective, 1%, 2%, 3%, 5%, and 8% defective. Then sketch a plot of the probability of acceptance versus the percent defective.

(c) About what percent defective leads to a 50–50 chance of acceptance?

(d) Repeat part (b) but use a different plan where 100 items are checked and the lot is accepted if 0, 1, or 2 are found defective, and rejected if 3 or more are found defective.

(e) Sketch both graphs from (b) and (d) on the same plot. Discuss the advantages and disadvantages of the two different plans.

6–21 (a) Make plots, as in Exhibit 6.13, but use $p = .4$ instead of $p = .5$. Use $n = 4$ and 16.

(b) Repeat part (a) using $p = .2$.

(c) What can you say about the normal approximation to the binomial? For what values of n and p does it seem to work best?

6–22 Suppose X has a binomial distribution with $p = .8$ and $n = 25$. Use Minitab to calculate each of the probabilities below exactly. Also compute the normal approximation to these probabilities. Remember to go .5 below and above when computing the normal approximation. Compare the binomial results with the normal approximations.

(a) Prob $(X = 21)$;

(b) Prob $(X \leq 21)$;

(c) Prob $(X \geq 24)$;

(d) Prob $(21 \leq X \leq 24)$.

6.4 The Poisson Distribution

A famous statistician once called the Poisson distribution "the distribution to read newspapers by." He did this because so many times we read facts like "crime increases 25% in year," or "543 fatalities expected this weekend." Such figures frequently can be assessed with the Poisson distribution.

The Poisson distribution arises when we count the number of occurrences of an event that happens relatively infrequently, given the number of times it could happen. For example, the number of automobile accidents that will occur in a given county next weekend could be very large. Any motorist in the area might have an accident. But the chance that any given motorist will have an accident is very small, so the actual number of accidents probably will not be too large.

Calculating Poisson Probabilities with PDF

A Poisson distribution is specified by just one parameter, the mean, μ. For example, if we know that there are, on the average, six accidents

Exhibit 6.14 Example of Probabilities for a Poisson Distribution

```
SET C1
 10     20      0      5
END
 PDF C1;
   POISSON 6.

POISSON WITH MEAN = 6.0
    K    P(X = K)
   10     0.0413
   20     0.0000
    0     0.0025
    5     0.1606
```

per weekend, then we can calculate the probability there will be, say, ten accidents next weekend, or 20 accidents, or no accidents. The PDF command will calculate probabilities for any values you specify. Exhibit 6.14 gives an example.

PDF for values in E [put results into E]

POISSON mean = K

Calculates probabilities for the Poisson distribution with the specified mean. If you do not store the results, they are printed.

If no arguments are given on the PDF line, a table is printed that gives all pdf values greater than .0005.

The mean, μ, must not exceed 100.

Cumulative Distribution Function

Often we are not interested in individual probabilities but in cumulative probabilities, such as the probability of ten or fewer accidents. We then use the CDF command. Exhibit 6.15 gives an example.

From this we see that if the average number of accidents per weekend is six, then the probability of no accidents next weekend is .0025, the

Exhibit 6.15 Cumulative Probabilities for a Poisson

```
CDF;
  POISSON 6.

POISSON WITH MEAN = 6.0
      K   P(X LESS OR = K)
      0            0.0025
      1            0.0174
      2            0.0620
      3            0.1512
      4            0.2851
      5            0.4457
      6            0.6063
      7            0.7440
      8            0.8472
      9            0.9161
     10            0.9574
     11            0.9799
     12            0.9912
     13            0.9964
     14            0.9986
     15            0.9995
     16            0.9998
     17            0.9999
     18            1.0000
```

probability of six or fewer is .6063, and the probability of ten or more accidents is 1 − .9161 = .0839. What is the probability of having 20 or fewer accidents? This value is not in the table. Minitab prints the cumulative probabilities only as far as 18. From there on, the cumulative probabilities all are essentially 1.

Suppose the police decide to crack down on speeders and that next weekend there are only three accidents. We can imagine a headline: "Police crackdown leads to 50% reduction in accidents." What do you think? Suppose we find the probability of three or fewer accidents for $\mu = 6$. It is .1512. That is, there is a 15% chance of a 50% or better reduction in accidents even if the sheriff's crackdown has no real effect whatsoever.

CDF for values in E [put results into E]

POISSON mean = K

Calculates the cumulative distribution function (cumulative probabilities) for the Poisson. If you do not store the results, they will be printed.

If you use no arguments on the CDF line, a table of cumulative probabilities is printed. Values that are essentially 0 (less than .00005) and values that are essentially 1 (over .99995) are omitted from the table. All other probabilities are printed.

Inverse CDF

Suppose, in another community, we know that the average number of accidents is 50. We want to get some idea how many accidents we can expect to occur in a given weekend. Exhibit 6.16 gives an example.

As with the binomial distribution, the Poisson does not have an exact answer for every probability we might specify. One way we could summarize this output is as follows: Since P(54 or fewer accidents) = .7423 and P(44 or fewer accidents) = .2210, then P(from 45 to 54 accidents) = .7423 − .2210 = .5213. Thus over 50% of the time there will be from 45 to 54 accidents. Similarly P(from 33 to 69 accidents) = P(69 or fewer) − P(32 or fewer) = .9957 − .0044 = .9917. Thus over 99% of the time

Exhibit 6.16 INVCDF for Poisson with $\mu = 50$

```
SET C1
.25 .75 .005 .995
END
INVCDF C1;
  POISSON 50.

POISSON WITH MEAN = 50.0
   K  P(X LESS OR = K)          K  P(X LESS OR = K)
  44            0.2210         45            0.2669
  54            0.7423         55            0.7845
  32            0.0044         33            0.0070
  68            0.9938         69            0.9957
```

there will be from 33 to 69 accidents. Rarely (about .5% of the time) will there be under 33 accidents and rarely (about .5% of the time) will there be over 69.

INVCDF for probabilities in E [put results into E]

POISSON mean = K

Computes the inverse cumulative distribution function. Answers are printed if they are not stored. If there is no exact answer for a probability you specify, both the value below and the value above are printed. If you request storage, the value above is stored.

Simulating Poisson Data

Poisson data may be simulated with the RANDOM command. For example, to simulate the number of accidents for the next ten weekends, given that $\mu = 6$, we could use

```
RANDOM 10 C1;
  POISSON 6.
PRINT C1
```

This instruction gave the following counts: 11 7 7 8 5 4 2 3 5 7

RANDOM K observations [into each of C, ..., C]

POISSON mean = K

Simulates random observations from a Poisson distribution. The BASE command (p. 123) also applies.

Exercises

6–23 A typist makes an average of only one error every two pages, or .5 errors per page. Suppose these mistakes follow a Poisson distribution. Use the PDF and CDF commands to answer each of the following.

(a) What is the probability the typist will make no errors on the next page?

(b) One error?

(c) Fewer than two errors?

(d) One or more errors?

(e) Sketch a histogram of this distribution (the PDF).

(f) Comment on reasons why the Poisson distribution might or might not be a good approximation for the number of errors made by a typist.

6–24 In high energy physics, the rate at which some particles are emitted has a Poisson distribution. Suppose, on the average, 15 particles are emitted per second.

(a) What is the probability that exactly 15 will be emitted in the next second?

(b) That 15 or fewer will be emitted?

(c) That 15 or more will be emitted?

(d) Find a number, x, such that approximately 95% of the time fewer than x particles will be emitted.

(e) Sketch a histogram of this distribution (the PDF).

6–25 In the fall of 1971, testimony was presented before the Atomic Energy Commission that the nuclear reactor used in teaching at Penn State University was causing increased infant mortality in the surrounding town of State College. This reactor had been installed in 1965. The following data for State College were presented as part of the testimony. In addition, data for Lebanon, Pennsylvania, a city of similar size and rural character, were presented.

	State College			*Lebanon*		
Year	*Live Births*	*Infant Deaths*	*Infant Deaths per 1000 Births*	*Live Births*	*Infant Deaths*	*Infant Deaths per 1000 Births*
1962	369	4	10.8	666	16	24.0
1963	403	4	9.9	464	15	23.2
1964	365	5	13.7	668	9	13.5
1965	365	6	16.4	582	10	17.2
1966	327	4	12.2	538	8	14.9
1967	385	6	15.6	501	8	16.0
1968	405	10	24.7	439	5	11.4
1969	441	6	13.6	434	3	6.9
1970	452	8	17.7	500	8	16.0

One of us (Brian Joiner) was asked to examine the evidence in detail and find out whether there was cause for concern or not. A wide variety of statistical procedures were used but one important discussion centered on whether there had been an abnormal peak in infant mortality in State College in 1968.

(a) Suppose we assume that infant deaths follow a Poisson distribution. Over the nine-year period presented in the testimony, there were 53 deaths in State College. On the average, this is $53/9 = 5.9$ per year. Simulate nine observations from a Poisson distribution with $\mu = 5.9$. Repeat this simulation ten times. Are peaks such as the one in State College in 1968 unusual?

(b) One of the strongest critics of Penn State's reactor drew the following conclusion from the data:

"Following the end of atmospheric testing by the U.S., U.S.S.R., and Britain in 1962, infant mortality declined steadily for Lebanon, while it rose sharply for State College. Using 1962 as a reference equal to 100, State College rose to $(24.7/10.8) \times 100 = 229$ by 1968 while Lebanon declined to $(11.4/24.0) \times 100 = 48$. Furthermore, not only is there an anomalous rise above the 1962 levels in State College after 1963, but there are two clear peaks of infant mortality rates in 1965 and 1968. Especially the high peak in 1968 has no parallel in Lebanon."

Do you think the data support his conclusions that Penn State's reactor has led to a significant rise in infant mortality? What criticisms of his argument can you make?

6–26 If you have a Poisson distribution with a large mean μ, its probabilities can be closely approximated by a normal distribution with the same mean μ and with a standard deviation equal to σ. For each of the following conditions compute the Poisson and normal pdfs and plot them on the same graph. Comment on the quality of the approximation.

(a) Use $\mu = 25$. The following program can be used. In this program, we used a special feature of SET (described on p. 332) that allows us to use 0:40 as an abbreviation for the integers 0, 1, 2, ..., 40.

```
SET C1
  0:40
END
PDF C1 INTO C2;
  POISSON MU = 25.
PDF C1 INTO C3;
  NORMAL MU = 25, SIGMA = 5.
MPLOT C2 VS C1, C3 VS C1
```

(b) Repeat (a) for $\mu = 4$.

(c) Repeat (a) for $\mu = 1$.

6–27 If you have a binomial distribution with a large value of n and a small value of p, its probabilities can be closely approximated by a Poisson distribution with mean equal to np.

(a) Use $n = 30$ and $p = .01$ and compute the corresponding binomial and Poisson probabilities (pdfs). Do they seem to be pretty close?
(b) Plot both pdfs on the same plot, using the MPLOT command. Do the pdfs agree pretty well everywhere?
(c) Compare the binomial and Poisson cdfs. How well do they agree?
(d) Repeat (a) but use $n = 30$ and $p = .5$. How good is the approximation now?
(e) Plot both pdfs from part (d) on the same plot. How well do they agree?
(f) Compare the binomial and Poisson cdfs. How well do they agree?

6.5 Sampling Finite Populations

So far in this chapter we have talked about random samples from theoretical distributions. But sometimes we are interested in a random sample from an actual finite population.

Sometimes we do this to study the properties of various statistical procedures when we are sampling from a finite population without replacement. In other cases we may have collected a very large set of data and want to do some preliminary analyses on a random subset. If we use a portion of the data, then our work will be faster and cheaper. Once we have some idea about the structure of the data, then we might want to analyze the full set.

SAMPLE K observations from C, ..., C put into C, ..., C

Takes a random sample of size K from the data in the first group of columns and stores the sample in the second group of columns. The sampling is done without replacement.

6.6 Summary of Distributions in Minitab

The commands PDF, CDF, INVCDF, and RANDOM all have the same collections of subcommands to specify different distributions. In this section we list these distributions and give the corresponding subcommands.

Continuous Distributions

NORMAL mu = K sigma = K

Normal distribution with mean μ and standard deviation σ. The pdf is

$$f(x) = \frac{1}{\sqrt{2\pi}\sigma} e^{-(x-\mu)^2/2\sigma^2}, \qquad \sigma > 0$$

UNIFORM a = K b = K

Uniform distribution on a to b. The pdf is

$$f(x) = \frac{1}{b - a}, \qquad a < x < b$$

CAUCHY a = K b = K

Cauchy distribution. The pdf is

$$f(x) = \frac{1}{(\pi b)[1 + \{(x - a)/b\}^2]}, \qquad b > 0$$

LAPLACE a = K b = K

Laplace or double exponential distribution. The pdf is

$$f(x) = \frac{1}{2b} e^{-|x-a|/b}, \qquad b > 0$$

LOGISTIC a = K b = K

Logistic distribution. The pdf is

$$f(x) = \frac{e^{-(x-a)/b}}{b[1 + e^{-(x-a)/b}]^2}, \qquad b > 0$$

LOGNORMAL mu = K sigma = K

Lognormal distribution. A variable x has a lognormal distribution if log x has a normal distribution with mean μ and standard deviation σ. The pdf of the lognormal is

$$f(x) = \frac{1}{x\sqrt{2\pi}\,\sigma} e^{-\{(\log_e x) - \mu\}^2/2\sigma^2}, \qquad x > 0$$

T with v = K

Student's t distribution with v degrees of freedom. The pdf is

$$f(x) = \frac{\Gamma[(v+1)/2]}{\Gamma[v/2]\sqrt{v\pi}} \frac{1}{(1 + x^2/v)^{(v+1/2)}}, \quad v > 0$$

F with u = K v = K

F distribution with u degrees of freedom for the numerator and v degrees of freedom for the denominator. The pdf is

$$f(x) = \frac{\Gamma[(u+v)/2]}{\Gamma[u/2]\Gamma[v/2]}\left(\frac{u}{v}\right)^{u/2} \frac{x^{(u-2)/2}}{[1 + (u/v)x]^{(u+v)/2}}, \quad x > 0, u > 0, v > 0$$

CHISQUARE v = K

χ^2 distribution with v degrees of freedom. The pdf is

$$f(x) = \frac{x^{(v-2)/2}\, e^{-x/2}}{2^{v/2}\, \Gamma(v/2)}, \quad x > 0, v > 0$$

EXPONENTIAL b = K

Exponential distribution. (Caution: Some books use $1/b$ where we have used b.) The pdf is

$$f(x) = \frac{1}{b} e^{-x/b}, \quad x > 0, b > 0$$

GAMMA a = K b = K

Gamma distribution. (Note: Some books use $1/b$ where we have used b.) The pdf is

$$f(x) = \frac{x^{a-1} e^{-x/b}}{\Gamma(a) b^a}, \quad x > 0, a > 0, b > 0$$

WEIBULL a = K b = K

Weibull distribution. The pdf is

$$f(x) = \frac{a x^{a-1} e^{-(x/b)^a}}{b^a}, \quad x > 0, a > 0, b > 0$$

BETA a = K b = K

Beta distribution. The pdf is

$$f(x) = \frac{\Gamma(a + b)x^{a-1}(1 - x)^{b-1}}{\Gamma(a)\,\Gamma(b)}, \qquad 0 \le x \le 1, a > 0, b > 0$$

Discrete Distributions

INTEGER a = K b = K

Each integer a, $a + 1$, ..., b has equal probability. (Sometimes called "discrete uniform.")

BERNOULLI p = K

The probability of a 1 is p and the probability of a 0 is 1 − p.

BINOMIAL n = K p = K

Binomial distribution. The probability of x is

$$f(x) = \binom{n}{x} p^x(1 - p)^{n-x}, \qquad x = 0, 1, \ldots, n$$

POISSON mu = K

Poisson distribution. The probability of x is

$$f(x) = \frac{e^{-\mu}\mu^x}{x!}, \qquad x = 0, 1, 2, 3, \ldots$$

DISCRETE Values in C, Probabilities in C

Arbitrary discrete distribution. You put the values and the corresponding probabilities in two columns beforehand. For example, to simulate 40 observations from a distribution which gives the three values −1, 0, +1 with probabilities 1/4, 1/2, 1/4, respectively, you can use:

```
READ C1 C2
  -1 .25
   0 .50
   1 .25
END
RANDOM 40 OBSERVATIONS INTO C6;
  DISCRETE C1,C2.
```

Exercises

6–28 *Continuous Uniform Distribution.* The uniform distribution and the t distribution with 2 degrees of freedom are examples of very nonnormal distributions. Their histograms should look quite different from those for normal data sets.

(a) Simulate 50 observations from a uniform distribution and make a histogram. Repeat five times. Comment on the general shape of the histograms.

(b) Do (a) but simulate data from a t distribution with 2 degrees of freedom.

6–29 *Discrete Uniform Distribution.* Suppose there are 30 people at a party. Do you think it is very likely that at least two of the 30 people have the same birthday? Try estimating this probability by a simulation. To simplify things ignore leap years (thus, assume all years have 365 days) and assume all days of the year are equally likely to be birthdays. Use the following commands to simulate ten sets of 30 birthdays:

```
RANDOM 30 observations into C1-C10;
  INTEGERS 1 to 365.
TALLY C1-C10
```

How many of your ten sets of 30 people had no matching birthdays?

6–30 *Chi-square Distribution.* The mean, standard deviation, and shape of the chi-square distribution change as the number of degrees of freedom changes. For a chi-square distribution with n degrees of freedom, the mean is n and the standard deviation is $\sqrt{2n}$.

(a) Simulate 200 observations from a chi-square distribution with 1 degree of freedom into C1. Use DESCRIBE to compute the sample mean and standard deviation. Are they approximately 1 and $\sqrt{2} = 1.414$, respectively? Now make a histogram and sketch the shape of the distribution. For all histograms in this exercise, use the subcommands START = .1 and INCREMENT = .2.

(b) Repeat (a) but use 2 degrees of freedom.

(c) Repeat (a) but use 5 degrees of freedom.

(d) Repeat (a) but use 10 degrees of freedom.

6–31 *Chi-square Distribution.* The chi-square distribution arises in several ways in statistics. Here we will use simulation to illustrate three ways.

(a) If X is from a standard normal distribution (with mean 0 and standard deviation 1) then X^2 has a chi-square distribution with 1 degree of freedom. Simulate 200 standard normals into a column, then square them. Then simulate 200 chi-square observations with 1 degree of freedom. Now make a histogram of the X^2s and a histogram of the chi-squares. The two histograms should be very similar.

(b) Another way the chi-square arises is from the variance of samples from a normal distribution. Simulate 200 rows of standard normal observations into C1–C4. Use RSTDEV to compute the standard deviations of the 200 rows. Square the results to get variances. Make a histogram of the variances. Compare this to a histogram of 200 chi-square observations with 3 degrees of freedom.

(c) Still another source of the chi-square is as follows: Simulate 200 Poisson observations with $\mu = 5$ into C1–C3. Now compute the "chi-square test," which we will discuss in Chapter 11, using the following commands:

```
LET C4 = (C1+C2+C3)/3
LET C6 = (C1-C4)**2+(C2-C4)**2+(C3-C4)**2
LET C7 = C6/C4
```

Now make a histogram of C7 and compare it to a histogram of a chi-square with 2 degrees of freedom.

6–32 *The t distribution.* Student's t distribution arises most naturally as $(\bar{x} - \mu)/(s/\sqrt{n})$. To illustrate the development of a t distribution, simulate 200 rows of normal data into C1–C3. You choose μ and σ. Use RMEAN and RSTDEV to compute $\bar{x}$ and s for C1–C3. Use LET to compute $t = (\bar{x} - \mu)/(s/\sqrt{3})$. Make a histogram of t. Compare this histogram to one you obtain by using RANDOM to simulate directly 200 observations for a t distribution with 2 degrees of freedom. Describe the shapes of the two histograms.

7

One-Sample Confidence Intervals and Tests for Population Means

We often want to know something about a population. The population could be all ten-year-old girls in Philadelphia and we want to know the mean blood pressure. The population could be all the light bulbs produced by a plant last month and we want to know what proportion are defective. In many cases, we cannot afford to study the entire population. So we take a small sample—perhaps 200 girls or 80 light bulbs—and we use information about the sample to estimate what we want to know about the population. If our sample is representative of the population, then our estimate may be quite good.

In the example with ten-year-old girls, we want to know the mean blood pressure of the population. This mean is denoted by the letter μ. We can use the mean of the sample, $\bar{x}$, as a guess or estimate of μ. In the light bulb example, we want to know the proportion of defective bulbs in the population. This population proportion is usually denoted by p. We can estimate p by the proportion of defectives in the sample of 80 bulbs. This sample proportion is often denoted by $\hat{p}$.

But how do we get a representative sample? One way is to take what is called a simple random sample. A simple random sample has two important properties: each member of the population has an equal chance of being chosen for the sample, and the observations form a random sequence, as discussed in Section 6.1. The command RANDOM, discussed in Chapter 6, simulates drawing a simple random sample from a population with a specified distribution.

One straightforward way to get a simple random sample of 200 girls is to first write the name of each ten-year-old girl in Philadelphia on a slip of paper. We might get a good list from school records. Then put all the slips in a large box. Mix them up and draw one slip. Mix them up again and draw a second slip, and so on until we have 200 slips. We could then contact these girls and measure their blood pressure. In this example, everything seems fairly simple. Unfortunately, in most studies it is very difficult to get a truly random sample. In Section 7.6 we will briefly discuss some ways to spot nonrandomness.

In this chapter we study different methods of learning about the mean of a population. These methods require two things: We must have a random sample and we must have a normal population. The first requirement is very important. The second, however, can be relaxed somewhat. In Section 7.6 we'll briefly discuss how a nonnormal population might affect your conclusions and how you can spot nonnormality using a sample of data.

7.1 How Sample Means Vary

Imagine drawing a random sample from a population with a normal distribution and calculating the mean of the sample. Imagine doing this 100 times so we have 100 values of $\bar{x}$. Exhibit 7.1 shows the results using 100 samples each with nine observations.

Most of the sample means are fairly close to the population mean, $\mu = 42$. All of them are between 37.5 and 47.5, and half are between 40.5 and 43.5.

In real life we usually have just one sample, not 100. That one sample would give just one $\bar{x}$, not 100 of them. We did this simulation of 100 $\bar{x}$s so we could study how $\bar{x}$ varies from sample to sample.

On the histogram of the 100 $\bar{x}$s, we sketched a histogram of the population of individual observations. This underlying population is normal, so its histogram is bell-shaped. What about the histogram of the $\bar{x}$s? It

Exhibit 7.1 Means for 100 Samples, Each with Nine Observations, from a Normal Distribution with $\mu = 42$ and $\sigma = 6$

```
RANDOM 100 C1-C9;
  NORMAL MU=42 SIGMA=6.
RMEAN C1-C9 INTO C10
HISTOGRAM  C10

Histogram of C10    N = 100

Midpoint   Count
      38       2  **
      39       4  ****
      40      21  *********************
      41      12  ************
      42      18  ******************
      43      20  ********************
      44       9  *********
      45      11  ***********
      46       1  *
      47       2  **

DESCRIBE C10

                N      MEAN    MEDIAN    TRMEAN     STDEV    SEMEAN
C10           100    42.101    41.852    42.069     1.936     0.194

              MIN       MAX        Q1        Q3
C10        38.114    47.427    40.328    43.352
```

seems to be fairly bell-shaped too, although taller and less spread out. If we simulated many more $\bar{x}$s, the histogram would be even closer to bell-shaped. This illustrates the important fact that if the underlying population is normal, the distribution of the $\bar{x}$s also will be normal.

Also notice that the mean of the 100 $\bar{x}$s is 42.101, which is very close to 42, the population mean. Even if we take samples from nonnormal distributions, the mean of the population of $\bar{x}$s will always be equal to the mean of the population of individual observations. In symbols, we write

$$\mu_{\bar{x}} = \mu$$

The distribution of $\bar{x}$ does differ from that of the individual observations in one important respect—it is less spread out. In fact, for random samples of size n from any population,

$$\sigma_{\bar{x}} = \frac{\sigma}{\sqrt{n}}$$

In our simulation, $\sigma = 6$ and $\sqrt{n} = \sqrt{9} = 3$, so $\sigma_{\bar{x}} = 2$. This is the standard deviation for the population of $\bar{x}$s. The standard deviation of our particular collection of 100 $\bar{x}$s is 1.936, which is fairly close.

Notice in Exhibit 7.1 that rarely did a sample mean $\bar{x}$ deviate from μ by more than $2\sigma_{\bar{x}} = 4$. This fact forms the basis for the confidence intervals described in Section 7.2 and for the test described in Section 7.4.

Suppose we form a new quantity

$$z = \frac{\bar{x} - \mu}{\sigma_{\bar{x}}} = \frac{\bar{x} - \mu}{\sigma/\sqrt{n}} = \frac{\bar{x} - 42}{2}$$

Exhibit 7.2 Histogram of the z Values for the Data of Exhibit 7.1

```
LET C12 = (C10-42)/2
HISTOGRAM C12

Histogram of C12   N = 100

Midpoint   Count
   -2.0        2  **
   -1.5        4  ****
   -1.0       21  *********************
   -0.5       12  ************
    0.0       18  ******************
    0.5       20  ********************
    1.0        9  *********
    1.5       11  ***********
    2.0        1  *
    2.5        2  **

DESCRIBE C12

                N      MEAN    MEDIAN    TRMEAN     STDEV    SEMEAN
C12           100    0.0506   -0.0740    0.0345    0.9681    0.0968

              MIN       MAX        Q1        Q3
C12       -1.9431    2.7134   -0.8359    0.6762
```

This quantity is called the standardized version of $\bar{x}$. Just as $\bar{x}$ has a distribution if we imagine taking repeated samples, so does z. The distribution of z is normal with a mean of 0 and a standard deviation of 1. Exhibit 7.2 contains a histogram of the z values for the data in Exhibit 7.1. Notice that most of these values are between -2 and 2 and all are between -2.75 and 2.75. It would be unusual to find a z value as large as 3 or as small as -3 since a standard normal distribution has 99% of its area between -2.576 and 2.579. The fact that very large and very small values of z are quite unusual forms the basis for the hypothesis tests in Section 7.4.

What if σ is not known? In practice we rarely know σ, the standard deviation of the population of xs. In most cases, the best we can do is estimate σ by s, the standard deviation of the sample. We can use this estimate to form the quantity

$$t = \frac{\bar{x} - \mu}{s/\sqrt{n}} = \frac{\bar{x} - 42}{s/\sqrt{9}} = \frac{\bar{x} - 42}{s/\sqrt{3}}$$

Exhibit 7.3 Histogram of the t Values for the Data of Exhibit 7.1

```
RSTDEV C1-C9 INTO C11
LET C13 = (C10-42)/(C11/SQRT(3))
HISTOGRAM C13

Histogram of C13   N = 100

Midpoint   Count
      -4       1  *
      -3       0
      -2       4  ****
      -1      28  ****************************
       0      33  *********************************
       1      24  ************************
       2       6  ******
       3       3  ***
       4       1  *

DESCRIBE C13

                 N      MEAN    MEDIAN    TRMEAN     STDEV    SEMEAN
C13            100     0.081    -0.070     0.038     1.201     0.120

               MIN       MAX        Q1        Q3
C13         -4.027     4.280    -0.731     0.860
```

This t statistic is called the Studentized version of $\bar{x}$. Notice that the formula for t is just like that for z except we have used s in place of σ. The population of t values does not have a normal distribution; it has a Student's t distribution. The t distribution is bell-shaped like the normal and has mean = 0. There are important differences, however: The t distribution is more spread out than a normal distribution. How much more depends on the sample size, n. When n is small, the t is much more spread out than a normal. When n is large, say over 30, the t is almost identical to the standard normal.

Exhibit 7.3 contains a histogram of the t values for the data from Exhibit 7.1. The t distribution is used as the basis for the confidence intervals described in Section 7.3 and the tests described in Section 7.5.

Exercises

7–1 (a) Run the program in Exhibit 7.1, but now use a sample of size $n = 2$ instead of 9. Discuss the output.

(b) Run the same program using $n = 9$ and $n = 25$. Compare the output for $n = 2$, 9, and 25. You might put the three histograms on the same scale, to facilitate comparisons.

7–2 Exhibit 7.1 helped illustrate the fact that means of random samples from normal distributions are normal. An equally important fact is that the means of samples of at least moderate size from most nonnormal distributions also are approximately normal. This approximation gets better as the sample size, n, gets larger. Simulate 200 observations from a uniform distribution into C1–C6. Then compute the means for samples of size $n = 1$, 2, 3, 4, 5, and 6. Commands to do this are given below.

```
RANDOM 200 C1-C6;
  UNIFORM 0, 1.
RMEAN C1     into C11
RMEAN C1-C2 into C12
RMEAN C1-C3 into C13
RMEAN C1-C4 into C14
RMEAN C1-C5 into C15
RMEAN C1-C6 into C16
HISTOGRAM C11-C16;
  SAME.
```

The last two lines make histograms of C11–C16. The SAME subcommand ensures they will all have the same scales. Compare the shapes of the histograms.

7–3 In this exercise we will take a look at how the distribution of the sample mean relates to the distribution of the individual observations in samples from some nonnormal distributions.

(a) Use RANDOM with the subcommand UNIFORM 0, 1 to simulate 100 observations from a distribution that is uniform on the interval 0 to 1. Put the observations into C10. Construct displays of these data with the following commands:

```
DOTPLOT C10
HISTOGRAM C10
HISTOGRAM C10;
  START .05;
  INCREMENT .1.
```

(b) How do these displays convey the feature that leads to the descriptive name "uniform distribution"? How do the results of the two HISTOGRAM commands differ? If particular bars of the two histograms look different, try to give an explanation. Discuss why we need to be careful when choosing cell boundaries in histograms, especially when the data are constrained to lie in a particular interval.

(c) Again use the uniform distribution on 0 to 1, but now simulate 100 samples, each of size 5, into C1–C5. Then use the command RMEAN (p. 42) to put the means of these 100 samples into C6. Now simulate 100 observations, uniform on 0 to 1, into C10. How does the distribution of the 100 means in C6 relate to the distribution of the 100 individual observations in C10? Use DESCRIBE C6,C10 and DOTPLOT C6,C10 to look at the differences. Discuss the shapes of the distributions. Does the shape of the distribution of the sample means look familiar? Record the mean and standard deviation of C6 and C10. Theory indicates that the mean of C6 should be approximately equal to the mean of C10. Is it? Theory also says that the standard deviation of C6 should be approximately equal to the standard deviation of C10 divided by $\sqrt{5}$. Is it? Why did we divide by $\sqrt{5}$?

7–4 In Exercise 7–3 above, we looked at data from a uniform distribution. Now we will look at another nonnormal distribution, the chi-square distribution with 3 degrees of freedom. (We will use this distribution again in Chapter 11.)

(a) Simulate 100 observations from the chi-square distribution and construct displays as follows:

```
RANDOM 100 C10;
  CHISQUARE 3.
DOTPLOT C10
HISTOGRAM C10
```

(b) How does the shape of this chi-square distribution compare to the normal curve?

(c) Again use the chi-square distribution with 3 degrees of freedom, but now simulate 100 samples, each of size 5, into C1–C5. Then use RMEAN to put the means of these 100 samples into C6. Now answer all the questions in part (c) of Exercise 7–3.

7.2 Confidence Interval for μ When σ Is Known

Suppose we want to estimate the mean μ of a population. We might take a sample, calculate the sample mean, $\bar{x}$, and use this as an estimate of μ. Exhibit 7.1 showed how close the population mean and sample mean are likely to be. A confidence interval uses this idea in a more formal way.

A 95% confidence interval is an interval, calculated from the sample, which is very likely to cover the unknown mean, μ. To be more precise, if we were to use the formula for a 95% confidence interval on many, many sets of data, then 95% of our intervals would cover the unknown mean we are trying to estimate and 5% would not cover it. Unfortunately, in practice we can never know which intervals are the successful ones and which are the failures. The value 95% is called the confidence level of the interval. The confidence levels most commonly used in practice are 95%, 99%, and 90%.

Suppose we happen to know the standard deviation of the population. Then we can use the ZINTERVAL command to find a confidence interval for the population mean.

As an example, suppose we want to find a 95% confidence interval for μ based on a sample of eight observations: 4.9, 4.7, 5.1, 5.4, 4.7, 5.2, 4.8, and 5.1. If we knew that $\sigma = 0.2$ then we could use the following commands:

```
SET C1
  4.9,4.7,5.1,5.4,4.7,5.2,4.8,5.1
END
ZINTERVAL SIGMA = 0.2, C1
```

The ZINTERVAL command gave the following results:

```
THE ASSUMED SIGMA = 0.200
     N  MEAN   STDEV SE MEAN   95.0 PERCENT C.I.
C1   8 4.987   0.253   0.071 (   4.849,     5.126)
```

The 95% confidence interval goes from 4.849 to 5.126. This set of data happens to be one we generated ourselves using the procedures in

Chapter 6, so we know $\mu = 5$. Since 5 is inside our confidence interval, we have a successful interval. We know that, in the long run, 95% of our intervals will be successful and 5% will be failures. Here we got one of the successful ones.

The value .071, labeled SE MEAN, is the standard error of the sample mean. This is just another name for the standard deviation of $\bar{x}$. Recall it is calculated as $\sigma_{\bar{x}} = \sigma/\sqrt{n}$. In the preceding section, we saw $\bar{x}$ and μ rarely differed by more than $2\sigma_{\bar{x}}$. This is the essence of the 95% confidence interval computed above. If we go up and down from $\bar{x}$ by $2\sigma_{\bar{x}}$, we get

$$\bar{x} - 2\sigma_{\bar{x}} = 4.987 - 2(.071) = 4.845$$
$$\bar{x} + 2\sigma_{\bar{x}} = 4.987 + 2(.071) = 5.129$$

The interval printed by ZINTERVAL differs slightly from this. It uses 1.96 instead of 2 as a multiplier for $\sigma_{\bar{x}}$. The 1.96 is the proper value for a 95% confidence interval since 95% of the area under the standard normal curve is located between -1.96 and $+1.96$.

ZINTERVAL [K percent confidence] sigma = K, for C, ..., C

For each column, ZINTERVAL calculates and prints a confidence interval for the population mean, μ, using the formula

$$\bar{x} - z(\sigma/\sqrt{n}) \quad \text{to} \quad \bar{x} + z(\sigma/\sqrt{n})$$

Here σ is the known value of the population standard deviation, $\bar{x}$ is the mean of the sample, n is the number of observations in the sample, and z is the value from the standard normal distribution corresponding to K percent confidence.

If the confidence level is not specified, 95% is used.

Exercises

7–5 The output on page 164 gives a 95% confidence interval for μ. Use this output to calculate by hand a 90% confidence interval.

7–6 Suppose $n = 9$ men are selected at random from a large population. Assume the heights of the men in this population are normal, with $\mu =$ 69 inches and $\sigma = 3$ inches. Simulate the results of this selection 20 times and in each case find a 90% confidence interval for μ. The following commands may be used:

```
RANDOM 9 C1-C20;
  NORMAL 69 3.
ZINTERVAL .90, SIGMA = 3, C1-C20
```

(a) How many of the intervals contain μ?

(b) Would you expect all 20 of the intervals to contain μ? Explain.

(c) Do all the intervals have the same width? Why or why not?

(d) Suppose you took 95% intervals instead of 90%. Would they be narrower or wider?

(e) How many of your intervals contain the value 72? The value 70? The value 69?

(f) Suppose you took samples of size $n = 100$ instead of $n = 9$. Would you expect more or fewer intervals to cover 72? 70? 69? What about the width of the intervals for $n = 100$? Would they be narrower or wider than for $n = 9$?

(g) Suppose you calculated 90% confidence intervals for 20 sets of real data. About how many of these intervals would you expect to contain μ? Could you tell which intervals were successful and which were not? Why or why not?

7–7 In this exercise we will simulate random samples into rows rather than columns. Using rows will make it easier to plot the confidence intervals for the samples, even though we will not be able to use the ZINTERVAL command since it expects data in columns.

(a) The following program first simulates 50 people into C1–C5. Each row contains one sample of five observations from a normal distribution with $\mu = 30$ and $\sigma = 2$. Next, a 90% confidence interval is calculated for each row: The lower endpoint of the interval is in C21 and the upper endpoint is in C22. The program then plots the 50 intervals. Run this program.

```
RANDOM 50 samples into C1-C5;
  NORMAL 30  2.
RMEAN C1-C5 into C10
LET K1 = 1.645 * 2/SQRT(5)
LET C21 = C10-K1
LET C22 = C10+K1
SET C23
1:50
END
MPLOT C21,C23 C22,C23
```

(b) Examine the plot. Are the intervals all of the same width? How many intervals missed the value of $\mu = 30$? With 90% confidence intervals about how many of the 50 intervals would you expect to miss?

7–8 In a z interval you must assume you know σ. This exercise takes a brief look at what happens if you specify an incorrect value.

(a) Repeat Exercise 7–7 above, but use 1 for σ instead of 2 when you calculate the endpoints of the interval. (Use $\sigma = 2$, however, in the NORMAL subcommand to RANDOM.) Did you still have about 90% of your intervals covering μ?

(b) Repeat exercise 7–7, but now use 4 for σ. Now what percent of your intervals cover μ?

(c) Would you say that knowing σ precisely was important or unimportant to the proper use of a z confidence interval?

7–9 The theory used in developing z confidence intervals assumes your data come from a normal distribution. In this exercise we will look at what happens when that assumption is violated; that is, when the data come from some other distribution.

(a) Simulate 50 random observations from a uniform distribution on 0 to 1 into C1–C5. This distribution has a standard deviation of $1/\sqrt{12} = .2887$ and a mean of .5.

(b) Compute and plot the 90% confidence intervals as in Exercise 7–7.

(c) What percent of your intervals were successful? The uniform distribution is a fairly extreme departure from normality. What can you conclude about the sensitivity of z intervals to nonnormality? That is, does the 90% hold up pretty well or not?

7–10 *Approximate binomial confidence intervals.* The normal approximation to the binomial distribution (see Section 6.3) can be used as the basis for computing approximate confidence intervals. For binomial data, the observed proportion of successes $\hat{p}$ provides an estimate of p. The standard deviation of $\hat{p}$ is $\sqrt{p(1-p)/n}$ which can be estimated by $\sqrt{\hat{p}(1-\hat{p})/n}$. Thus the normal approximation indicates that we can compute a 90% confidence interval for p with the formula $\hat{p} \pm 1.645 \sqrt{\hat{p}(1-\hat{p})/n}$.

(a) The following commands simulate 50 binomial observations with $n = 20$ and $p = .3$, compute $\hat{p}$, put the upper and lower confidence limits into C3 and C4, and then plot these intervals. Run this program.

```
RANDOM 50 C1;
  BINOMIAL 20  .3.
```

```
LET C2=C1/20
LET C10 = 1.645*SQRT(C2*(1-C2)/20)
LET C3 = C2 - C10
LET C4 = C2 + C10
SET C5
1:50
END
MPLOT C3 C5, C4 C5
```

(b) Examine the plot. Do all the intervals have the same width? Explain how many intervals were successful in catching the correct value of p.

7.3 Confidence Interval for μ When σ Is Not Known

Usually we do not know the standard deviation of the population and must estimate it from the data on hand. Then we can use Student's t confidence interval procedure and the TINTERVAL command.

As an example, suppose we want to find a 90% confidence interval based on the random sample: 4.9, 4.7, 5.1, 5.4, 4.7, 5.2, 4.8, 5.1. We could use the following commands:

```
SET C2
  4.9,4.7,5.1,5.4,4.7,5.2,4.8,5.1
END
TINTERVAL 90 PERCENT C2
```

The TINTERVAL command gave the following results:

```
          N   MEAN STDEV SE MEAN    90.0 PERCENT C.I.
C2        8  4.987 0.253   0.090 (    4.818,     5.157)
```

Based upon this output we might say, "We estimate the mean to be about 4.99, and we are 90% confident that it is somewhere between 4.82 and 5.16." Since we simulated these data from a population with μ = 5, we know this interval was a success: It caught the population mean.

Here, SE MEAN is the estimated standard error (or standard deviation) of $\bar{x}$. It is calculated by $s/\sqrt{n}$, where s is the sample standard deviation.

TINTERVAL [K percent confidence] for data in C, ..., C

For each column, TINTERVAL calculates and prints a t confidence interval for the population mean, using the formula

$$\bar{x} - t(s/\sqrt{n}) \quad \text{to} \quad \bar{x} + t(s/\sqrt{n})$$

Here $\bar{x}$ is the sample mean, s is the sample standard deviation, n is the sample size, and t is the value from the t distribution, using $(n - 1)$ degrees of freedom and K percent confidence.

If the confidence level is not specified, 95% is used.

Exercises

7–11 (a) Get a histogram for the SAT verbal scores in sample A of the Grades data (p. 309). Calculate $\bar{x}$ and get a 95% confidence interval for μ = mean score. How does the confidence interval compare to the histogram?

(b) Also get a 90% and then a 99% confidence interval for mean SAT verbal scores. How do they compare with the 95% confidence interval in part (a)? Do they have the same centers? Do they have the same widths?

7–12 Repeat the simulation of Exercise 7–6 but now assume σ is unknown and use the TINTERVAL command. Get a total of 20 90% intervals.

(a) How many of the 20 intervals contain μ?

(b) Would you expect all of the intervals to contain μ? Explain.

(c) Do all of the intervals have the same width?

(d) Compare the t intervals of this exercise to the z intervals of Exercise 7–6. On the average, which kind of interval seems to be wider?

(e) Suppose you calculated 95% t intervals instead of 90%. Would they be narrower or wider?

(f) How many of your intervals contain 72? 70? 69?

(g) Suppose you took samples of size $n = 100$ instead of $n = 9$. Would you expect more or fewer intervals to cover 72? 70? 69? What about the size of the intervals for $n = 100$? Would they be longer or shorter than those for $n = 9$?

(h) Suppose you calculated 20 90% t intervals for real data. About how many would you expect to contain the true μ? Could you tell which?

7–13 Repeat Exercise 7–7, except use t confidence intervals rather than z confidence intervals. You will need to use RSTDEV to calculate the standard deviation for each row and use the appropriate value from a t table. The command INVCDF could be used to get the t table value.

7–14 The theory used in developing t confidence intervals assumes your data come from a normal distribution. In this exercise we will look at what happens when that assumption is violated; that is, when the data come from some other distribution.

(a) Simulate 50 random observations from a uniform distribution on 0 to 1 into C1–C5.

(b) Compute and plot the 90% t confidence intervals as in Exercise 7–13 above. What is the appropriate value to use from the t table?

(c) What percent of your intervals were successful? The uniform distribution is a fairly extreme departure from normality. What can you conclude about the sensitivity of t intervals to nonnormality? That is, did the 90% hold up pretty well in this simulation?

7.4 Test of Hypothesis for μ When σ Is Known

Suppose a machine is set to roll aluminum into sheets that are 40.0 thousandths of an inch thick. To check that the machine stays in adjustment, the operator periodically takes five measurements of sheet thickness. On the latest occasion, the results were:

40.1 39.2 39.4 39.8 39.0

This sample has a mean of 39.5, which is below the desired mean of 40.0. Is this discrepancy just due to random fluctuation or does it indicate the machine is not adjusted correctly? To answer this, we must have some idea of how much the thickness naturally varies from sheet to sheet. Suppose we know from past experience that the standard deviation of sheet thickness is $\sigma = .4$. Then we can use Minitab's ZTEST command to test the null hypothesis that the machine is properly adjusted, H_0: $\mu = 40.0$, versus the alternative hypothesis that the machine is misadjusted, H_1: $\mu \neq 40.0$.

The following commands do the test for the machine data:

```
SET C1
  40.1 39.2 39.4 39.8 39.0
END
ZTEST MU = 40, SIGMA = .4 C1
```

The ZTEST command gave the following results:

```
TEST OF MU = 40.0 VS MU N.E. 40.0
THE ASSUMED SIGMA = 0.400

           N    MEAN   STDEV  SE MEAN       Z    P VALUE
C1         5  39.500   0.447     0.18   -2.80     0.0053
```

The value of $\bar{x}$ differs from 40 by 5. It has a z value of

$$z = \frac{\bar{x} - \mu}{\sigma/\sqrt{n}} = \frac{39.5 - 40}{.4/\sqrt{5}} = -2.80$$

As we saw in Section 7.1, values of z bigger than 2 rarely occur. Therefore, we should check to see if we have misjudged something. Perhaps μ is not equal to 40, perhaps μ is really 39.7, or even some lower value, implying the machine may be misadjusted.

The p-value (also called significance level) says how unusual such a discrepancy is. Here p-value $= .0053$ or .53%. This means that if $\mu = 40$, then .53% of the time $\bar{x}$ will be this far or further from μ. The rest of the time, $\bar{x}$ will be closer than this to μ. Therefore, $\bar{x} = 39.5$ is quite unusual for a perfectly adjusted machine.

The value of the test statistic, z, and the p-value are related. Here we are using a two-sided test; that is, a test where the alternative hypothesis is H_1: $\mu \neq$ K. In such cases the p-value equals twice the area beyond z, under a standard normal curve.

In most textbooks, statistical tests are done by specifying what is called an α level. Once we know the p-value, however, we can determine the results of a test for any value of α. If the p-value is less than α, we reject H_0; if it is greater than α, we don't reject H_0. For example, if we did a test here using $\alpha = .01$, we would reject H_0 since .0053 is less than .01. If we used $\alpha = .001$, we would not reject.

Practical Significance Versus Statistical Significance

It is rare that practical significance and statistical significance coincide precisely. Here our best estimate is that the process mean is about 39.5. The ZTEST showed this result was statistically significant if $\alpha = .01$. This in itself gives us no clue whatsoever about whether the process mean has shifted seriously off target. Is 39.5 far from 40.0 in a practical sense? It depends on the use of the rolled aluminum, the costs of production, and so on.

In some applications, tolerances are very tight. Even results as close as 39.9 may still be too far off to be acceptable. In other situations, it may not make much practical difference as long as the sheets are at least as thick as 39.3. In other words, a statistical test of significance tells us only whether or not the observed data are unusual under the hypothesized situation. It tells us nothing about what practical course of action we should take. This is the heart of the reason why we prefer confidence intervals over tests of significance. A 95% confidence interval for these data goes from 39.09 to 39.91. Thus, we are fairly confident that the true mean thickness is between 39.09 and 39.91. If all we really need is a mean thickness of 39.3 or more, then, for practical purposes, the machine may be okay.

ZTEST [of mu = K] sigma = K on data in C, ..., C

For each column, ZTEST tests the null hypothesis, H_0: μ = K, against the alternative hypothesis, H_1: $\mu \neq$ K, where μ is the hypothesized population mean. If μ is not specified on the ZTEST command, H_0: $\mu = 0$ is tested. The command calculates the test statistic

$$z = \frac{\bar{x} - K}{\sigma/\sqrt{n}}$$

Here $\bar{x}$ is the sample mean, n is the sample size, σ is the known population standard deviation, and K is the hypothesized value of the population mean.

If you want to do a one-sided test, use the subcommand

ALTERNATIVE = K

where K = -1 corresponds to H_1: $\mu <$ K and K = $+1$ corresponds to H_1: $\mu >$ K.

Exercises

7–15 Imagine choosing $n = 16$ women at random from a large population and measuring their heights. Assume that the heights of the women in this population are normal, with $\mu = 64$ inches and $\sigma = 3$ inches. Suppose you then test the null hypothesis H_0: $\mu = 64$ versus the alternative that H_1: $\mu \neq 64$, using $\alpha = .10$. Assume σ is known. Simulate the results of doing this test 20 times as follows:

```
RANDOM 16 C1-C20;
  NORMAL 64 3.
ZTEST MU = 64 SIGMA = 3 on C1-C20
```

(a) In how many tests did you fail to reject H_0? That is, how many times did you make the "correct decision"?

(b) How many times did you make an "incorrect decision" (that is, reject H_0)? On the average, how many times out of 20 would you expect to make the wrong decision?

(c) What is the attained significance level (p-value) of each test? Are they all the same?

(d) Suppose you used $\alpha = .05$ instead of $\alpha = .10$. Does this change any of your decisions to reject or not? Should it in some cases?

7–16 As in Exercise 7–15 above, simulate choosing 16 women at random, measuring their heights, and testing H_0: $\mu = 64$ versus H_1: $\mu \neq 64$, but this time assume that the population really has a mean of $\mu = 63$, instead of 64. Thus, use the subcommand NORMAL with $\mu = 63$ and $\sigma = 3$

to simulate the samples. Use α = .10 and assume σ is known. Do this for a total of 20 tests.

(a) In how many tests did you reject H_0? That is, how many times did you make the "correct decision"? How many times did you make an "incorrect decision"?

(b) What we are investigating is what's called the power of a test; that is, how well the test procedure does in detecting that the null hypothesis is wrong, when indeed it is wrong. Repeat the above simulation, but now assume the true population mean is μ = 62 (still use σ = 3 and the same null hypothesis). How often did you make the correct decision (reject) in these 20 tests? On the average, would you expect to make more "correct" decisions if the true mean were 62 or if the true mean were 63?

7–17 In this exercise, we will use simulation to study hypothesis tests. We will do our arithmetic with rows of data rather than columns because it is easier this way. We will not be able to use the ZTEST command because ZTEST expects the data in columns.

(a) The following program first simulates 200 samples into C1–C5. Each row contains one sample of five observations from a normal distribution with μ = 30 and σ = 2. Next, the z statistic for testing H_0: μ = 30 versus H_1: $\mu \neq$ 30 is calculated for each row. The program then makes a dotplot of these 200 z values. Run this program.

```
RANDOM 200 into C1-C5;
  NORMAL 30,2.
RMEAN C1-C5 into C10
LET C11 = (C10-30)/(2/SQRT(5))
DOTPLOT C11
```

(b) Suppose we do the test using α = .05. Then we reject H_0 if the z value is less than −1.96 or greater than 1.96. For (approximately) what percentage of your 200 samples do you reject H_0? (It should be roughly 5%). This illustrates the fact that when you do hypothesis tests, you can make the wrong decision. In fact, if you do tests using α = .05, and the null hypothesis is true, you will make the wrong decision 5% of the time, on the average.

7.5 Test of Hypothesis for μ When σ Is Not Known

Suppose we once again test whether or not the aluminum sheet machine mentioned in the previous section is correctly adjusted. But this time let's not assume we know the value of σ. The following commands can be used:

```
SET C1
  40.1 39.2 39.4 39.8 39.0
END
TTEST MU = 40.0, C1
```

The TTEST command gave the following output:

```
TEST OF MU = 40.0 VS MU N.E. 40.0

            N    MEAN   STDEV SE MEAN       T      P VALUE
C1          5  39.500   0.447    0.20   -2.50        0.067
```

Again, the p-value is small. Therefore, we suspect that the rolling machine is not perfectly adjusted. This test, called Student's t-test, is based on the Studentized value of $\bar{x}$, which is labeled T in this output.

TTEST [of mu = K] on data in C, ..., C

For each column, TTEST tests the null hypothesis H_0: μ = K against the alternative hypothesis H_1: $\mu \neq$ K. If μ is not specified on the TTEST command, H_0: μ = 0 is used.

The command calculates Student's t-test statistic

$$t = \frac{\bar{x} - \mathrm{K}}{s/\sqrt{n}}$$

Here, $\bar{x}$ is the sample mean, n is the sample size, s is the sample standard deviation, and K is the hypothesized value of the population mean.

If you want to do a one-sided test, use the subcommand

ALTERNATIVE = K

where K = -1 corresponds to H_1: $\mu <$ K and K = $+1$ corresponds to H_1: $\mu >$ K.

Exercise

7–18 Do a test to see if there is evidence that the preprofessionals who participated in the Cartoon experiment (p. 303) have OTIS scores that differ significantly from the national norm of 100. First, use COPY (fully described on p. 335) to select the data for the preprofessionals. Then get a histogram. What do you think? Now do the appropriate test, using α = .05. The following commands may be used:

```
RETRIEVE 'CARTOON'
COPY C1-C9 INTO C1-C9;
  USE C3 = 0.
HISTOGRAM 'OTIS'
TTEST MU=100 'OTIS'
```

7.6 Departures from Assumptions

Many procedures in statistical inference, including those described in this chapter, are based on the assumption that the data are a random sample from a normal distribution. Here we take a quick look at how you can spot departures from these assumptions. We treat the most serious problem first—lack of randomness.

Nonrandomness. Most data are not a random sample from any population. In particular, observations taken close together are often more alike than those taken further apart. For example,

People from the same part of the country tend to be more alike than those in different regions.

Light bulbs manufactured on the same day tend to be more alike than those manufactured on different days.

Adjacent pieces from a roll of tape tend to have more nearly the same degree of stickiness than do pieces of tape made on different days or by different machines.

The measurements made by one inspector are often more alike than measurements made by different inspectors.

Observations taken in close proximity, either in time or in space, often are correlated. Observations that are correlated do not form a simple random sample. One of the best ways to look for correlations is to make plots. For example, if measurements are made one after the other, plot them in that order. If they are made in bunches, plot them so the bunches are readily identifiable. If you see a clear pattern, then you probably do not have a random sample.

Exhibit 7.4 gives six examples. In each case, the observations, *y*,

Exhibit 7.4 Plots of Normal Data with Various Departures from Randomness

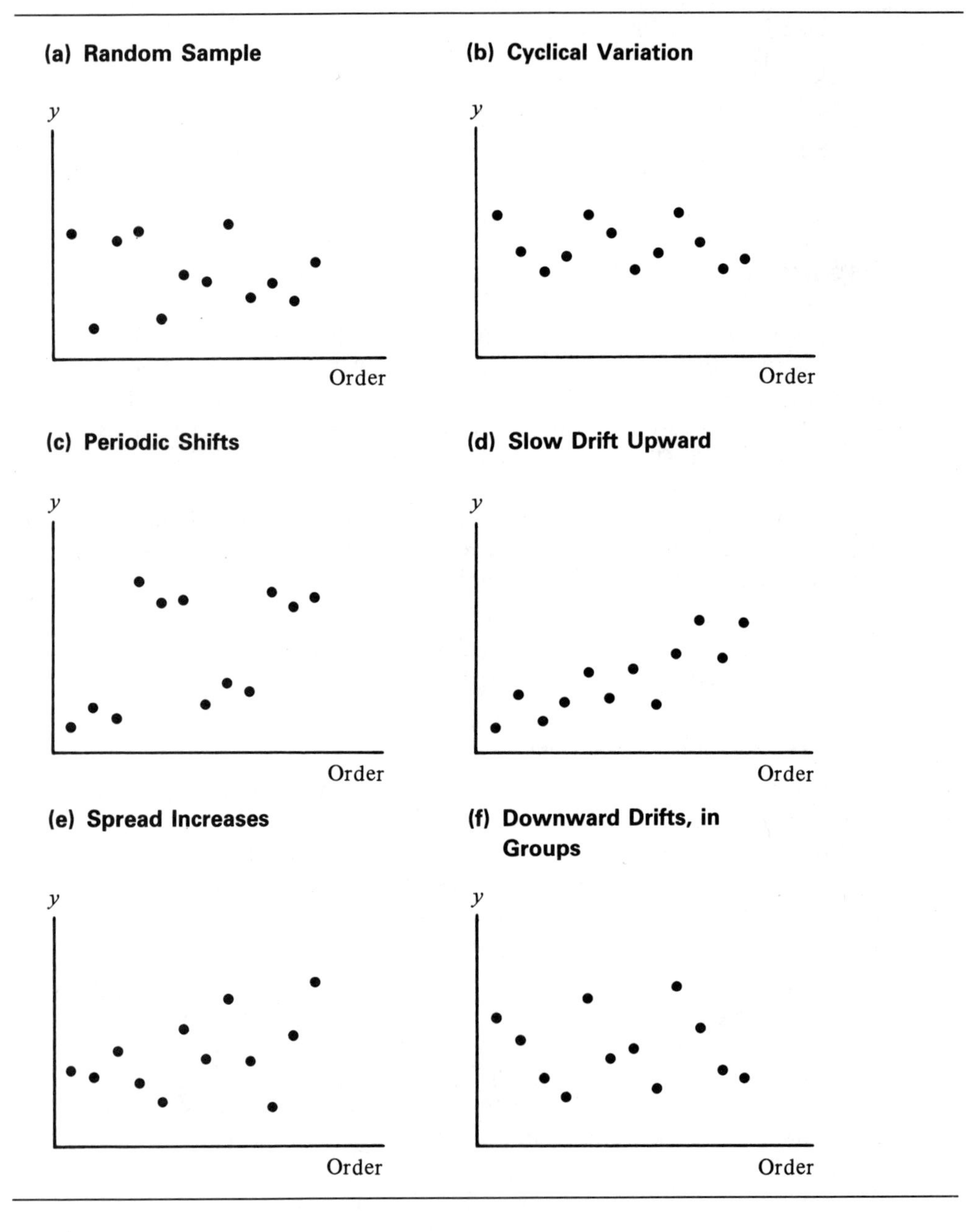

are plotted versus "order." Here order could be time order, spatial order, or some other order that helps us see the nonrandom nature of the data. The first plot shows no evidence of nonrandomness, but all the other plots have very clear patterns.

Nonnormality. The techniques in this chapter, and Chapters 8–10, all require one more condition if they are to be exactly valid: The population from which our sample is drawn must have a normal distribution. This would seem to rule out most practical applications of these techniques since no real population is ever exactly normal. But, in fact, it does not. Techniques based on the normal distribution give very good, though approximate, answers even if the production is nonnormal.

For example, suppose we construct a 95% t confidence interval for μ using 20 observations. If the population is nonnormal, then the true confidence will not be 95%. It might be 94% or 90% or 96%, depending on what the actual shape of the population is; but it is not likely to be very far from 95%. Similarly, suppose we do a t test using these data and get a p-value of .031, say. If the population is nonnormal the correct p-value is not .031. It might be .038 or .047, or .026, but it will not be very far from .031.

We should mention that even though these procedures usually work well for nonnormal populations, in some cases the nonparametric methods in Chapter 12 are better (or more powerful, in statistical terminology). The confidence interval constructed by a nonparametric procedure may be shorter than the t confidence interval. A shorter interval gives us a more precise estimate of μ. A nonparametric test may be more likely to reject the null hypothesis when the null hypothesis is indeed false.

There are many methods that can help us decide if a population is normal. Some of the sophisticated methods use formal tests. Here we will look at two simple graphical techniques.

Suppose we take a sample from the population and get a histogram. If the population is normal, the sample histogram should have approximately the shape of a normal curve. If the sample size is large, this approximation should be very good. Of course, if we have just a few observations, say 10 or 15, then it will be difficult to see a clear shape in the sample histogram; but we still may be able to spot gross departures from normality.

Another plot, called a normal probability plot, is a useful supplement to histograms in checking for nonnormality. It plots the sample versus the values we would get, on the average, if the sample came from a normal population. This plot is approximately a straight line if the sample is from a normal population, but exhibits curvature if the population is not normal.

Minitab can make a normal probability plot. If C1 contains the data, use the two commands

```
NSCORES OF C1 PUT IN C2
PLOT C1 VS C2
```

Exhibit 7.5 gives a probability plot of the OTIS scores from the Cartoon experiment (p. 303). Again we are faced with the problem of judging a picture. Does this plot look straight? There certainly is some curvature. But is it curved enough to doubt the normality of the OTIS scores? Good judgment of graphs comes from experience. Some find it easier to learn how to judge probability plots than histograms. Some of the exercises at the end of this section show how you can use simulation to improve your judgment of the nonnormality of histograms and normal probability plots.

Exhibit 7.5 Normal Probability Plot of OTIS Scores from the Cartoon Data

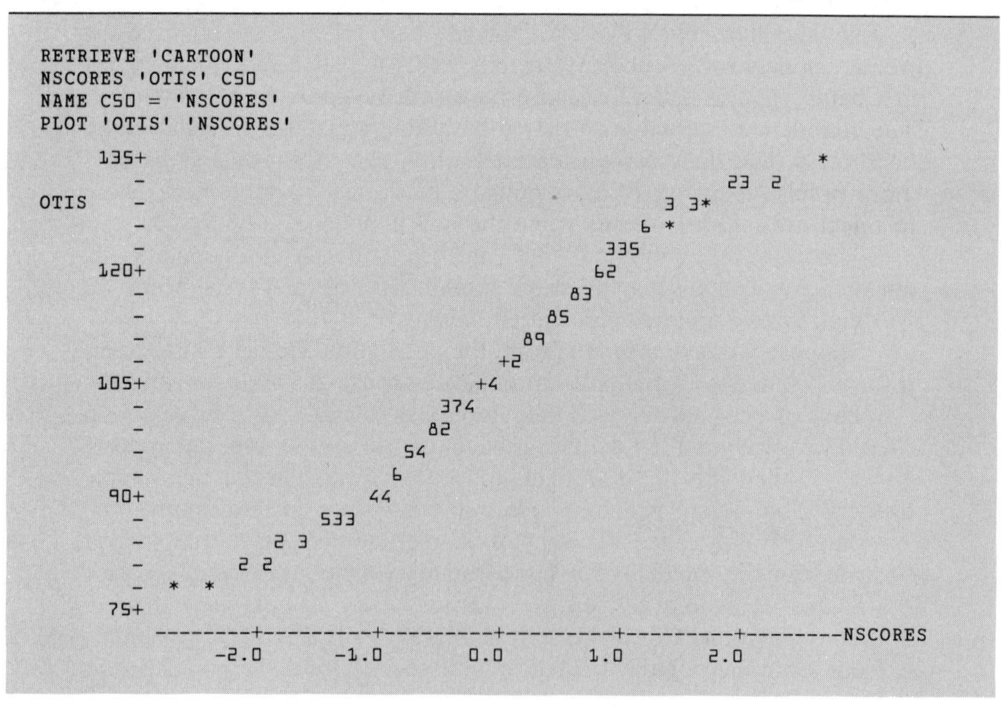

NSCORES of C put in C

Calculates the normal scores of a set of data.

Example

```
NSCORES OF C1 PUT IN C2
```

C1	*C2*
1.1	− .20
1.9	1.28
.8	−1.28
1.3	.64
1.2	.20
.9	− .64

Loosely speaking, the normal scores in the example above can be defined as follows: The number −1.28 in C2 is the smallest value you would get, on the average, if you took samples of size $n = 6$ from a standard normal population (with $\mu = 0$, $\sigma = 1$). This number is placed next to the smallest value in C1. The number −.64 in C2 is the second smallest value you would get, on the average, in samples of size 6 from a standard normal. It is placed next to the second smallest value in C1. This is continued for all six values. (More precisely, the ith smallest normal score is found by computing the inverse cdf of $(i - 3/8)/(n + 1/4)$.)

Exercises

7–19 This exercise is designed to give you some idea what a normal probability plot can look like. It will help you learn to decide if a particular plot looks definitely nonnormal.

(a) Use the RANDOM command to simulate 20 observations from a normal distribution with mean 0 and standard deviation 1. Make a normal probability plot. Repeat this a total of five times.

(b) Repeat (a) but use $n = 50$ observations.

7–20 (a) Simulate 50 observations from a normal population with $\mu = 0$, $\sigma = 1$. Get a histogram and a probability plot.

(b) Repeat part (a) until you have five histograms and five probability plots. How do the histograms and plots compare?

7–21 Simulate five normal probability plots to compare with the OTIS scores. (Choose appropriate values for n, μ, and σ.) Compare these with the plot in Exhibit 7.5. Is there any evidence that the OTIS scores may not be a random sample from a normal population?

7–22 Simulate 100 observations from a normal distribution with mean 0 and standard deviation 1 into C1. Then make a normal probability plot and comment on the appearance of the plot under each of the following circumstances:

(a) Use the entire set of 100 simulated observations.

(b) Use just the observations between -1 and 1, selected with the following instruction (COPY is described on p. 335):

```
COPY C1 INTO C2;
  USE C1 = -1:1.
```

(c) Repeat (b) but select the observations between -2 and 2.

(d) Append two "outliers" with the following command (STACK is described on p. 337):

```
STACK C1 ON -4, +4, PUT IN C3.
```

(e) Repeat (d) but put the outliers at -5 and 5. Then do (d) again with the outliers at -6 and 6.

8 Comparing Two Means: Confidence Intervals and Tests

Often we are interested in the question, "How different are the means of two populations?" This chapter shows how Minitab can be used to answer this question in some common situations. The techniques discussed depend heavily on the concepts treated in Sections 5.1, 5.2, and 5.3. In particular, it is important to be able to distinguish between paired data and independent samples.

The remarks in Chapter 7 concerning random samples and normal populations apply here also. That is, although the methods we present are "exact" only if the populations have normal distributions, they still work quite well for most populations. However, the nonparametric methods of Chapter 12 are more powerful for some nonnormal populations.

8.1 Difference Between Two Means: Paired Data

In Section 5.1 (p. 92), we described some data that were collected in a study of the blood cholesterol levels of heart attack patients. A total of 28 heart attack patients had their cholesterol levels measured two days after the attack, four days after, and 14 days after. In addition, the cholesterol levels were recorded for a control group of 30 people who had not had a heart attack.

How did cholesterol level change between the second and fourth days following the heart attack? As we mentioned in Section 5.1, this is an example of paired data—each 2-DAY value is paired with a 4-DAY value. We will pretend these patients are a random sample of heart attack patients from that hospital. We are interested in estimating how much difference on the average there is between the day 2 and day 4 cholesterol levels of all heart attack patients of this hospital.

Confidence Interval

To get a confidence interval for the average difference, we first find the change in each individual's cholesterol level. The mean of the changes in this sample of 28 patients gives us an estimate of the mean change in the population. To get some idea of the uncertainty in this estimate, we compute a *t* confidence interval. Exhibit 8.1 shows how to do this in Minitab.

Exhibit 8.1 Analysis of the Change in Cholesterol Level, from Day 2 to Day 4 Following a Heart Attack

```
RETRIEVE 'CHOLEST'
NAME C5='CHANGE'
LET 'CHANGE' = '4-DAY' - '2-DAY'
STEM-AND-LEAF 'CHANGE'

Stem-and-leaf of CHANGE     N  =   28
Leaf Unit = 10

    1    -1   0
    3    -0   98
    5    -0   66
    9    -0   5554
   14    -0   33322
   14    -0   11100
    9     0   000011
    3     0   23
    1     0   4

TINTERVAL 95 'CHANGE'

                 N        MEAN      STDEV   SE MEAN     95.0 PERCENT C.I.
CHANGE          28       -23.3       38.3       7.2   (  -38.1,      -8.4)
```

TINTERVAL gives the average change in cholesterol for these 28 patients, which was −23.3. The minus sign indicates that this change was negative, so cholesterol decreased on the average. From the stem-and-leaf display, we see that about two-thirds of the patients had a decrease. But, on the other hand, about a third experienced an increase. At this point we have a good description of what happened to these 28 patients. But suppose we are interested in how cholesterol changes in some larger population of heart attack patients. The confidence interval gives us some indication of what might happen there.

The confidence interval means, "If these 28 patients were a random sample from a large population of patients, then a 95% confidence interval for the average change in that population would be −38.1 to −8.4." Thus, we would be fairly certain that cholesterol goes down, on the average, by at least eight and maybe by as much as 38 units.

These 28 patients, however, were not a random sample from any population. They were just the 28 heart attack patients who happened to appear in that hospital during the time when these data were being collected. There are some who would argue that without a random sample, the confidence interval is useless. There is some merit to this position, but we think it goes too far. We feel the confidence interval gives us flawed but still useful information. Our rationale is as follows: Even if the sample of patients were not random, the changes they experienced might well act much like a random sample. In other words, if the changes were a random sample, the confidence interval still would be appropriate. In this study the doctors saw no evidence that this group of patients was different in any important way from other heart attack patients they knew. If the doctors were right, the confidence interval might well be a good indicator for the population of patients at that hospital.

The next level of extrapolation, to the population of all heart attack victims in the state, nation, or world, is probably even more risky. There may well be differences in factors such as age, race, and climate that make patients at this hospital different from the larger population.

To sum up, confidence intervals sometimes are useful indicators even when based on a nonrandom sample. When a confidence interval is not based on a random sample, and they seldom are, then the real uncertainty is usually somewhat larger, and often much larger, than indicated by the confidence interval. Thus, one use of confidence intervals in nonrandom samples is to express a lower bound to the real uncertainty in the estimate. Here, as in all other cases, statistics need to be used with caution and intelligence.

Paired *t*-Test

We just saw that a confidence interval for paired data could be done by first computing the changes, then computing a t interval based on the

Exhibit 8.2 Paired *t* Test

```
RETRIEVE 'CHOLEST'
NAME C5 = 'CHANGE'
LET 'CHANGE' = '4-DAY' - '2-DAY'
TTEST 0 'CHANGE'

TEST OF MU = 0 VS MU N.E. 0

                N       MEAN      STDEV    SE MEAN         T     P VALUE
CHANGE         28      -23.3       38.3        7.2     -3.22      0.0033
```

changes. The t-test for paired data can be done in a similar manner. For example, Exhibit 8.2 shows how to test the null hypothesis that the average difference in cholesterol levels in the "population" was zero.

We see, as before, that the estimate of the mean change is -23.3. The p-value for the test is .0033. With such a small p-value, we have strong evidence that the mean change in the population is not zero. We would reject our null hypothesis if we had used $\alpha = .05$ or $\alpha = .01$, or any value of α down to .0033.

Exercises

8–1 Use the output in Exhibit 8.1 to calculate by hand a 90% confidence interval for CHANGE. Also calculate a 99% confidence interval by hand.

8–2 Table 5.2 (p. 102) contains data to compare two materials for making shoes.

(a) Do an appropriate test to see if there is statistically significant evidence of a difference between these two materials, using $\alpha = .05$.

(b) Estimate the difference with a 95% confidence interval.

8–3 In this section, we saw that the cholesterol level of heart attack patients decreases by about 23 units between the second and fourth day after a heart attack. Estimate how much change there is between the fourth and fourteenth day following the attack. Notice there are some missing observations in the data. Minitab omits these cases from the analysis in TTEST and TINT. What effect might this have on your conclusions?

8–4 The 179 participants in the Cartoon experiment (p. 303) each saw cartoon and realistic slides.

(a) Do an appropriate test to see if there is any difference between the two types of slides.

(b) Estimate the difference between the two types of slides with a 90% confidence interval.

8.2 Difference Between Two Means: Independent Samples

This section builds on the concepts developed in Section 5.2 (p. 95). In particular, slightly different steps are used in the analysis of "stacked" and "unstacked" data. We recommend here, as in most other analyses, that some plots precede calculations.

Unstacked Data

The Cholesterol data set (p. 93) contains two independent groups: the heart attack patients and the controls. As an example, let us compare the day 2 readings with the control group. In Section 5.2, we did appropriate plots. Now, suppose we wanted to compare these two groups with a confidence interval and *t*-test. Exhibit 8.3 contains an example. Notice that TWOSAMPLE-T does both a confidence interval and a *t*-test.

The procedure used by TWOSAMPLE-T is slightly different from that described in many textbooks. These textbooks use the pooled *t*, a procedure which assumes that both populations have the same standard deviation. The name *pooled* is used because the standard deviations from the two samples are pooled to get an estimate of the common standard deviation. If you use the pooled procedure when it is not appropriate—that is, when the standard deviations of the two populations are not

Exhibit 8.3 Two-sample *t* Procedure Using Cholesterol Data

```
RETRIEVE 'CHOLEST'
TWOSAMPLE-T  '2-DAY' VS 'CONTROL'

TWOSAMPLE T FOR 2-DAY VS CONTROL
          N       MEAN      STDEV    SE MEAN
2-DAY    28      253.9       47.7        9.0
CONTROL  30      193.1       22.3        4.1

95 PCT CI FOR MU 2-DAY - MU CONTROL: (40.7, 80.8)
TTEST MU 2-DAY = MU CONTROL (VS NE): T=6.15 P=0.0000 DF=37.7
```

equal—you could be seriously misled. For example, you might falsely claim to have evidence that the two populations differ when they really do not. Of course, you always have a chance of making such an error (called a Type I error) whenever you do a statistical test. In fact, this is exactly what α measures. When you do a test at $\alpha = .05$, you are supposed to have a 5% chance of claiming there is a difference when in actuality there isn't any. But if you use the pooled procedure when the population standard deviations are not equal, your chances of a Type I error may be very different from 5%. How different depends on how unequal the standard deviations and the sample sizes are.

Minitab has a subcommand that allows you to do the pooled analysis. This subcommand, POOLED, is described on page 188. If you do not use POOLED when you safely could have (i.e., when the standard deviations of the two populations are equal) then, on the average, your analysis will be slightly conservative. That is, you will get a slightly larger confidence interval and you'll be slightly less likely to reject a true null hypothesis. This conservatism essentially disappears with moderately large sample sizes (say, if both n_1 and n_2 are greater than 30). So, in most cases, and especially for large sample sizes, it's better not to use the POOLED subcommand. If the standard deviations are equal, you've lost little and if they're unequal, you may have gained a lot.

TWOSAMPLE-T first sample in C, second sample in C

Prints the results of a t-test and confidence interval to compare two independent samples. Suppose there are n_1 observations in the first sample, with mean $\bar{x}_1$ and standard deviation s_1. Suppose n_2, $\bar{x}_2$, and s_2 are the corresponding values for the second sample. Then the confidence interval goes from

$$(\bar{x}_1 - \bar{x}_2) - t\sqrt{\frac{s_1^2}{n_1} + \frac{s_2^2}{n_2}} \quad \text{to} \quad (\bar{x}_1 - \bar{x}_2) + t\sqrt{\frac{s_1^2}{n_1} + \frac{s_2^2}{n_2}}$$

where t is the value from a t-table corresponding to 95% confidence and degrees of freedom defined below. The test statistic is

$$t = \frac{(\bar{x}_1 - \bar{x}_2)}{\sqrt{\frac{s_1^2}{n_1} + \frac{s_2^2}{\mathrm{n}_2}}}$$

The degrees of freedom is based on the following approximation:

$$\text{d.f.} = \frac{((s_1^2/n_1) + (s_2^2/n_2))^2}{\dfrac{(s_1^2/n_1)^2}{(n_1 - 1)} + \dfrac{(s_2^2/n_2)^2}{(n_2 - 1)}}$$

A confidence level other than 95% can be specified as follows:

```
TWOSAMPLE-T K percent confidence, samples in C, C
```

TWOSAMPLE-T has the two subcommands: POOLED, to do the pooled procedure, and ALTERNATIVE, to do one sided tests. These are described on pages 188–90.

Stacked Data

A convenient example of stacked data comes from the PULSE experiment (p. 318). Suppose we were interested in knowing how much difference, on the average, there is between pulse rates for males and females. We will use PULSE1 since these pulse rates were taken before anyone exercised.

There are two independent samples: one of 57 males and the other of 35 females. There is no pairing between the two samples. The first male is not associated with any particular female, nor is the second, and so on. Similar data would be paired if, for example, we had male-female pairs who were brother and sister or husband and wife.

Minitab's TWOT command can be used to compare two independent samples that are stacked. Exhibit 8.4 contains an example.

TWOT [K percent confidence] data in C, groups in C

Prints the results of a *t*-test and confidence interval to compare two independent samples. The observations from the two samples are in the first column. The second column specifies which sample each observation belongs to. The formulas for the test and confidence interval are the same as those used by TWOSAMPLE-T.

TWOT has two subcommands, POOLED and ALTERNATIVE, described on pages 188–190.

Exhibit 8.4 Comparison of Male and Female Pulse Rates

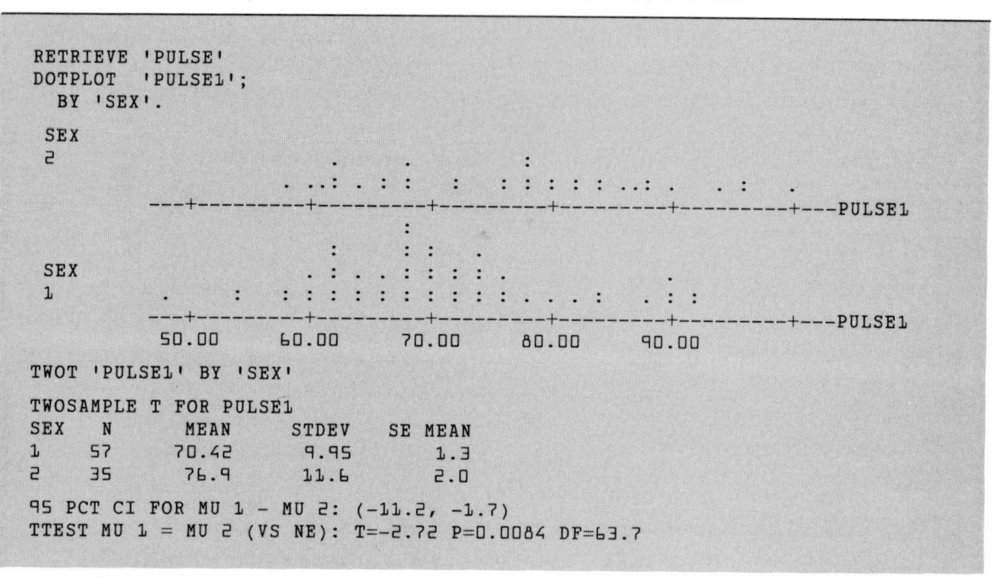

```
RETRIEVE 'PULSE'
DOTPLOT  'PULSE1';
  BY 'SEX'.

 SEX
 2                        .     .         :
                    . ..: . : :   :   : : : : : ..: .   . :   .
          ---+---------+---------+---------+---------+---------+---PULSE1
                                :
                          :     : :   .
 SEX                    . : . . : : : : .             .
 1              .     :   : : : : : : : : : . . . :   . : :
          ---+---------+---------+---------+---------+---------+---PULSE1
            50.00     60.00     70.00     80.00     90.00

TWOT 'PULSE1' BY 'SEX'

TWOSAMPLE T FOR PULSE1
SEX    N       MEAN      STDEV   SE MEAN
1     57      70.42       9.95       1.3
2     35       76.9       11.6       2.0

95 PCT CI FOR MU 1 - MU 2: (-11.2, -1.7)
TTEST MU 1 = MU 2 (VS NE): T=-2.72 P=0.0084 DF=63.7
```

The POOLED and ALTERNATIVE Subcommands

The commands TWOSAMPLE-T and TWOT have the same two subcommands, POOLED and ALTERNATIVE.

In the PULSE data, the assumption of equal standard deviations seems plausible, so we used POOLED to do the pooled t procedure. Exhibit 8.5 gives the results. The test statistic is -2.82, which is quite similar to the unpooled t-value of -2.72, given in Exhibit 8.4. The p-value for the pooled test is .0058, which again is similar to the unpooled value .0084. In both cases, there is statistically significant evidence that the pulse rates of the males and females are different even if we use an α as small as .01. We also note that the observed standard deviations of the males and females are reasonably close: 9.95 and 11.6. Thus the data do not contradict our assumption of equal standard deviations.

POOLED

This subcommand for TWOSAMPLE-T and TWOT tells Minitab to use the pooled procedure. The following pooled estimate of the common variance is calculated:

Exhibit 8.5 Pooled *t*-Test to Compare Male and Female Pulse Rates

```
RETRIEVE 'PULSE'
TWOT 'PULSE1' BY 'SEX';
   POOLED.

TWOSAMPLE T FOR PULSE1 USING POOLED STDEV
SEX    N        MEAN      STDEV    SE MEAN   POOLED STDEV
1     57       70.42       9.95        1.3          10.75
2     35        76.9       11.6        2.0          10.75

95 PCT CI FOR MU 1 - MU 2: (-11.0, -1.9)
TTEST MU 1 = MU 2 (VS NE): T=-2.82 P=0.0058 DF=90.0
```

$$s_p^2 = \frac{(n_1 - 1)s_1^2 + (n_2 - 1)s_2^2}{n_1 + n_2 - 2}$$

The confidence interval goes from

$$(\bar{x}_1 - \bar{x}_2) - ts_p\sqrt{\frac{1}{n_1} + \frac{1}{n_2}} \quad \text{to} \quad (\bar{x}_1 - \bar{x}_2) + ts_p\sqrt{\frac{1}{n_1} + \frac{1}{n_2}}$$

where t is the value from a t-table corresponding to $n_1 + n_2 - 2$ degrees of freedom. The test statistic is

$$t = \frac{(\bar{x}_1 - \bar{x}_2)}{s_p\sqrt{\frac{1}{n_1} + \frac{1}{n_2}}}$$

ALTERNATIVE = K

TWOSAMPLE-T and TWOT both allow one-sided as well as two-sided tests. (The confidence intervals, however, are always two-sided.) The subcommand ALTERNATIVE tells Minitab to do a one-sided test. It uses the following codes:

K = −1 means μ_1 is less than μ_2
K = +1 means μ_1 is greater than μ_2

Example

Here, $H_1 : \mu_1 < \mu_2$ and a 90% confidence interval, using the pooled procedure, is requested.

```
TWOT 90 'PULSE' 'SEX';
  POOLED;
  ALTERNATIVE -1.
```

Exercises

8–5 The following questions refer to the output in Exhibit 8.3.

(a) Use this output to calculate by hand a 95% confidence interval for the mean cholesterol level of people who did not have a heart attack.

(b) Using the formula for t in the box for TWOSAMPLE, verify (by hand or use a calculator) that $t = 6.15$.

(c) Using the formulas in the box for TWOSAMPLE, find a 90% confidence interval for the mean difference between the 2-DAY and CONTROL scores. Note: Round the degrees of freedom to the nearest value that is in your t-table.

8–6 The following questions all refer to the output in Exhibit 8.5, using pooled procedures.

(a) Using the formula for s_p^2, in the box for the POOLED subcommand, verify that the POOLED STDEV = 10.75.

(b) Using the formula for t in the box for the POOLED subcommand, verify that $t = -2.82$.

(c) Calculate a 90% confidence interval for the difference between male and female pulse rates.

8–7 The data given below were obtained in a study of tool life. Ten specimens were untreated and ten were treated by a new process thought to improve wear resistance. Tool wear was measured as volume loss in millionths of a cubic inch.

Untreated	.56	.50	.69	.59	.47	.42	.45	.47	.50	.50
Treated	.13	.13	.18	.23	.18	.31	.35	.23	.31	.33

These data were first given in Exercise 5–6 (p. 99), where you were asked to compare treated and untreated tools using various displays and DESCRIBE. Now do an appropriate test to see if there is a difference between treated and untreated tools. Estimate the difference with a confidence interval.

8–8 Physicists constantly are trying to obtain more accurate values of the fundamental physical constants, such as the mean distance from the earth to the sun and the force exerted on us by gravity. These are very difficult measurements and require the utmost ingenuity and care. Here are results obtained in a Canadian experiment to measure the force of gravity. The first group of 32 measurements was made in August and the second group in December of the following year. (Note: The force of gravity, measured in centimeters per second squared, can be obtained from these numbers by first dividing each number by 1000 then adding 980.61; thus, the first measurement converts to 980.615.)

Measurements Made in August 1958

5	20	20	25	25	30	35
15	20	20	25	25	30	35
15	20	25	25	30	30	
15	20	25	25	30	30	
20	20	25	25	30	30	

Measurements Made in December 1959

20	30	35	40	40	45	55
25	30	35	40	40	45	60
25	30	35	40	45	50	
30	30	35	40	45	50	
30	35	40	40	45	50	

(a) Use the data from August to estimate the force of gravity. Also find a 95% confidence interval.

(b) Repeat (a) for the December measurements.

(c) Now compare the measurements made in August to those made in December. Do a dotplot and an appropriate test using $\alpha = .05$. Is there a statistically significant difference between the two groups? What is the practical significance of this result?

(d) In between these two sets of measurements, it was necessary to change a few key components of the apparatus. Might this have made a difference in the measurements? If so, can you draw any guidelines for sound scientific experimentation? What if the experimenters had done all their measurements at one time? Might they have misled themselves about the accuracy of their results? What if still other parts of the apparatus were changed? Might the measurements change even more?

8–9 A small study was done to compare how well students with different majors do in an introductory statistics course. Seven majors were tested:

biology, psychology, sociology, business, education, meteorology, and economics. At the end of the course, the students were given a special test to measure their understanding of basic statistics. Then a series of t-tests were performed to compare every pair of majors. Thus, biology and psychology majors were compared, biology and sociology majors, psychology and sociology majors, and so on, for a total of 21 t-tests. Simulate this study assuming that all majors do about the same. Assume there are 20 students in each major and that scores on the test have a normal distribution with $\mu = 12$ and $\sigma = 2$. Use

```
RANDOM 20 C1-C7;
  NORMAL 12 2.
```

This puts a sample for each of the seven majors into a separate column.

(a) What is the null hypothesis?

(b) What are the 21 pairs of majors for the 21 t-tests?

(c) Do the 21 t-tests. Unfortunately, you will need to type TWOSAMPLE 21 times.

(d) In how many of the tests did you reject the null hypothesis at $\alpha = .10$?

(e) Since this study was simulated, the true situation is known—there aren't any differences. But you probably did find at least one pair of majors where there was a significant difference. This illustrates the "hazards" of doing a lot of comparisons without making proper adjustments in the procedure. Try to think of some other situations where one might do a lot of statistical tests. For example, suppose a pharmaceutical firm had 16 possible new drugs which they wanted to try out in hopes that at least one was better than the present best competing brand. What are some of the consequences of doing a lot of statistical tests?

9

Analysis of Variance

9.1 One-way Analysis of Variance

In Section 5.4 we showed how to use displays and summary statistics to compare data from several populations. One-way analysis of variance provides a test to see if these populations have different means. We will start with an example.

The flammability of children's sleepwear has received a lot of attention over the years. There are standards to ensure that manufacturers don't sell children's pajamas that burn easily. But a problem always arises in cases like this. How do you test the flammability of a particular garment? The shape of the garment might make a difference. How tightly it fits also might be important. Of course, the flammability of clothing can't be tested on children. Perhaps a metal manikin could be used. But how should the material be lighted? Different people putting a match to identical cloth will probably get different answers. The list of difficulties is endless, but the problem is real and important.

One procedure that has been developed to test flammability is called the Vertical Semirestrained Test. There are a lot of details to the test, but basically it involves holding a flame under a standard-size piece of cloth which is loosely held in a metal frame. The dryness of the fabric, its temperature, the height of the flame, how long the flame is held under the fabric, and so on are all carefully controlled. After the flame is removed and the fabric stops burning, the length of the charred portion of the fabric is measured and recorded.

Once we have a proposed way to test flammability, one important question is, "Will the laboratories of the different garment manufacturers all be able to get about the same results if they apply the same test to

Table 9.1 Data from an Interlaboratory Study of Fabric Flammability

	Laboratory				
	1	*2*	*3*	*4*	*5*
	2.9	2.7	3.3	3.3	4.1
	3.1	3.4	3.3	3.2	4.1
	3.1	3.6	3.5	3.4	3.7
	3.7	3.2	3.5	2.7	4.2
Length of	3.1	4.0	2.8	2.7	3.1
Charred Portion	4.2	4.1	2.8	3.3	3.5
of Fabric	3.7	3.8	3.2	2.9	2.8
	3.9	3.8	2.8	3.2	3.5
	3.1	4.3	3.8	2.9	3.7
	3.0	3.4	3.5	2.6	3.5
	2.9	3.3	3.8	2.8	3.9
Sample Means	3.34	3.60	3.30	3.00	3.65

Data are in the saved worksheet called FABRIC.

the same fabric?'' A study was conducted to answer this question. A small part of the data is in Table 9.1.

Dotplots of these data are given in Exhibit 9.1. This display tells us a lot by itself. Even within a given laboratory, the char-length varies from specimen to specimen. All of the specimens supposedly are identical since they were cut from the same bolt of cloth. But there probably still were some differences between them. In addition, the test conditions surely differed slightly from specimen to specimen—the tautness of the material, the exact time the flame was held under the specimen, and so on. We must expect some variation in results even when all measurements are made in the same laboratory, under apparently identical conditions. The variation within a laboratory can be thought of as random error. There are several other names for this variation: within-group variation, unexplained variation, residual variation, or variation due to error.

There is a second source of variation in the data—this variation is due to differences among labs. This variation is called among-group variation, variation due to the factor, or variation explained by the factor. Here the factor under study is laboratory. In statistics, the word *treatment* sometimes is used instead of factor. If we look at the dotplot again, we see some variation among the five labs. Lab 4 got mostly low values whereas labs 2 and 5 got mostly high values. Instead of using a dotplot, we could compare the sample means for the five labs. These are given in Table 9.1. The five sample means do vary from lab to lab, but again some variation is to be expected. The question is, ''Do the five sample means differ any more than we'd expect from just random variation?''

Exhibit 9.1 Dotplots of the Flammability Data

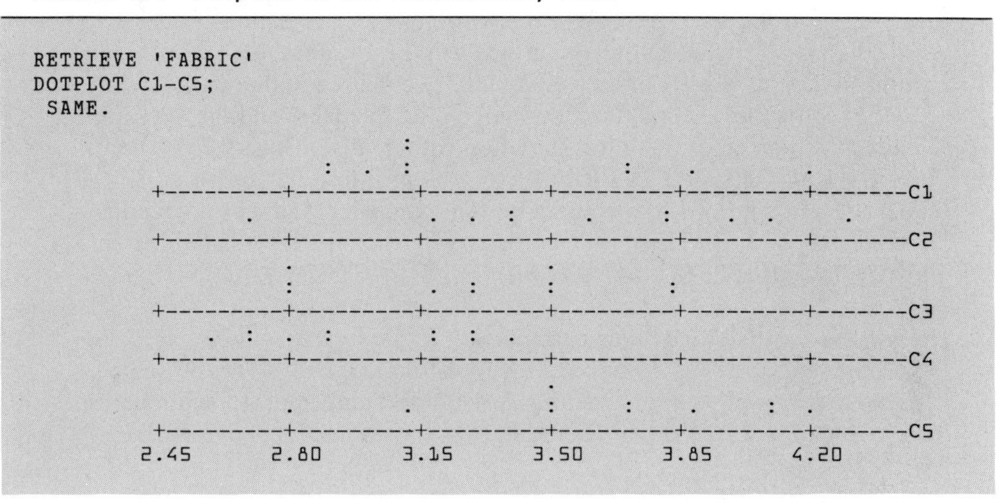

Put another way, "Is the variation we see among the groups significantly greater than the variation we would expect to see given the amount of variation within groups?" Analysis of variance is a statistical procedure that gives an answer to this question.

The One-way Analysis of Variance Procedure

In one-way analysis of variance we want to compare the means of several populations. We assume that we have a random sample from each population, that each population has a normal distribution, and that all of the populations have the same variance, σ^2. In practice, the normality assumption is not too important, the equal variances assumption is not important (provided the numbers of observations in each group was about the same), but the assumption of a random sample is very important. If we do not have a random sample from the population, or something that is very close to it, our conclusion can be far from the truth. The main question is, "Do all of the populations have the same mean?" Suppose a is the number of populations we have, and that μ_1 is the mean of the first population, μ_2 is the mean of the second population, μ_3 is the mean of the third, and so on. Then the null hypothesis of no differences is

$$H_0\colon \mu_1 = \mu_2 = \mu_3 = \cdots = \mu_a$$

To test this null hypothesis, we can use the Minitab command AOVONEWAY.

Exhibit 9.2 uses AOVONEWAY to analyze the data in Table 9.1. The first part of the output is an analysis of variance table. In it, the total sum of squares is broken down into two sources—the variation due to the factor (here it is differences among the five labs) and the variation due to random error (here it is variation within labs). Thus, (SS TOTAL) = (SS FACTOR) + (SS ERROR). In this example, 11.219 = 2.987 + 8.233. Each sum of squares has a certain number of degrees of freedom associated with it. These will be used when we do tests. The degrees of freedom also add up, (DF TOTAL) = (DF FACTOR) + (DF ERROR).

Formulas for the Three Sums of Squares. In the following table, x_{ij} is the jth observation in the sample from population i, $\bar{x}_i$ and n_i are the sample mean and sample size for sample i, a is the number of populations, n is the total number of observations, and $\bar{x}$ is the mean of all n observations.

Source	*DF*	*SS*
Factor	$a - 1$	$\Sigma_i n_i(\bar{x}_i - \bar{x})^2$
Error	$n - a$	$\Sigma_i\Sigma_j(x_{ij} - \bar{x}_i)^2$
Total	$n - 1$	$\Sigma_i\Sigma_j(x_{ij} - \bar{x})^2$

Exhibit 9.2 Output from AOVONEWAY Using the Flammability Data

```
RETRIEVE 'FABRIC'
AOVONEWAY C1-C5

ANALYSIS OF VARIANCE
SOURCE      DF          SS          MS          F
FACTOR       4       2.987       0.747       4.53
ERROR       50       8.233       0.165
TOTAL       54      11.219
                                          INDIVIDUAL 95 PCT CI'S FOR MEAN
                                          BASED ON POOLED STDEV
LEVEL        N        MEAN       STDEV   ----+---------+---------+---------+--
C1          11      3.3364      0.4523            (------*-------)
C2          11      3.6000      0.4604                   (-------*------)
C3          11      3.3000      0.3715          (-------*-------)
C4          11      3.0000      0.2864   (-------*------)
C5          11      3.6455      0.4321                    (-------*-------)
                                         ----+---------+---------+---------+--
POOLED STDEV =      0.4058                 2.88      3.20      3.52      3.84
```

Exhibit 9.2 gives mean squares (abbreviated MS). Each mean square is just the corresponding sum of squares divided by its degrees of freedom. The last column gives the quotient: (F-RATIO) = (MS FACTOR)/(MS ERROR). This F-RATIO is a useful test statistic. It is very large when MS FACTOR is much larger than MS ERROR, that is, when the variation among the labs is much greater than the variation due to random error. In such cases, we would reject the null hypothesis that the labs all have the same average char-length. How large the F-RATIO must be in order to reject is determined by an *F*-table. To use this table, we need the degrees of freedom for the numerator of the *F*-ratio and the degrees of freedom for the denominator of the *F*-ratio. Here the numerator has 4 degrees of freedom and the denominator has 50. The corresponding value from an *F*-table, using $\alpha = .05$, is about 2.6. Since 4.53 is greater than 2.6, we reject the null hypothesis and conclude we have evidence that there are some differences among the five labs. (Note: You can use Minitab's INVCDF command (p. 153) instead of an *F*-table.)

The next table in Exhibit 9.2 summarizes the results separately for each lab. The sample size, sample mean, sample standard deviation, and a 95% confidence interval are given for each lab. Each confidence interval is calculated by the formula:

$$\bar{x}_i - ts_p/\sqrt{n_i} \quad \text{to} \quad \bar{x}_i + ts_p/\sqrt{n_i}$$

Here $\bar{x}_i$ and n_i are the sample mean and sample size for level i, s_p = POOLED STDEV = $\sqrt{\text{MS ERROR}}$ is the pooled estimate of the common standard deviation, σ, and t is the value from a t-table corresponding to 95% confidence and the degrees of freedom associated with MS ERROR. These intervals give us some idea of how the population means differ.

AOVONEWAY C, . . . , C

Performs a one-way analysis of variance. The first column contains the sample from the first population (sometimes called the first group or level), the second column contains the sample from the second population, the third column from the third population, and so on. The sample sizes need not be equal.

The ONEWAY Command for Stacked Data

In Section 5.4 we discussed two ways to arrange data in the worksheet: stacked and unstacked. The AOVONEWAY does an analysis of variance for unstacked data, whereas ONEWAY analyzes stacked data. Both commands do the same analysis and give the same output.

Suppose we want to compare the OTIS scores for the three levels of education (preprofessional, professional, student) in the Cartoon data set (p. 303). The ONEWAY command could be used as follows:

```
RETRIEVE 'CARTOON'
ONEWAY 'OTIS' 'ED'
```

ONEWAY on data in C, levels in C

Does the same analysis and prints the same output as AOVONEWAY. The difference is the form of the input: For ONEWAY, all data for all levels must be put in one column. A second column indicates what level each observation belongs to. Note: The numbers used for levels must be integers. (More in Section 9.4.)

Exercises

9–1 Here are some data from the Plywood study, discussed in Section 3.2, pages 59–62.

Temperature	*Torque*									
60°F	17.5	16.5	17.0	17.0	15.0	18.0	15.0	22.0	17.5	17.5
120°F	15.5	17.0	18.5	15.5	17.0	18.5	16.0	14.0	17.5	17.5
150°F	16.5	17.5	15.5	16.5	16.0	13.5	16.0	13.5	16.0	16.5

Thirty logs were used. A chuck was inserted into each end of a log. The log then was turned and a sharp blade was used to cut off a thin layer of wood. Three levels of temperature were used: Ten logs were tested at 60°F, ten at 120°F, and ten at 150°F. The torque that could be applied to the log before the chuck spun out was measured.

Use DOTPLOT and AOVONEWAY to see how temperature affects the amount of torque that can be applied.

9–2 In a simple pendulum experiment, a weight (bob) is suspended at the end of a length of string. The top of the string is supported by some sort of stable frame. Under ideal conditions the time (T) required for a single cycle of the pendulum is related to the length (L) of the string by the equation

$$T = \frac{2\pi}{\sqrt{g}}\sqrt{L}$$

where, as usual, $\pi = 3.1415 \ldots$ and g is the pull (acceleration) of gravity. If the time required for a cycle and the length of the pendulum are measured, then this equation can be solved to give an estimate of g. In theory, neither the length of the pendulum nor the type of bob should have any effect on the estimate of g. In practice, however, things that are not supposed to make any difference often do. So it is often a good policy, when doing an experiment, to vary, in a carefully balanced manner, things that are not supposed to matter. Therefore, the experiment was run using four different lengths and two types of bobs.

The following estimates of g, in centimeters per second^2, were obtained by college professors during a short course on the use of statistics in physics and chemistry courses.

	Length of Pendulum (cm)			
	60	*70*	*80*	*90*
Heavy Bob	924	994	970	1000
	973	969	975	1017
	955	968	970	1055
Light Bob	966	973	985	960
	949	997	999	1041
	955	988	994	962

(a) Use DOTPLOT to display the data and do a one-way analysis of variance to see if length affects the estimate of g. Use only the data obtained with the heavy bob.

(b) Repeat part (a), using only the data obtained with the light bob.

9–3 Table 5.3 (p. 107) contains data from a radiation experiment. Use AOVONEWAY to see if there is statistical evidence of a difference among the four types of devices. Compare your results to the DOTPLOT in Exhibit 5.8.

9–4 Very small amounts of manganese are important to a good diet. Unfortunately, measurement of the small amounts is quite difficult. To help researchers evaluate their ability to measure such small amounts, the National Bureau of Standards sells samples of cow liver together with an accurate chemical analysis of the amount of manganese in the sample.

The data given here are from one part of the evaluation. Eleven pieces were taken from one cow's liver. The experimenters wanted to know if the amount of manganese varied from piece to piece. Of course, even if the pieces were exactly the same, there would still be some differences in the recorded amounts just due to errors in making the measurements. Therefore, the question posed was, "Do the 11 recorded amounts vary more than you'd expect from measurement error alone?'' To get some idea how large measurement error was, the NBS experimenters measured each piece twice. The amount of manganese (in parts per million) is given in the table below. Use AOVONEWAY to see if there is a statistically significant difference among the 11 pieces.

Piece										
1	*2*	*3*	*4*	*5*	*6*	*7*	*8*	*9*	*10*	*11*
10.02	10.41	10.25	9.41	9.73	10.07	10.09	9.85	10.02	9.92	9.7
10.03	9.79	9.80	10.17	10.75	9.76	9.38	9.99	9.51	10.01	10.0

9–5 Use DOTPLOT and ONEWAY to see if the OTIS scores differed by a statistically significant amount for the three education groups—pre-professional, professional, student—in the Cartoon experiment.

9–6 In the Restaurant survey (p. 321), the various restaurants were classified by the variable TYPEFOOD as fast food, supper club, and other. Use appropriate displays and ONEWAY to determine the following:

(a) Do these three groups spend approximately the same percent of their sales on advertising? (Use the variable ADS.)

(b) Do they spend the same percent on wages? (Use the variable WAGES.)

(c) Do they spend the same percent on the cost of goods? (Use the variable COSTGOOD.)

9.2 Randomized Block Designs

In Section 5.6 we discussed two experiments—billiard balls and alfalfa—that were conducted according to what is called a randomized block design (abbreviated RBD). In a RBD, the material, people, locations, time periods, or whatever within a block are relatively homogeneous. For example, the three portions of plastic within one block of the billiard ball experiment all came from the same batch of plastic. Similarly, the six plots within one block of the alfalfa experiment all were located in the same field. The treatments we wish to compare then are assigned at random within each block, with each treatment appearing exactly once in each block. In Section 5.6 we used MPLOT and LPLOT to display data from a RBD. We can do a formal test using an analysis of variance procedure.

Table 9.2 contains data from a RBD used to compare different methods of freezing meatloaf. Meatloaf was to be baked, then frozen for a time, and finally compared by expert tasters. Eight loaves could be baked in

Table 9.2 Drip Loss in Meatloaves

Oven Position	Batch 1	2	3
1	7.33	8.11	8.06
2	3.22	3.72	4.28
3	3.28	5.11	4.56
4	6.44	5.78	8.61
5	3.83	6.50	7.72
6	3.28	5.11	5.56
7	5.06	5.11	7.83
8	4.44	4.28	6.33

Data are in the saved worksheet called MEATLOAF, in the form used in Exhibit 9.3.

Oven Position on the Shelf

5	6	7	8
4	3	2	1

Front of Shelf

Note: Thermometers to measure the temperature of the loaves were inserted into the four corner loaves (1, 4, 5, 8) and in one center loaf (7).

the oven at one time. But some parts of an oven are usually hotter or differ in some other important way from other parts. If this is the case, loaves baked in one part of the oven might taste better than those baked in another part, and differences in freezing methods might be masked. So a preliminary test was conducted to see if there were any noticeable differences among the eight oven positions used in the study.

The data in Table 9.2 came from this preliminary experiment. Eight loaves were mixed. One loaf was assigned at random to each oven position; the loaves then were baked and analyzed. A second batch of eight loaves was mixed, assigned at random to the eight oven positions, baked and analyzed, and then a third batch was tested in the same manner. Each batch is a block and there are eight treatments (oven positions) within each block. Each loaf was analyzed by measuring the percentage of drip loss (i.e., the amount of liquid which dripped out of the meatloaf during cooking divided by the original weight of the loaf).

Exhibit 9.3 shows commands to READ the data, to print out the data in a table with row and column means, to do a plot, and to use Minitab's TWOWAY command to do a test.

Exhibit 9.4 contains the output from TABLE. The body of the table contains the original data. This output is in the same form as the data in Table 9.2, so we easily can check for typing errors. The margins of the table give batch and position means. The batch means allow us to compare the three batches. On the average, batch 1 had the lowest drip loss, batch 2 was in the middle, and batch 3 had the highest. The means for oven positions indicate how the eight positions compared. For example, position 1 had an average drip loss of 7.83. This was more than twice that of position 2, which had a drip loss of just 3.74.

Exhibit 9.3 Commands to Analyze the Meatloaf Data of Table 9.2. Output is in Exhibits 9.4, 9.5, and 9.6.

```
NAME C1='DRIPLOSS' C2='BATCH' C3='POSITION'
READ 'DRIPLOSS' 'BATCH' 'POSITION'
7.33    1   1
3.22    1   2
3.28    1   3
6.44    1   4
3.83    1   5
3.28    1   6
5.06    1   7
4.44    1   8
8.11    2   1
3.72    2   2
5.11    2   3
5.78    2   4
6.50    2   5
5.11    2   6
5.11    2   7
4.28    2   8
8.06    3   1
4.28    3   2
4.56    3   3
8.61    3   4
7.72    3   5
5.56    3   6
7.83    3   7
6.33    3   8
END
TABLE 'POSITION' BY 'BATCH';
  MEAN 'DRIPLOSS'.
LPLOT 'DRIPLOSS' BY 'POSITION', CODE FOR 'BATCH'
TWOWAY 'DRIPLOSS' 'BATCH' 'POSITION'
```

The LPLOT in Exhibit 9.5 allows us to compare individual observations. We connected the letters by hand to help us see better. Here it seems obvious that there were some systematic differences among the three batches. The readings for batch 1 were the lowest at almost every

Exhibit 9.4 Output from TABLE for the Meat Loaf Data

```
TABLE 'POSITION' BY 'BATCH';
  MEAN 'DRIPLOSS'.

ROWS: POSITION        COLUMNS: BATCH
            1           2           3        ALL

 1     7.3300      8.1100      8.0600     7.8333
 2     3.2200      3.7200      4.2800     3.7400
 3     3.2800      5.1100      4.5600     4.3167
 4     6.4400      5.7800      8.6100     6.9433
 5     3.8300      6.5000      7.7200     6.0167
 6     3.2800      5.1100      5.5600     4.6500
 7     5.0600      5.1100      7.8300     6.0000
 8     4.4400      4.2800      6.3300     5.0167
ALL    4.6100      5.4650      6.6187     5.5646

 CELL CONTENTS --
          DRIPLOSS:MEAN
```

Exhibit 9.5 Output from LPLOT for the Meat Loaf Data

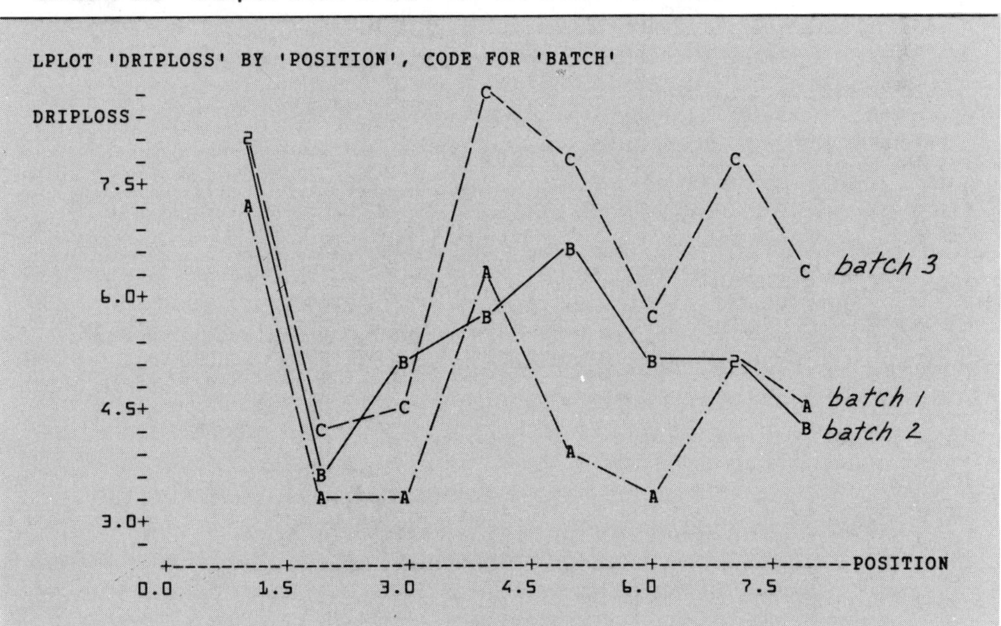

position, while those for batch 3 were almost always the highest. Thus, part of the variation in the data seems to be due to differences among batches. It also appears that oven position made a difference. All the readings for position 1 were fairly high; those for 2 and 3 were low, and so on. So, part of the variation seems to be due to differences among oven positions.

Now, just as we did in one-way analysis of variance, we can express the total variation in the data as the sum of the variation from several sources:

(Total variation in data) = (variation due to batch)
+ (variation due to oven position)
+ (variation due to random error)

If the variation due to oven position is much greater than the variation due to random error, we will have statistically significant evidence of a difference among the eight oven positions.

The TWOWAY output in Exhibit 9.6 gives the breakdown of the total variation. In the table below, a is the number of treatments, b is the number of blocks, x_{ij} is the observation in block i given treatment j, $\bar{x}_{i.}$ is the mean of all r observations in block i, $\bar{x}_{.j}$ is the mean of all b observations given treatment j, and $\bar{x}$ is the mean of all ab observations.

Source	*DF*	*SS*
Blocks	$b - 1$	$a\Sigma_i(\bar{x}_{i.} - \bar{x})^2$
Treatments	$r - 1$	$b\Sigma_j(\bar{x}_{.j} - \bar{x})^2$
Error	$(b - 1)(r - 1)$	$\Sigma_i\Sigma_j(x_{ij} - \bar{x}_{i.} - \bar{x}_{.j} + \bar{x})^2$
Total	$br - 1$	$\Sigma_i\Sigma_j(x_{ij} - \bar{x})^2$

Exhibit 9.6 Output from TWOWAY for the Meat Loaf Data

```
TWOWAY 'DRIPLOSS' 'BATCH' 'POSITION'

ANALYSIS OF VARIANCE  DRIPLOSS

SOURCE          DF        SS          MS
BATCH            2     16.259       8.130
POSITION         7     40.396       5.771
ERROR           14      9.290       0.664
TOTAL           23     65.945
```

In the meatloaf experiment, the sum of squares for blocks, SS BATCH, is 16.259 and has 3 − 1 = 2 degrees of freedom. The sum of squares for treatments, SS POSITION, is 40.396 and has 8 − 1 = 7 degrees of freedom. SS ERROR is 9.290 and has (2)(7) = 14 degrees of freedom. Notice that 16.259 + 40.396 + 9.290 = 65.945, the SS TOTAL. As in one-way analysis of variance, each mean square is just the corresponding SS divided by its DF.

To see if there is statistically significant evidence of a difference among the eight oven positions, we form the *F*-ratio: (MS POSITION)/(MS ERROR) = 5.771/0.664 = 8.74. We compare the result to the value from an *F*-table corresponding to 7 degrees of freedom in the numerator and 14 in the denominator. Here the *F*-table value is 2.76, for $\alpha = .05$. Since 8.74 is greater than 2.76, we have statistically significant evidence that drip loss varies among the eight oven positions. This agrees with what we observed in Exhibits 9.4 and 9.5.

A Further Look

In many cases an analysis of variance would stop at this point, but whenever possible it will pay us to look deeper to learn as much as we can. In Table 9.2, we notice that the three positions with the lowest drip loss are the three without thermometers. Perhaps the hole made by the thermometer allowed some juices to escape. One obvious solution would have been to put a thermometer in every loaf or in no loaves. Then all eight loaves would be comparable. It's too late to improve this experiment, but perhaps the next experiment will benefit from our careful work and attention to details.

TWOWAY analysis, obs in C, blocks in C, treatments in C

Does both a two-way analysis of variance (see Section 9.3) and a randomized block design analysis. In a RBD, the first column contains all the observations, the second column denotes which block each observation is in, and the third column denotes which treatment each observation was given. (More in Section 9.4.)

Exercises

9–7 Table 5.5 (p. 115) contains data from a randomized block design to study the elasticity of billiard balls. Use TWOWAY to analyze these data. Compare your results to the MPLOT in Exhibit 5.12.

9–8 Table 5.6 (p. 119) contains data from a randomized block design to compare six varieties of alfalfa. Use TWOWAY to analyze these data. Compare your results to the LPLOTS in Exhibit 5.13.

9.3 Analysis of Variance with Two Factors

In Section 5.5 we looked at ways to plot data which were classified by two variables or factors. Here we will show how to use two-way analysis of variance to do formal tests.

Table 9.3 contains data from an experiment designed to study the effect of two factors on the quality of pancakes. The two factors were the amount of whey and whether or not a supplement was used. There are four levels of whey (0%, 10%, 20%, and 30%) and two levels of supplement (used and not used), giving a total of 4 × 2 = 8 treatment combinations or cells. Three pancakes were baked using each treatment combination. Each pancake then was rated by an expert and the three ratings averaged to give one overall quality rating. The higher the quality rating, the better the pancake. This was done three times for each treatment combination giving a total of 3 × 8 = 24 overall quality ratings.

Exhibit 9.7 contains commands to analyze these data. The quality ratings were all put in one column. A second column was used for supplement (0 = no, 1 = yes) and a third column for the amount of whey (0, 10, 20, 30). The TABLE output in Exhibit 9.8 gives the data in a format similar to the one in Table 9.3. This makes it easy to check for typing errors. Exhibit 9.9 contains a TABLE of cell means and a hand-drawn plot. This exhibit gives us an idea of how supplement and whey affect pancakes. First, let's look at the average effect of each factor. Pancake quality increases, on the average, as the percentage of whey increases. The six pancakes with 0% whey have a mean quality of 3.8, the six with 10% whey have a mean quality of about 4.2; for 20%

Table 9.3 Quality of Pancakes

	Amount of Whey			
	0%	*10%*	*20%*	*30%*
No Supplement	4.4	4.6	4.5	4.6
	4.5	4.5	4.8	4.7
	4.3	4.8	4.8	5.1
Supplement	3.3	3.8	5.0	5.4
	3.2	3.7	5.3	5.6
	3.1	3.6	4.8	5.3

Exhibit 9.7 Commands to Analyze the Pancake Data. Output is in Exhibits 9.8, 9.9 and 9.10.

```
NAME C1 = 'QUALITY' C2 = 'SUPPLMNT' C3 = 'WHEY'
SET 'QUALITY'
4.4 4.5 4.3  4.6 4.5 4.8  4.5 4.8 4.8  4.6 4.7 5.1
3.3 3.2 3.1  3.8 3.7 3.6  5.0 5.3 4.8  5.4 5.6 5.3
END
SET 'SUPPLMNT'
0 0 0  0 0 0  0 0 0  0 0 0  1 1 1  1 1 1  1 1 1  1 1 1
END
SET 'WHEY'
0 0 0  10 10 10  20 20 20  30 30 30
0 0 0  10 10 10  20 20 20  30 30 30
END
TABLE 'SUPPLMNT' 'WHEY';
  DATA 'QUALITY'.
TABLE 'SUPPLMNT' 'WHEY';
  MEAN 'QUALITY'.
TWOWAY 'QUALITY' 'SUPPLMNT' 'WHEY'
```

Exhibit 9.8 TABLE Output for Pancake Data

```
TABLE 'SUPPLMNT' 'WHEY';
  DATA 'QUALITY'.

ROWS: SUPPLMNT      COLUMNS: WHEY
            0         10         20         30

 0     4.4000     4.6000     4.5000     4.6000
       4.5000     4.5000     4.8000     4.7000
       4.3000     4.8000     4.8000     5.1000

 1     3.3000     3.8000     5.0000     5.4000
       3.2000     3.7000     5.3000     5.6000
       3.1000     3.6000     4.8000     5.3000

 CELL CONTENTS --
          QUALITY:DATA
```

Exhibit 9.9 Table and Plot of Cell Means for Pancake Data

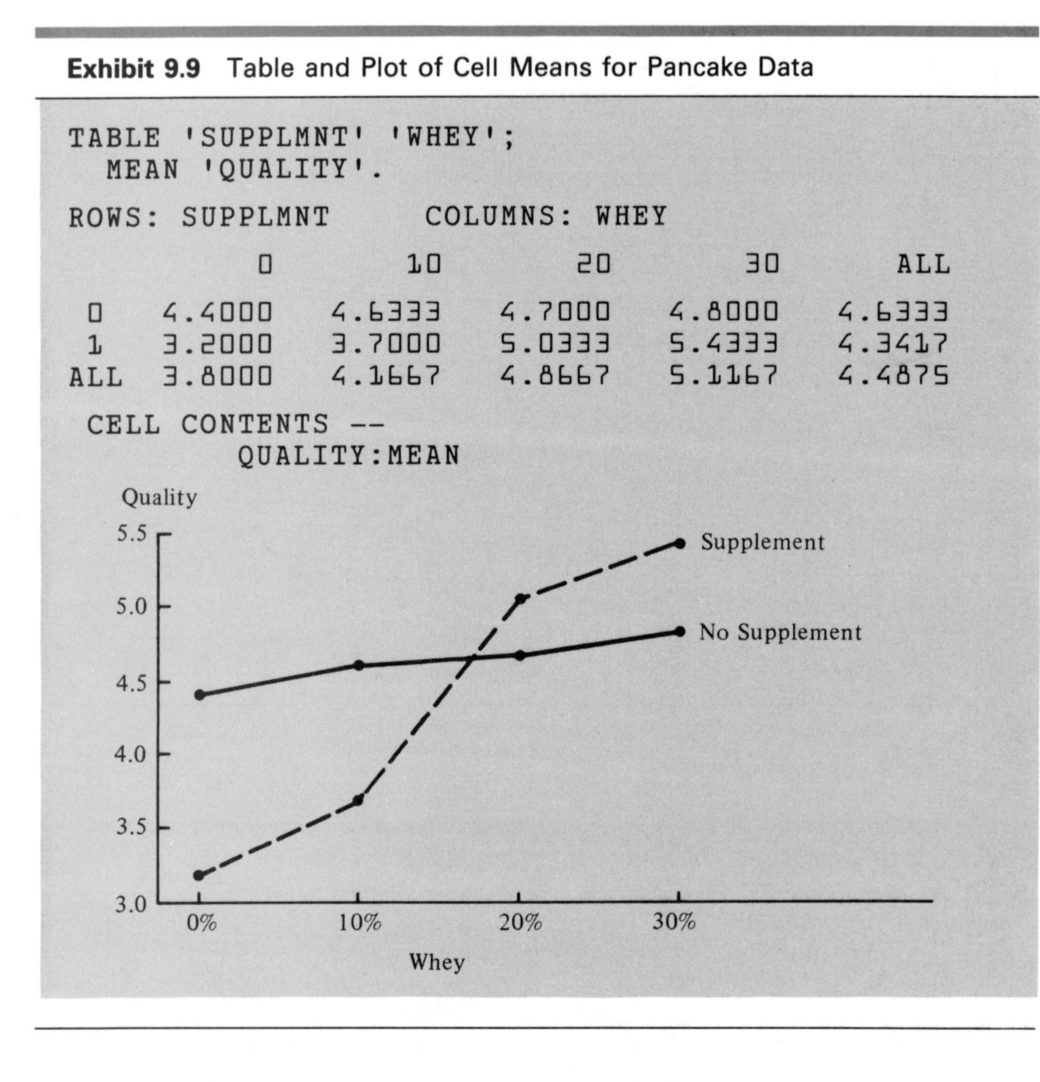

```
TABLE 'SUPPLMNT' 'WHEY';
  MEAN 'QUALITY'.

ROWS: SUPPLMNT        COLUMNS: WHEY

            0         10         20         30        ALL

 0     4.4000     4.6333     4.7000     4.8000     4.6333
 1     3.2000     3.7000     5.0333     5.4333     4.3417
ALL    3.8000     4.1667     4.8667     5.1167     4.4875

 CELL CONTENTS --
         QUALITY:MEAN
```


the mean is 4.9, and for 30% it's 5.1. When we use the TWOWAY command, we test whether these means differ by more than we would expect if the amount of whey made no difference in quality.

The average effect of using the supplement is not as great as the effect of whey: the mean quality of the 12 pancakes with no supplement is just a little higher than the mean quality for the 12 with the supplement. This difference forms the basis of the test we do on supplement when we use the TWOWAY command.

We also can look at the individual cells using the TABLE and the hand-drawn plot in Exhibit 9.9. Notice that both supplement and no supplement show a similar overall pattern—as percentage of whey increases, quality also increases. There is a difference, however: The increase in quality is much greater when the supplement is used than when it is not.

The effect of the supplement is not clear-cut: Supplement increases quality when 20% or 30% whey is used but seems to lower quality when 0% or 10% is used. Thus, whether or not the supplement increases quality *depends* on the amount of whey we use. When the effect of one factor depends on the level of another we say the two factors interact. Here supplement and whey appear to interact. If the two factors did not interact, the two lines in the plot of Exhibit 9.9 would be roughly parallel. The TWOWAY command also does a test for interaction.

Exhibit 9.10 contains the output from TWOWAY. This is the same command we used for the RBD of Section 9.2 and the formulas are similar. We will use the following notation: a is the number of levels of the first factor (that is, the number of rows in the table of data); b is the number of levels of the second factor (that is, the number of columns in the table of data); n is the number of observations in each cell; x_{ijk} is the kth observation in cell (i, j) (i.e., in the cell in row i and column j); $\bar{x}_{ij.}$ is the mean of the n observations in cell (i, j); $\bar{x}_{i..}$ is the mean of the bn observations in row i; $\bar{x}_{.j.}$ is the mean of the an observations in column j; and $\bar{x}$ is the mean of all abn observations in the table.

Source	*DF*	*SS*
Rows	$a - 1$	$bn\Sigma_i(\bar{x}_{i..} - \bar{x})^2$
Columns	$b - 1$	$an\Sigma_j(\bar{x}_{.j.} - \bar{x})^2$
Interaction	$(a - 1)(b - 1)$	$n\Sigma_i\Sigma_j(\bar{x}_{ij.} - \bar{x}_{i..} - \bar{x}_{.j.} + \bar{x})^2$
Error	$ab(n - 1)$	$\Sigma_i\Sigma_j\Sigma_k(x_{ijk} - \bar{x}_{ij.})$
Total	$abn - 1$	$\Sigma_i\Sigma_j\Sigma_k(x_{ijk} - \bar{x})$

Exhibit 9.10 Output from TWOWAY for Pancake Data

```
TWOWAY 'QUALITY' 'SUPPLMNT' 'WHEY'

ANALYSIS OF VARIANCE  QUALITY

SOURCE          DF        SS          MS
SUPPLMNT         1    0.5104      0.5104
WHEY             3    6.6912      2.2304
INTERACTION      3    3.7246      1.2415
ERROR           16    0.4800      0.0300
TOTAL           23   11.4062
```

The total variation is now partitioned into four sources:

$$\begin{aligned}(\text{Total variation}) = {} & (\text{variation due to rows}) \\ & + (\text{variation due to column}) \\ & + (\text{variation due to interaction}) \\ & + (\text{variation due to error}).\end{aligned}$$

Notice that the sums of squares (and the degrees of freedom) in Exhibit 9.10 also add up as in this formula.

Tests are based on mean squares. A mean square is the corresponding sum of squares divided by its degrees of freedom. There are three tests we can do: the effect of rows (here supplement), the effect of columns (here whey), and whether these two factors interact. We always test for interaction first. If interaction is present, then we must be careful how we interpret the other two tests.

To see if there is statistically significant evidence of interaction, we form the F-ratio, (MS interaction)/(MS error) = 1.2415/.0300 = 41.38. Then we compare this value to the value from an F-table with 3 degrees of freedom in the numerator and 16 in the denominator. The table value is 3.24 for $\alpha = .05$ and 5.29 for $\alpha = .01$. Since 41.38 exceeds both these values, there is statistical evidence of an interaction. That is, there is evidence that the effect of the supplement depends on how much whey is used. Or, put another way, the effect of whey depends on whether or not the supplement is used.

Now let's look at the two other tests. These often are called tests for the average or main effects of the two factors, here they are supplement and whey. To test for the effect of supplement, we form the F-ratio (MS supplement)/(MS error) = 17.01. The F-table value with 2 degrees of freedom in the numerator and 16 degrees of freedom in the denominator is 8.53 for $\alpha = .01$. Since 17.01 exceeds 8.53, there is significant evidence of an overall effect of supplement.

Suppose we look at the table in Exhibit 9.9. Notice that the mean for no supplement is 4.6333, which is larger than the mean of 4.3417 for supplement. This is what our test has found. But the test does not tell the whole story. In fact, it tends to distort the real story. If we look at the individual cells, we see that sometimes supplement is better than no supplement and sometimes it is worse. On the average, no supplement is better than supplement. In general, whenever there is a significant interaction, you should look at the individual cells to see what's really going on. Otherwise, the tests on the main effects may be very misleading.

The other main effect test is the test on whey. Here the F-ratio, (MS whey)/(MS error) = 74.35, is also much larger than the F-table value of 5.29 for $\alpha = .01$. Thus, the test says there is significant evidence of differences due to whey.

Another Example: Potato Rot A substantial percentage of the potatoes raised in this country never have a chance to reach the table. Instead they fall victim to potato rot while being stored for later use. To find out what could be done to reduce this loss, an experiment was carried out at the University of Wisconsin. Potatoes were injected with a bacteria known to cause rot, then stored under a variety of conditions. After five days the diameter of the rotted portion on each potato was measured.

Three factors were varied in this experiment: (1) the amount of bacteria injected into the potato (1 = low amount, 2 = medium amount, 3 = high amount); (2) the temperature during storage (10°C, 16°C); (3) the amount of oxygen during storage (2%, 6%, 10%).

Table 9.4 contains the data for those potatoes that were injected with a high bacteria amount. (We will analyze the rest of the data in the exercises.) Exhibit 9.11 contains a table of means, a plot of those means, and output from TWOWAY. The F-ratio for interaction is 8.2/29.1 = .28, much smaller than the value from an F-table. Therefore, there is no statistical evidence of any interaction between temperature and oxygen. The two lines in the plot are very close to being parallel. They are not exactly parallel, but we always have a certain amount of variation due just to random error. The test for interaction compares the amount of interaction (as measured by MS Interaction) to the amount of random variation in the data as measured by MS Error. If MS Interaction is much larger than MS Error, then we have statistically significant evidence that the two factors interact.

Since there is no statistical evidence of interaction here, the two tests on main effects can be interpreted more easily. The F-ratio for temperature is 600.9/29.1 = 20.65, which is much larger than the F-table value of 9.33 for α = .01. Thus there is statistical evidence of a difference between storing potatoes at 10°C and at 16°C. The plot also shows this: At each level of oxygen, there is less rot at 10°C than at 16°C, and the difference is always about the same—about 11 to 14 millimeters.

Table 9.4 Potato Rot Data (diameter of rotted area is in millimeters)

	Oxygen		
Temperature	*2%*	*6%*	*10%*
10°C	13	10	15
	11	4	2
	3	7	7
16°C	26	15	20
	19	22	24
	24	18	8

Exhibit 9.11 Table of Cell Means, a Hand-drawn Plot of Cell Means, and Analysis of Variance for Potato Rot Data

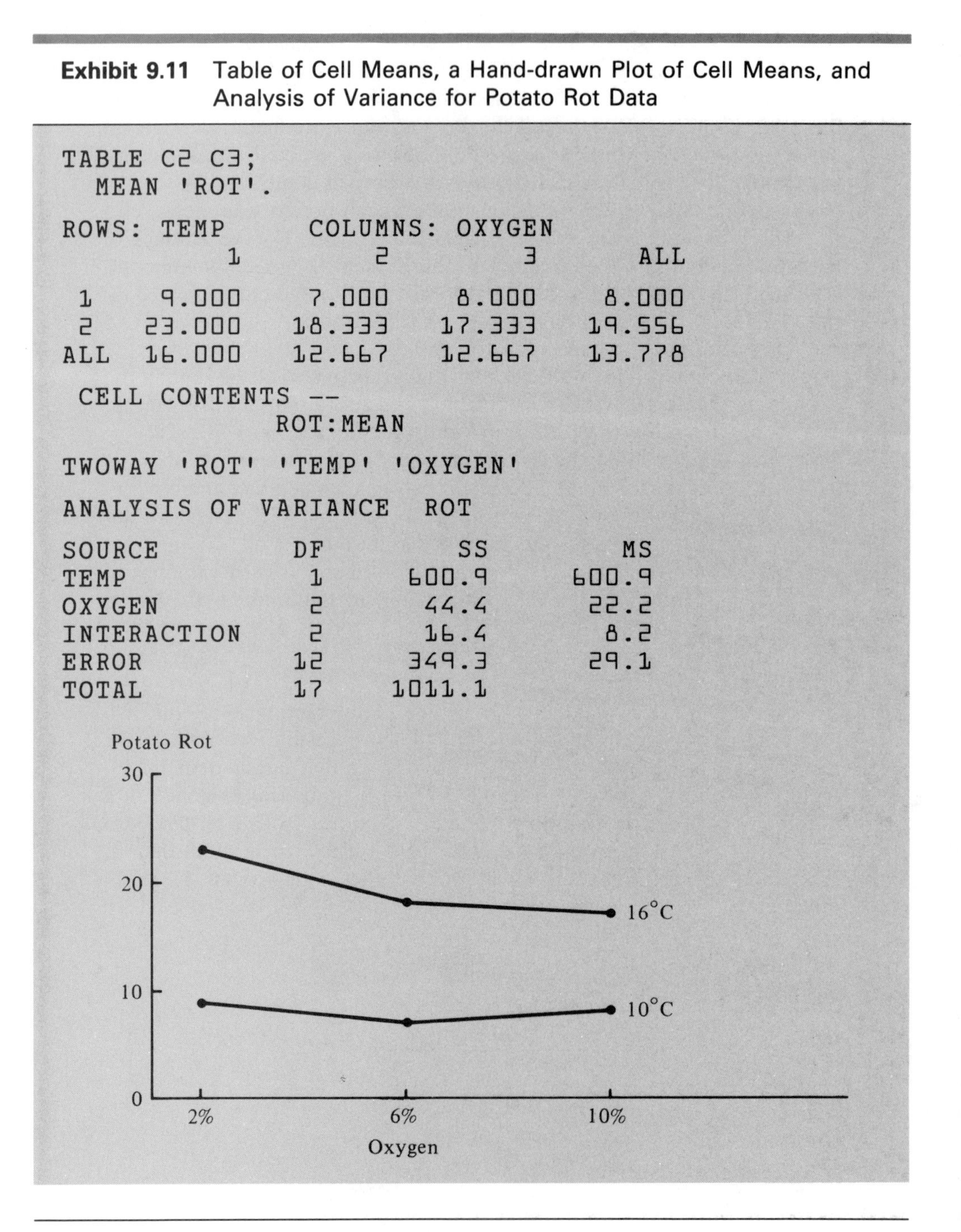

```
TABLE C2 C3;
  MEAN 'ROT'.

ROWS: TEMP       COLUMNS: OXYGEN
            1          2          3        ALL

 1      9.000      7.000      8.000      8.000
 2     23.000     18.333     17.333     19.556
ALL    16.000     12.667     12.667     13.778

 CELL CONTENTS --
               ROT:MEAN

TWOWAY 'ROT' 'TEMP' 'OXYGEN'

ANALYSIS OF VARIANCE  ROT

SOURCE          DF          SS          MS
TEMP             1       600.9       600.9
OXYGEN           2        44.4        22.2
INTERACTION      2        16.4         8.2
ERROR           12       349.3        29.1
TOTAL           17      1011.1
```

The F-ratio for oxygen is $22.2/29.1 = .76$, much smaller than the F-table value of 2.81 for $\alpha = .10$. Thus, this experiment gives no statistical evidence that oxygen influences the amount of rot in potatoes. The small amount of difference we observe in these means is no greater than the variation we must expect from data with an inherent variability indicated by a MS Error of 29.1.

TWOWAY analysis, obs in C, factors levels in C and C

Does a two-way analysis of variance. The design must be balanced, that is, each cell must contain the same number of observations. If each cell contains just one observation, then SS Interaction cannot be calculated. Note: The level numbers must be integers. (More in Section 9.4.)

Exercises

9–9 An experiment was done to see whether the damage to corn, wheat, dried milk, and other stored crops due to a beetle, *Trogoderma glabrum,* could be reduced. The idea was to lure male adult beetles to inoculation sites where they would pick up disease spores which they would then carry back to the remaining population for infection. The experiment consisted of two factors, "sex attractant or none" and "disease pellet present or not." The dependent variable was the fraction of larvae failing to reach adulthood.

Analyze the data presented in the table below. Use TABLE to get cell means, and plot these means by hand. Use TWOWAY to do an analysis of variance. Interpret your results.

	Disease	
Attractant	*Yes*	*No*
Yes	.979	.222
	.743	.189
	.775	.143
	.885	.262
No	.312	.239
	.293	.253
	.188	.159
	.388	.241

9–10 The pendulum data of Exercise 9–2 (p. 199) involve two factors, length and bob. Do a two-way analysis of variance on the data. Use TABLE to get cell means. Plot these by hand.

(a) Does length seem to affect the estimate of g? Does the type of bob used affect the estimate? Is there a significant interaction between length and bob?

(b) Suppose you wanted to estimate the acceleration of gravity using a pendulum. What do the results in part (a) tell you about designing such an experiment?

9–11 The entire potato rot data set, discussed in this section, is given here and is in the saved worksheet called POTATO.

Bacteria	*Temperature*	*Oxygen*	*Rot*	*Bacteria*	*Temperature*	*Oxygen*	*Rot*
1	1	1	7	2	2	1	17
1	1	1	7	2	2	1	18
1	1	1	9	2	2	1	8
1	1	2	0	2	2	2	3
1	1	2	0	2	2	2	23
1	1	2	0	2	2	2	7
1	1	3	9	2	2	3	15
1	1	3	0	2	2	3	14
1	1	3	0	2	2	3	17
1	2	1	10	3	1	1	13
1	2	1	6	3	1	1	11
1	2	1	10	3	1	1	3
1	2	2	4	3	1	2	10
1	2	2	10	3	1	2	4
1	2	2	5	3	1	2	7
1	2	3	8	3	1	3	15
1	2	3	0	3	1	3	2
1	2	3	10	3	1	3	7
2	1	1	2	3	2	1	26
2	1	1	4	3	2	1	19
2	1	1	9	3	2	1	24
2	1	2	4	3	2	2	15
2	1	2	5	3	2	2	22
2	1	2	10	3	2	2	18
2	1	3	4	3	2	3	20
2	1	3	5	3	2	3	24
2	1	3	0	3	2	3	8

These data probably should be analyzed using a three-way analysis of variance; but you can discover most of what is going on by using several two-way analyses and some plots and tables.

(a) Do a two-way analysis of variance on all the data, using bacteria and temperature as the two factors (ignore oxygen for the moment). Use TABLE to get cell means. Plot these means by hand.
(b) Next, do a two-way analysis of variance on all the data, using bacteria and oxygen as the two factors (ignore temperature in this analysis). Use TABLE to get cell means. Plot these means by hand.
(c) How do each of the three factors influence potato rot? Do any of the factors appear to interact?
(d) The factors that can be controlled, to some extent, by the food supplier are temperature and oxygen. Analyze these two factors at each of the three levels of bacteria. The analysis for BACTERIA = 3 was done earlier in this section. You do the analysis for BACTERIA = 1 and the analysis for BACTERIA = 2. Do temperature and oxygen seem to have the same effect at each level?
(e) What recommendations would you give to someone who was storing potatoes?

9.4 Residuals and Additive Models

The commands ONEWAY and TWOWAY have some additional capabilities for more advanced work in statistics. Here we will briefly describe how the commands work.

Residuals and Fitted Values

Analysis of variance can be viewed in terms of fitting a model to the data. This leads to a fitted value and a residual for each observation. A fitted value is our best estimate of the underlying population mean value corresponding to that observation. The residual is how much our observation differs from its fitted value. (The concepts of fitted values and residuals are discussed in more depth in Section 10.7, on regression models.)

In one-way analysis of variance, the fitted values are just the group means. A residual is then the difference between an observed value and the corresponding group mean. Consider, for example, the Flammability data. The first few observations, fitted values, and residuals are as follows:

	Laboratory						
	1				*2*		
Observation	2.9	3.1	3.1	3.7 ...	2.7	3.4	3.6 ...
Fitted Value	3.34	3.34	3.34	3.34 ...	3.60	3.60	3.60 ...
Residual	−.44	−.24	−.24	.36 ...	−.90	−.20	.00 ...

When you use ONEWAY, Minitab will store the fitted values and residuals if you specify two extra columns on the ONEWAY command.

It is good practice to make plots of the residuals to check for unequal variances, nonnormality, dependence on other variables, and so on. We often make the following plots:

Plot of the residuals versus the fitted values.

Histogram of residuals.

Plot of the residuals versus the order in which the data were collected.

Plot of the residuals versus other variables (when data on other variables are available).

The command AOVONEWAY will not compute fitted values and residuals.

In ordinary two-way analysis of variance, the fitted values are the cell means. A residual is then the difference between the observed value and the corresponding cell mean. We will use the Pancake data in Exhibits 9.8 and 9.9 as an example. The observations, fitted values, and residuals from the two cells where whey = 0% are as follows:

	No Supplement			*Supplement*		
Observation	4.4	4.5	4.3	3.3	3.2	3.1
Fit	4.4	4.4	4.4	3.2	3.2	3.2
Residual	.0	.1	−.1	.1	.0	−.1

The TWOWAY command, like ONEWAY, will store the fitted values and residuals if you specify two extra columns.

Additive Models

In two-way analysis of variance it sometimes is desirable to try a model in which it is assumed there is no interaction. This can be done with the ADDITIVE subcommand. In an additive model, the fitted values are no longer the cell means. In the balanced data case treated by the TWOWAY command, the fitted values are computed as follows:

$$\begin{aligned}\text{Fitted value for cell } (i, j) &= (\text{mean of data in row } i)\\ &\quad + (\text{mean of data in column } j)\\ &\quad - (\text{mean of all data})\end{aligned}$$

Consider, for example, the Potato Rot data in Table 9.4 and Exhibit 9.11. The fitted value for cell (1, 1) can be computed as follows:

Fitted value for cell (1, 1) = 8.000 + 16.000 − 13.778 = 10.222

The residuals for this cell are thus

(13 − 10.222) = 2.778, (11 − 10.222) = .778, (3 − 10.222) = −7.222

ONEWAY data in C, levels in C, [store resids in C [fits in C]]

Does a one-way analysis of variance and stores both the fitted values and the residuals.

TWOWAY data in C, levels in C, C [store resids in C [fits in C]]

Does either a RBD or a two-way analysis of variance and stores the fitted values and residuals. To fit an additive model (i.e., a model which has no interaction) use the subcommand

ADDITIVE

10 Correlation and Regression

Some of the most interesting problems in statistics occur when we try to find a model for the relationship between several variables. The data in Table 10.1 illustrate a common situation. Two tests were given to 31 individuals. We often want to answer questions like the following:

What is the correlation between the scores on the two tests?

If you know someone's score on the first test, does that help you at all in predicting the score on the second test?

What is a good prediction of the second score for a person who scored 70 on the first test?

In this chapter we will see how questions like these can be answered.

10.1 Correlation

There are several ways to measure the association between two variables. The most common measure is the Pearson product moment correlation coefficient, or just *correlation* for short. This is usually designated by the letter r.

Table 10.1 contains some data and Exhibit 10.1 contains a plot and the output from the command CORRELATION. In this example, Minitab printed the result $r = .703$.

The correlation coefficient is always between -1 and $+1$. We will denote the two variables by x and y. Here x would be the first score and y the second. The correlation coefficient is positive if y tends to

Table 10.1 Two Scores for 31 People

Subject Number	*First Test Score*	*Second Test Score*
1	50	69
2	66	85
3	73	88
4	84	70
5	57	84
6	83	78
7	76	90
8	95	97
9	73	79
10	78	95
11	48	67
12	53	60
13	54	79
14	79	79
15	76	88
16	90	98
17	60	56
18	89	87
19	83	91
20	81	86
21	57	69
22	71	75
23	86	98
24	82	70
25	95	91
26	42	48
27	75	52
28	54	44
29	54	51
30	65	73
31	61	52

increase as x increases, that is, if a plot of y versus x slopes upward. Conversely, the correlation is negative if y tends to decrease as x increases, that is, if a plot of y versus x slopes downward. If the points fall exactly on a straight line, then $r = +1$ if the points slope upward and -1 if the trend is downward. The closer the points are to forming a straight line, the closer r is to $+1$ or -1. The closer r is to $+1$ or to -1, the easier it is to predict y from x.

If there is almost no association between x and y, then r will be near 0. The converse, however, is not true. There are cases where r is near 0 but there is still a clear association between x and y. Exhibit 10.2

Exhibit 10.1 Plot and Correlation of Data in Table 10.1

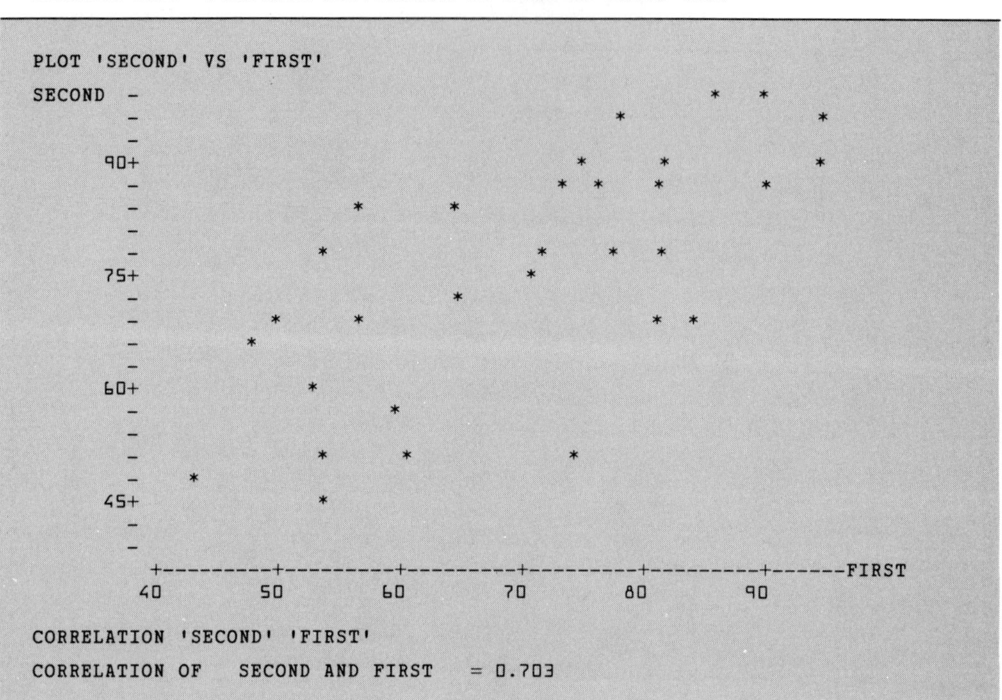

CORRELATION coefficient between C and C

Computes the correlation between the two columns of data. The usual Pearson product moment correlation coefficient is used.

$$r = \frac{\Sigma (x - \bar{x})(y - \bar{y})}{\sqrt{\Sigma (x - \bar{x})^2 \Sigma (y - \bar{y})^2}}$$

CORRELATION coefficients between C, ..., C

Minitab prints a table giving the correlations between all pairs of columns. Example:

```
CORRELATION C1, C2, C3
```

Three correlations are computed: the correlation between C1 and C2, between C1 and C3, and between C2 and C3.

Exhibit 10.2 Two Examples Where the Correlation Coefficient r Is Zero

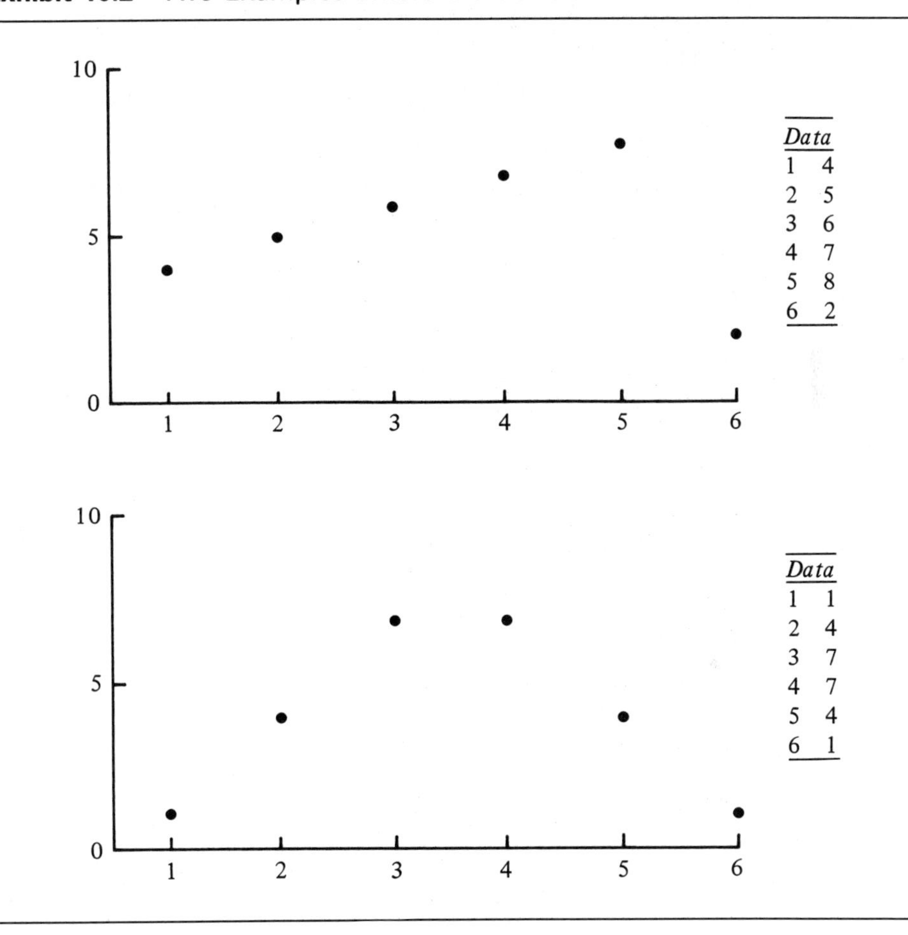

gives two examples. In the first plot, the correlation would be almost +1 if it were not for the point at $x = 6$. In the second plot, there is a strong association but it is not linear, that is, the points do not lie on a straight line. They do, however, lie on a curve. The correlation coefficient measures one type of association—how closely the points fall on a straight line. Fortunately, this is the most common type of association. But the two examples in Exhibit 10.2 show that a correlation coefficient can at times be quite misleading and that plots may give you a better view of what is really going on.

Exercises

10–1 There are 3 variables in the Trees data (p. 328): diameter, height, and volume. Plot volume versus height, volume versus diameter, and height versus diameter. Find the correlation associated with each of the three plots. Which relationship seems strongest? Which variable would probably be the best predictor of volume? Is there any evidence to make you doubt the reasonableness of the *linear* (straight line) correlation coefficients you computed?

10–2 Listed below are the price and number of pages for 15 books that were reviewed in the February 1982 issue of the journal *Technometrics*. Make a plot of price versus book length. Compute the correlation between book length and price.

Pages	*Price ($)*
302	30
425	24
526	35
532	42
145	25
556	27
426	64
359	59
465	55
246	25
143	15
557	29
372	30
320	25
178	26

10–3 In this exercise we first simulate data under various conditions. We then plot the data, guess the correlations, and check our guesses.

(a) Use the following command to simulate two independent random normal samples into columns C1 and C2:

```
RANDOM 50 C1 C2
```

Make a plot of C1 versus C2 and guess the correlation coefficient. Then compute the correlation coefficient and check your guess. Repeat this process four more times.

(b) A trick allows us to simulate correlated samples. Repeat (a), but this time use:

```
RANDOM 50 C1 C2
LET C3 = C1 + C2
```

Now plot C1 versus C3, guess the correlation, and check it. Repeat four more times.

(c) Repeat (b), but this time use:

```
RANDOM 50 C1 C2
LET C3 = C2 - C1
```

(d) Repeat (b) but use:

```
RANDOM 50 C1 C2
LET C3 = 5*C1 + C2
```

(e) Repeat (b) but use:

```
RANDOM 50 C1 C2
LET C3 = C2 - 5*C1
```

10–4 (a) Exhibit 3.1 is a plot of gas consumption per hour versus the outside temperature. Guess the correlation of these data. Compute the correlation coefficient and check your guess.

(b) Repeat (a) for the plot of cartoon scores versus OTIS scores in Exhibit 3.2.

10–5 Use SET to put the integers 1 to 50 into C1. Now compute

```
C2 = C1*C1
C3 = SQRT(C1)
C4 = 3*C1
C5 = (C1-25.5)**2
```

In each case, there is an exact relationship with C1. Plot C2 versus C1, C3 versus C1, and so on. Look at the relationships. Now compute the correlations among C1–C5. Explain why the correlation coefficients are not all 1.

10.2 Simple Regression: Fitting a Straight Line

Correlation tells us how much association there is between two variables; but regression goes further. It gives us an equation that uses one variable to help explain the variation in another variable. In this section we will show how to use the REGRESS command to fit straight lines.

Look at the data in Table 10.1 again. Suppose we wanted an equation that relates the second test scores to the first. From Exhibit 10.1 we see that if a person scored 70 on the first test, we might expect a score of about 75 on the second test. If another person scored 90 on the first test, we might expect about a 90 on the second test. If still another scored 55 on the first test, we might expect a second score of about 65.

In fact, it looks as if we might be able to draw a straight line on the plot and use it to predict scores on the second test. Suppose we decide to do that. How would we draw the line? Any straight line we draw will miss some of the points. Exhibit 10.3 gives another copy of this plot with several lines drawn. Which one looks best?

A method called *least squares* often is used to decide which line to choose. Suppose we look at the deviation between what a given line suggests and what actually happened for each person. If the line suggests a second score of 79 and the actual score was 88, then the deviation is $88 - 79 = 9$. Suppose we compute all the deviations for a line, square them, and add them up.

Intuitively, a line with large deviations is not as good as a line with smaller deviations. The least squares criterion says we should use the line that gives us the smallest sum of squared deviations. The surprising thing is that it is relatively easy to find the equation for that line.

If we are trying to use x to determine y, any straight line can be written in the form $y = a + bx$ where a and b are two numbers that tell us which line we are using. For example, $y = 3 + 2x$ and $y = -5$

Exhibit 10.3 Several Possible Lines for Predicting Second Scores from First Scores

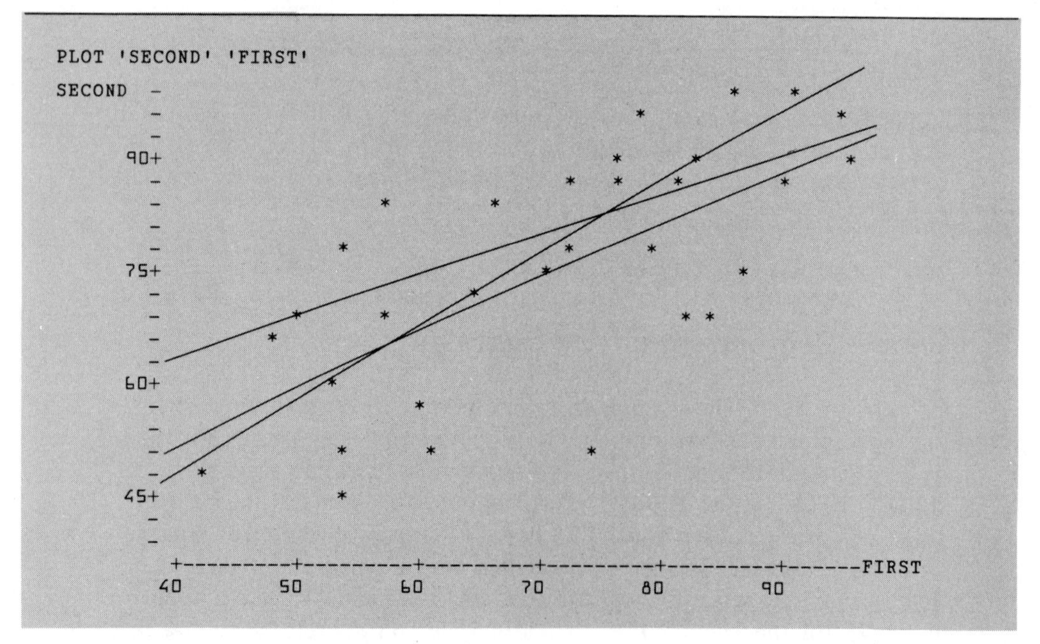

+ .8x specify two different straight lines. For the least squares line, a and b can be found with the formulas:

$$b = \frac{\Sigma (x - \bar{x})(y - \bar{y})}{\Sigma (x - \bar{x})^2} \text{ and } a = \bar{y} - b\bar{x}$$

Minitab will give values for a and b, and many other useful quantities, when the REGRESS command is used.

Suppose we want to use REGRESS to relate the second test scores to the first scores. All we need to do, once the data have been put in the worksheet, is type

```
REGRESS 'SECOND' on 1 predictor, 'FIRST'
```

Notice you must type the number of predictors. Here there is just 1 predictor, the score on the first test. Exhibit 10.4 shows the full output that may be obtained from the REGRESS command. Ordinarily an abbreviated version is given; we use the full output here to facilitate our explanation.*

Look at some of the simpler parts of this output. The regression equation is SECOND = 22.5 + .755 FIRST or $y = 22.5 + .755x$. This equation corresponds to one of the lines in Exhibit 10.3. (Can you determine which one?) Since it is the least squares line, it is impossible to find any straight line that gives a smaller sum of squared deviations than this one. The sum of squared deviations for this line is 3844. It is printed in the Analysis of Variance table in Exhibit 10.4 as SS Error. SS is an abbreviation for sum of squares and Error is another name for deviation.

The next block of output again gives the values of $a = 22.47$ and $b = .7546$, with some additional information. Next comes the Analysis of Variance table, which gives, as we will see later, a breakdown of the variation in the data.

The last block of output lists each x value and y value and then gives the fitted y value from the equation. For example, the first person's score on the first test was 50 and his score on the second test was 69. The fitted score was 60.20, which was computed by 22.47 + .7546 (50). The corresponding deviation or residual is thus y − (fitted y) = 69 − 60.2, which gives 8.8.

R-SQUARED (Coefficient of Determination)

The output also gives R-sq = 49.4%. Whenever a straight line is fit to a set of data, R^2 is just the square of the ordinary correlation coefficient

* To get the full output, type the command BRIEF 3 (p. 259). Then all REGRESS commands that follow will give the full output.

Exhibit 10.4 REGRESS Output for Predicting Second Score from the First Score

```
BRIEF 3
REGRESS 'SECOND' on 1 predictor 'FIRST'

The regression equation is
SECOND = 22.5 + 0.755 FIRST

Predictor         Coef      Stdev.Coef       t-ratio
Constant         22.47           10.22          2.20
FIRST           0.7546          0.1417          5.32

s = 11.51         R-sq = 49.4%    R-sq(adj) = 47.7%

Analysis of Variance

SOURCE         DF          SS            MS
Regression      1      3757.4        3757.4
Error          29      3844.0         132.6
Total          30      7601.4

Obs     FIRST    SECOND        Fit   Stdev.Fit   Residual   St.Resid
  1      50.0     69.00      60.20        3.58       8.80       0.80
  2      66.0     85.00      72.27        2.17      12.73       1.13
  3      73.0     88.00      77.55        2.09      10.45       0.92
  4      84.0     70.00      85.85        2.80     -15.85      -1.42
  5      57.0     84.00      65.48        2.83      18.52       1.66
  6      83.0     78.00      85.10        2.71      -7.10      -0.63
  7      76.0     90.00      79.81        2.20      10.19       0.90
  8      95.0     97.00      94.15        4.02       2.85       0.26
  9      73.0     79.00      77.55        2.09       1.45       0.13
 10      78.0     95.00      81.32        2.32      13.68       1.21
 11      48.0     67.00      58.69        3.82       8.31       0.77
 12      53.0     60.00      62.46        3.24      -2.46      -0.22
 13      54.0     79.00      63.21        3.14      15.79       1.43
 14      79.0     79.00      82.08        2.38      -3.08      -0.27
 15      76.0     88.00      79.81        2.20       8.19       0.72
 16      90.0     98.00      90.38        3.44       7.62       0.69
 17      60.0     56.00      67.74        2.56     -11.74      -1.05
 18      89.0     87.00      89.62        3.32      -2.62      -0.24
 19      83.0     91.00      85.10        2.71       5.90       0.53
 20      81.0     86.00      83.59        2.54       2.41       0.21
 21      57.0     69.00      65.48        2.83       3.52       0.32
 22      71.0     75.00      76.04        2.07      -1.04      -0.09
 23      86.0     98.00      87.36        3.00      10.64       0.96
 24      82.0     70.00      84.34        2.62     -14.34      -1.28
 25      95.0     91.00      94.15        4.02      -3.15      -0.29
 26      42.0     48.00      54.16        4.56      -6.16      -0.58
 27      75.0     52.00      79.06        2.16     -27.06      -2.39R
 28      54.0     44.00      63.21        3.14     -19.21      -1.73
 29      54.0     51.00      63.21        3.14     -12.21      -1.10
 30      65.0     73.00      71.51        2.22       1.49       0.13
 31      61.0     52.00      68.50        2.48     -16.50      -1.47

R denotes an obs. with a large st. resid.
```

we discussed in Section 10.1. There we found $r = .703$, which, when squared, gives .494 or 49.4%, the same value given here in the REGRESS output.

R^2 also has two other, more general, interpretations. It is the square of the correlation between the observed y values and the fitted y values. It also is the fraction of the variation in y that is explained by the fitted equation. Look at the Analysis of Variance table in Exhibit 10.4. The total sum of squared deviations, written Total SS, is a measure of the variation of y about its mean. Here it is 7601. The Regression SS is the amount of this variation which is explained by the regression line. Here it is 3757. The fraction of variation explained is 3757/7601 which is again .494. It sometimes is more convenient to convert this to a percentage and say the regression equation explains 49.4% of the variation in y.

REGRESS the *y* values in C on 1 explanatory variable in C

Computes the regression equation for determining y from x. Output includes the equation and other information.

More details about REGRESS are given in the remainder of this chapter.

Exercises

10–6 Refer to the output from REGRESS in Exhibit 10.4.

(a) Based on the fitted equation, if a person scored 54 on the first test, what score would you predict for the second test?

(b) How many individuals got 54 on the first test? What did they each get on the second test? Find the residual (deviation) for each.

(c) If a person got 100 on the first test, what score would you predict on the second test? (Always check your answers for reasonableness.)

10–7 The following are the mean Scholastic Aptitude Test scores (SAT scores) for the academic years 1967 through 1981.

Year	*Verbal*	*Math*
1967	466	492
1968	466	492
1969	463	493
1970	460	488
1971	455	488
1972	453	484
1973	445	481

Year	*Verbal*	*Math*
1974	444	480
1975	434	472
1976	431	472
1977	429	470
1978	429	468
1979	427	467
1980	424	466
1981	424	466

(a) Plot the verbal SAT scores versus year. Repeat for math scores.

(b) Fit a regression line using year to predict the verbal SAT scores.

(c) Fit a regression line using year to predict the math SAT scores.

(d) If you have a copy of the plot on paper, draw in the two regression lines (by hand).

(e) Do both verbal and math scores seem to be changing at about the same rate (same number of points per year)?

(f) What is the predicted average SAT score for math in 1970? In 1983? In 2000? Which of these seem to make sense? Repeat for verbal scores.

10–8 Given below are the winning times for the men's 1500-meter run in the Olympics from 1900–1980. (Note: No Olympics were held in 1916, 1940, and 1944 due to wars; in 1980 the United States and some other countries did not participate.)

Year	1900	1904	1908	1912	1920	1924	1928	1932	1936
Time (sec)	246.0	245.4	243.4	236.8	241.8	233.6	233.2	231.2	227.8
Year	1948	1952	1956	1960	1964	1968	1972	1976	1980
Time (sec)	229.8	225.2	221.2	215.6	218.1	214.9	216.3	219.2	218.4

(a) Plot winning time versus year. Does winning time seem to be changing over the years? Is it changing according to a straight line?

(b) Fit a regression line for predicting winning time from year. If you have a plot of the data on paper, draw the regression line on the plot.

(c) What winning time would you predict for the 1988 Olympics? The year 2000 Olympics?

(d) On the average, how much has the winning time decreased in the four years between Olympics? Use the fitted equation.

(e) Are there any appreciable departures from the straight line that you might have anticipated from outside information? Explain.

10–9 The following data were collected to study the relationship between the temperature of a battery and its output voltage. The eight measurements were taken in the order listed.

	Reading							
	1	*2*	*3*	*4*	*5*	*6*	*7*	*8*
Temperature (Celsius)	10.0	10.0	23.1	23.1	34.0	34.0	45.6	45.6
Voltage	4290	4270	4470	4485	4723	4731	4920	4935

(a) Plot the data and use REGRESS to fit a straight line to estimate voltage as a function of temperature. Does the regression line seem to fit the data well?

(b) Do you spot any weakness in the order in which the readings were taken? Can you think of some better orders in which to make the eight readings if you had to do this experiment again? Discuss.

10–10 (a) How well can you predict the volume of a tree from its diameter? Use the Trees data (p. 328) to find an equation for black cherry trees in Pennsylvania.

(b) Plot the data (volume versus diameter) and, if the plot is on paper, draw your regression line on it. How well does the line seem to fit the data? Do you see any problems?

(c) Repeat (a) and (b) but use height as the predictor of volume.

(d) Which is the better predictor, height or diameter? Which equation would be easier to use in the woods to predict the volume of a given tree? Why?

10.3 Making Inferences from Straight-Line Fits

Sometimes we will be content to use REGRESS just to fit an equation to the set of data we happen to have. On other occasions we may want to generalize from the data in hand to some larger population. Sometimes our data will be a random sample from some population. In such cases we often will be more interested in the characteristics of the population than in the characteristics of the sample. Other times, our data will have resulted from a carefully designed and executed experiment. Then we usually will be more interested in the underlying relationship between the variables than in the values we happened to get in this particular experiment.

In other cases, our data will be more happenstance in nature. For example, we may have good estimates of the automobile fatality rate and the average highway speeds of vehicles in each of the 50 states. Sometimes we may want to pretend, for a moment, that such data are like a random sample from some population. We will know that in reality there are only 50 states in all, but pretending our 50 are a random sample allows us to use the procedures of statistical inference for general guidance. Then we might seek to answer questions like, "Is the relationship we observe between fatalities and speed one that could reasonably have occurred due to chance alone?"

Conditions for Inference

There are several conditions that must be met, at least approximately, before we can make reasonable statistical inferences:

1. In the underlying population, the relationship between x and y should be a straight line. Suppose for each value of x, we find the mean of all the corresponding y values. Then these means must, at least approximately, fall on a straight line. We will denote this straight line by $A + Bx$, where A and B are some fixed values. For example, suppose we could calculate the mean of all the ys in the population corresponding to $x = 1$, the mean of all the ys corresponding to $x = 2$, the mean of all the ys corresponding to $x = 3$, and so on. Then all these means should fall on a straight line. We will call this line the *population regression line*. The values a and b, discussed in Section 10.2, are our sample estimates of the population values A and B.
2. For each x, the amount of variation in the population of ys should be approximately the same. This variance is usually called the variance of y about the regression line, and is denoted by σ^2. Correspondingly, σ is called the standard deviation of y about the regression line.
3. For each value of x, the distribution of ys in the population should be approximately normal.
4. The ys that actually are obtained should be approximately independent. In other words, the amount by which a particular y value differs from its mean should not be related to the amount by which any other y value differs from its mean.

In Sections 10.7 and 10.8, we will discuss some procedures for checking whether these conditions hold. Condition 4 is the most important and the most difficult to check. The best advice when doing surveys or experiments is to try to do them properly in the first place. For example, try to make sure that no observations are unduly related to one another.

Whenever practical, use randomization to determine the order in which measurements are made. Try to avoid biases or systematic errors.

Interpreting the Output

The following all require that the conditions for inference be at least approximately satisfied.

Estimate of σ. We begin by looking at s in Exhibit 10.4. This often is called "the standard deviation of y about the regression line" or "the standard error of estimate." This quantity gives us an estimate of σ. It is computed using the formula

$$s = \sqrt{\frac{\Sigma(y - \text{fitted } y)^2}{n - 2}}$$

The value s can be thought of as a measure of how much the observed y values differ from the corresponding average y values as given by the least squares line. It has $(n - 2)$ degrees of freedom and is used in all of the formulas for standard deviations. All t-tests and confidence intervals will be based on this s, and all thus will have $(n - 2)$ degrees of freedom. In our example, the number of degrees of freedom is $(31 - 2) = 29$.

Standard Deviation of Coefficients. Under the Conditions for Inference, the estimated coefficients a and b each have an approximately normal distribution. The estimated standard deviations of these coefficients are given in the column headed Stdev. Coef. in Exhibit 10.4. The estimated standard deviation of a is 10.22, and the estimated standard deviation of b is 0.1417.

Confidence Intervals. The general formula for a t confidence interval is

$$(\text{quantity}) \pm (\text{value from } t\text{-table}) \times (\text{estimated stdev of quantity})$$

The t value corresponding to 29 degrees of freedom and 95% confidence is 2.045. Thus a 95% confidence interval for A (the population value of a) is

$$a \pm t\ (\text{stdev of } a)$$

or

$$22.47 \pm 2.045\ (10.22)$$

This gives the interval 1.57 to 43.37. A 95% confidence interval for B (the population value of b) is similarly given by

$$0.7546 \pm 2.045(0.1417)$$

This gives the interval .46 to 1.04.

Tests of Significance. We often want to know if there is any statistically significant evidence of an association between x and y. Thus, a hypothesis we frequently want to test is that B is 0. We use the general formula

$$t = \frac{b - \text{(hypothesized value)}}{\text{(estimated stdev of } b\text{)}}$$

Here we find that

$$t = \frac{.7546 - 0}{.1417} = 5.33$$

This is given in the column headed T-RATIO. With 29 degrees of freedom, this value of t is highly significant, giving us evidence that B is probably not 0. This in turn implies that the score on the first test is at least slightly useful as a predictor of the score on the second test. (Note: This t-test is equivalent to testing whether there is evidence that the population correlation coefficient is not 0.)

A similar t-test of the null hypothesis that $A = 0$ gives

$$t = \frac{22.47 - 0}{10.22} = 2.20$$

This also is statistically significant, although just barely. Thus, from a statistical standpoint, both a and b have been shown to be useful in the equation.

Standard Deviation of a Fitted y-Value. The estimated standard deviations of the fitted y values are given in the column headed Stdev.Fit in Exhibit 10.4. These can be used to get confidence intervals for the population mean of all y values corresponding to a given value of x. For example, the first line shows that the fitted value for $x = 50$ is 60.20. The 95% confidence interval for the mean of all ys corresponding to $x = 50$ is given by

$$\text{(fit)} \pm \text{t(stdev of fit)}$$

or

$$60.2 \pm 2.045\ (3.58).$$

This gives the interval 52.9 to 67.6. Thus, we can be 95% confident that the mean score on the second test, for all persons who score 50 on the first test, is between 52.9 and 67.6.

This tells us how well persons with a first test score of 50 will do on the average. But how about one particular individual who scores 50 on the first test?

Prediction Interval for a Single y Value. Since individuals are never as predictable as averages, we must expect more uncertainty in this prediction. The following calculations give us an interval which we can be 95% confident will contain the second test score of an *individual* who gets a 50 on the first test:

$$60.2 \pm 2.045 \sqrt{(3.58)^2 + (11.51)^2}$$

This gives the interval 35.5 to 84.9. The 3.58 is, again, the estimated standard deviation of the fitted value and $s = 11.51$ is the estimated standard deviation of an individual. Note that the interval for an individual is, as expected, larger than the interval for the average.

Most texts give the following, slightly different looking, formula for a prediction interval corresponding to a given x:

$$(\text{fitted } y) \pm ts \sqrt{\frac{1}{n} + \frac{(x_0 - \bar{x})^2}{\Sigma\,(x - \bar{x})^2} + 1}$$

We can rewrite this formula as

$$(\text{fitted } y) \pm t \sqrt{(\text{estimated stdev of fitted } y)^2 + s^2}$$

This is precisely the formula we used in the calculations above.

Predictions for New Values of x. The procedure we just used for $x = 50$ will work for any value of x that was in our set of data, such as $x = 66$, $x = 73$, or $x = 84$. But how about a prediction interval for a value of x that was not in our original set of data? For example, how could we find a prediction interval for $x = 68$? To get the predicted second score is not too difficult. All we have to do is substitute 68 into the regression equation. This gives

$$y = 22.47 + (.7546)\,(68) = 73.78$$

The easiest way to get the estimated standard deviation of this predicted y value is to use the PREDICT subcommand. To use PREDICT for $x = 68$, type

```
REGRESS 'SECOND' ON 1 'FIRST';
  PREDICT 68.
```

Minitab will then print the following extra lines.

```
  Fit   Stdev.Fit         95% C.I.              95% P.I.
73.78        2.10 (  69.49,   78.08)  (  49.84,   97.92)
```

REGRESS C on 1, C

PREDICT for E

The PREDICT subcommand computes estimates for any values of x. It prints out a table of fitted y, standard deviation of fitted y, a 95% confidence interval, and a 95% prediction interval.

E may be a constant such as 68 or K3 or it may be a column containing a list of x values.

Exercises

10–11 Refer to the output in Exhibit 10.4.

(a) Calculate a 90% confidence interval for B, the slope of the underlying regression line.

(b) Find a 90% confidence interval for A, the intercept of the underlying regression line.

(c) Find a 90% confidence interval for the average of the second test scores of all persons in the population who score a 90 on the first test.

10–12 Refer to the output in Exhibit 10.4. Suppose a person scored an 86 on the first test. What score should be expected on the second? Find an interval which you are 95% confident will cover the second score for that individual.

10–13 Refer to the output in Exhibit 10.4. Test the null hypothesis, H_0: $A = .5$ versus the alternative hypothesis, H_1: $A \neq .5$.

10–14 Refer to the test data in Table 10.1.

(a) Find a 90% confidence interval for the mean second test scores for all persons in the population who obtained a score of 77 on the first test. You will need to use the PREDICT subcommand since 77 is not in the data set.

(b) Find a 90% prediction interval for the second test score for an individual who achieved a score of 77 on the first test.

10–15 Refer to Exercise 10–8 (p. 228) for the 1500-meter race. Give 95% confidence intervals for the two regression coefficients.

10–16 (a) We can simulate data from a regression model as follows. To choose a model, we need to specify 3 things, A, B, and σ. Suppose we use $A = 3$, $B = 5$, and $\sigma = .5$. Next, we must specify values for x. Suppose we take two observations at each integer from 1 to 10. First, we use SET to put the 20 values of x into a column. Then use LET to calculate $A + Bx$. Now, simulate 20 observations from a normal distribution with $\mu = 0$ and $\sigma = .5$. Add these to $A + Bx$ to get the observed ys. Get a plot and do a regression for these simulated data. Record a, b, S, and R^2.

(b) Repeat part (a) using $\sigma = 2.00$. Compare the results with those of part (a).

(c) Repeat part (a) using $\sigma = 10.0$. Compare the results with those of (a) and (b).

10–17 Maple trees have winged fruit called samara which come spinning to the ground in the fall. A forest scientist was interested in the relationship between the velocity with which the samara fell and their "disk loading." The disk loading is a function of the size and weight of the fruit and is closely related to the aerodynamics of helicopters.

Tests were run on samara from three trees with the following results.

Tree 1		*Tree 2*		*Tree 3*	
Loading	*Velocity*	*Loading*	*Velocity*	*Loading*	*Velocity*
.239	1.34	.238	1.20	.192	.91
.208	1.06	.206	1.06	.200	1.13
.223	1.14	.172	.88	.175	1.00
.224	1.13	.235	1.24	.187	.98
.246	1.35	.247	1.37	.181	.96
.213	1.23	.239	1.37	.195	.88
.198	1.23	.233	1.43	.155	.81
.219	1.15	.234	1.32	.179	.91
.241	1.25	.189	.99	.184	1.00
.210	1.24	.192	1.00	.177	.87
.224	1.34	.209	1.12	.177	1.02
.269	1.35			.186	.94

Data are in the saved worksheet called MAPLE.

(a) Does velocity seem to be a straight line function of loading? Examine separately for each tree.
(b) A scientist hypothesizes that the straight lines will go through the origin (the point $x = 0$, $y = 0$). Do they seem to do this, at least approximately?
(c) Test H_0: $A = 0$ and compare to your answer in (b).
(d) Is there any difference in the relationship between velocity and loading for the three different trees?

10.4 Multiple Regression

So far we have described how one variable can be used to help explain the variation in another, for example, how the score on one test relates to that on another test. But, what if you have two or three, or even more variables that could help with the explanation? One technique you can use, and the one we will describe in this section, is multiple regression. We begin with an example.

Many universities use multiple regression to estimate how well the various applicants would do if they were admitted to that university. An equation used at one major university was

$$\text{(Freshman GPA)} = .61813\text{(H.S. GPA)} + .00137\text{ (SAT verbal)} + .00063\text{(SAT math)} - .19787$$

This equation shows that the estimated grade point average (GPA) at the end of the freshman year is equal to .61813 times the high-school grade point average plus .00137 times the Scholastic Aptitude Test verbal score, plus .00063 times the Scholastic Aptitude Test mathematics score, minus .19787. This equation was an important criterion in deciding who to admit to that university. The variables (H.S. GPA), (SAT verbal), and (SAT math) are often called *predictor variables*. Here they are used to predict Freshman GPA.

This equation was obtained by using multiple regression on the records of previous students. The university had the freshman year GPA for some past students, as well as their high school GPA and their two SAT scores. They asked, "Which equation best explains freshman GPA from these other variables?" The procedure they used is very similar to the one we used in Sections 10.2 and 10.3 for straight-line equations.

The Grades example (p. 309) gives some data from another university.

Suppose we want to develop a similar equation for freshman GPAs at that university using just the SAT verbal and SAT math scores. We used Minitab to find an equation based on a sample of 100 freshmen (sample A from p. 309). The results are shown in Exhibit 10.5. Notice you must type the number of predictor variables on the REGRESS command. Exhibit 10.5 gives the default output from REGRESS: Only those observations which are "unusual" because of their x values (marked by X on the output) or because of their residuals (marked by R on the output) are printed. If you want all 100 observations printed, type BRIEF 3 before you type REGRESS.

Exhibit 10.5 Multiple Regression Output for Predicting Freshman GPA from SAT Scores

```
RETRIEVE 'GA'
REGRESS 'GPA' 2 'VERBAL' 'MATH'

The regression equation is
GPA = 0.471 + 0.00356 VERBAL +0.000158 MATH

Predictor        Coef       Stdev.Coef      t-ratio
Constant       0.4706          0.5433          0.87
VERBAL      0.0035628       0.0007350          4.85
MATH        0.0001576       0.0008514          0.19

s = 0.5018      R-sq = 23.5%   R-sq(adj) = 22.0%

Analysis of Variance

SOURCE         DF           SS            MS
Regression      2       7.5137        3.7568
Error          97      24.4227        0.2518
Total          99      31.9364

SOURCE         DF       SEQ SS
VERBAL          1       7.5051
MATH            1       0.0086

Unusual Observations
Obs    VERBAL        GPA        Fit   Stdev.Fit   Residual   St.Resid
  2       454     2.3000     2.1624      0.1544     0.1376     0.29 X
 40       490     1.2000     2.3269      0.1149    -1.1269    -2.31R
 54       592     2.4000     2.6493      0.1863    -0.2493    -0.54 X
 89       361     2.4000     1.8517      0.1682     0.5483     1.16 X

R denotes an obs. with a large st. resid.
X denotes an obs. whose X value gives it large influence.
```

Interpreting the Output

The equation for Freshman GPA is

$$\text{(fitted GPA)} = .471 + (.00356)\,(\text{VERBAL}) + (.000158)\,(\text{MATH})$$

We can use this equation to estimate how well a student will do who made 500 on both SAT scores. We compute

$$\begin{aligned}\text{(fitted GPA)} &= .471 + (.00356)\,(500) + (.000158)\,(500)\\ &= .471 + 1.78 + 0.08\\ &= 2.33\end{aligned}$$

Thus we forecast the student's GPA as 2.33.

For a student who made 500 on the verbal test and 800 on the math test we would predict

$$\begin{aligned}\text{(Estimated GPA)} &= .471 + (.00356)\,(500) + (.000158)\,(800)\\ &= 2.38\end{aligned}$$

Thus a student with scores of 500 and 500 and another student with scores of 500 and 800 have estimated GPAs which differ by only .05. This seems to indicate that SAT math scores are not very useful estimators of GPA at that university.

How good is this estimation equation as a whole? Put another way, how much might the GPAs for individual students vary from what we predict? One way to answer this question is to look at the value of R^2. It is 23.5%, which means that our equation explains only 23.5% of the variation in GPAs. The remaining 76.5% of the variation in freshman GPAs is left unexplained.

From these results, it appears that SAT scores have only limited value in forecasting who will succeed in college. We can give several possible reasons for this. First, we do not have results for a random sample of all students. All we have is students who applied, were admitted, and attended that university. Second, perhaps those students with low SAT scores were advised to take easier courses and thus received higher grades than they ordinarily would have, while those with higher SAT scores were encouraged to take more difficult courses. Third, it is possible that tests like the Scholastic Aptitude Test simply do not do a very good job of measuring whatever it takes to make good grades in college.

Notation and Assumptions in Multiple Regression with Two Predictors

Suppose there are two explanatory variables—call them x_1 and x_2. We assume the following:

1. The underlying population regression line is approximately $y = B_0 + B_1x_1 + B_2x_2$.
2. For all values of x_1 and x_2, the ys have approximately the same variance, σ^2.
3. For each x_1 and x_2, the ys have approximately normal distributions.
4. The ys are approximately independent.
5. We use b_0, b_1, b_2, and s for the estimated values of B_0, B_1, B_2, and σ, respectively.

Confidence Intervals and Tests

All the confidence intervals we calculated for straight lines in Section 10.3 can be calculated here in essentially the same way. For example, in Exhibit 10.5, the estimated standard deviation of b_2 is .0008514. Since there are 97 degrees of freedom, a 95% confidence interval for B_2 is .000158 ± (1.99)(.0008514) which gives the interval −.00154 to .00185.

We can also do t-tests on the coefficients just as we did in simple regression. For example, to test H_0: $B_2 = .005$, we form t ratio $t = (b_2 - .005)/(\text{estimated stdev of } b_2)$. This ratio has t distribution with 97 degrees of freedom.

The t ratio for the hypothesis that $B_2 = 0$ is given on the output, and is just .19. Thus, b_2 was not statistically different from zero. This test indicates that SAT math scores were not statistically significant in explaining freshmen GPAs in this sample of students.

Suppose we want a 95% confidence interval for the population mean value of y corresponding to an SAT verbal score of 454 and an SAT math score of 471. This pair of scores happens to be in the data we used in the regression; it's the second observation (see p. 309) and it is in Exhibit 10.5. The fitted GPA is 2.1624, and the Stdev.Fit is .1544. This means that 2.1624 ± (1.99)(.1544) or 1.86 to 2.47 gives a 95% confidence interval for the mean GPA of all students in the population who had an SAT verbal score of 454 and an SAT math score of 471.

For a new *individual* with these same test scores the prediction interval can be computed, as on page 233, to be

$$2.1624 \pm (1.99)\sqrt{(.5018)^2 + (.1544)^2}$$

This gives the interval 1.66 to 2.67, emphasizing again our lack of ability to make precise forecasts of freshman GPAs.

Suppose we want to compute an estimate for scores that did not happen to be in the original set of data, say 600 and 750. We then can calculate

$$\begin{aligned}(\text{Pred GPA}) &= .471 + (.00356)(600) + (.000158)(750) \\ &= 2.73\end{aligned}$$

To obtain a confidence interval or prediction interval for values that are not in the original data set, we use the PREDICT subcommand. For example,

```
REGRESS 'GPA' 2 'VERBAL' 'MATH';
  PREDICT 600 750.
```

This gave the following output:

```
  Fit  Stdev.Fit         95% C.I.              95% P.I.
2.7265    0.0954   ( 2.5371, 2.9159)  ( 1.7125, 3.7405)
```

We see that a student with an SAT verbal score of 600 and an SAT math score of 750 has an expected freshman GPA of 2.73, as we calculated. In addition, we are 95% confident that the average GPA of all such freshmen at this university is between 2.5371 and 2.9159. The 95% prediction interval for an individual freshman is 1.7125 to 3.7405.

REGRESS C on K predictors C, ..., C

Calculates and prints a multiple regression equation.

REGRESS C on K predictors C, ..., C

PREDICT for E, ..., E

The PREDICT subcommand tells Minitab to compute predicted values, standard deviations, 95% confidence intervals and 95% prediction intervals. The arguments on PREDICT should be either all columns of the same length, or all constants. There should be as many arguments on the PREDICT subcommand as there are predictors on the REGRESS command.

Examples

```
REGRESS C8 on 4 C1-C4;
  PREDICT 16, 20, 4, 30.

REGRESS C8 on 4 C1-C4;
  PREDICT C11-C14.
```

Exercises

Refer to the output in Exhibit 10.5.

10–18 (a) Get a 95% confidence interval for B_1, the coefficient of SAT verbal score.

(b) Is B_1 significantly different from zero (use $\alpha = .05$)? What does this say about the relationship between SAT verbal score and GPA?

(c) Is your answer to (b) consistent with the low value of R^2? Explain.

10–19 (a) Use the data in sample B from the Grades data (p. 309) to develop another equation for predicting GPA from SAT scores.

(b) How does this equation compare to the equation in Exhibit 10.5? Compare the estimates of B_0, B_1, B_2, σ, and R^2 for the two equations.

10–20 In Exercise 10–10, we fitted an equation for estimating volume of a black cherry tree from its diameter. Suppose we use height as a second explanatory variable.

(a) Find an equation for estimating volume from diameter and height. How much extra help does height seem to give you when predicting volume?

(b) Use REGRESS and PREDICT to calculate a 95% confidence interval for estimating the average volume of trees with diameter = 11 and height = 70.

10.5 A Longer Example of a Multiple Regression Problem

The Peru data on page 317 were collected to study the long-term effects of a change in environment on blood pressure. Some questions we might ask of these data are as follows: "What happens to the Indians' blood pressure as they live longer periods in the new environment?" "Does it go up or down?" "Does it change by about the same amount each year?'' In this section we will show how multiple regression can be used to help answer these questions. We begin by showing what happens when we use just one explanatory variable.

First, we need to decide how we should measure ''how long'' the Indians have lived in their new environment. We could just use the ''number of years since migration.'' But that has problems. Younger people adapt to new surroundings more quickly than older people. A 25-year-old might be able to adapt as well in one year as a 50-year-old could in two years. In this analysis we chose to use the fraction of life in the new environment as a measure of ''how long'' someone has lived there. Thus, one year for a 25-year-old and two years for a 50-year-old will both be listed as 1/25 of their life. The data on page 318 were put into the worksheet, then fraction of life was computed using the following instructions.

```
RETRIEVE 'PERU'
NAME C15 'FRACTION'
LET 'FRACTION' = 'YEARS'/'AGE'
```

Now look at what happens when we plot systolic blood pressure versus fraction of life. Exhibit 10.6 shows this plot as well as the output from fitting a straight line to these data. The plot shows that there is very little relationship between blood pressure and the fraction of life that was spent in the new environment. The regression coefficient b_1 is negative (-15.8) which suggests that, if anything, there may be a tendency for blood pressure to decrease with increasing fraction of life in the modern society. However, the t ratio of -1.75 is not statistically significant. This lack of significance means that from this analysis it appears that it may well have been chance alone that gave the -15.8 coefficient in the sample. The coefficient in the population may well be zero, in which case there would be no linear association between blood pressure and fraction of life since migration.

The regression ouput lists two unusual observations. In both cases the individual had an unusually high blood pressure. Both points were checked to see if a blunder had been made in typing, or at some other place, but no errors were found. Of course, these may just have been people with abnormally high blood pressure, or they may have been excited when their blood pressure was taken. Perhaps an error was made when the anthropologist took the blood pressure. In any event, these data points seem to be different from the rest. Another regression was done with these points temporarily set aside. When this was done, there was even less indication of a significant relationship between the two variables.

Exhibit 10.6 Plot and Regression for Peru Data

```
RETRIEVE 'PERU'
NAME C15='FRACTION'
LET 'FRACTION'='YEARS'/'AGE'
PLOT C9 'FRACTION'

        -
SYSTOL  -    *
        -
        -
     160+
        -
        -                                              *
        -    *
        -                         *
     140+   * *                                * *
        -       *     *   *  **   *
        -                            *                         *
        -                 2                    *        *
        -           **  *                 *
     120+          *   * *        * *    *
        -  *                *     *  **
        -                            *          *
        -                                            *           *
        -
     100+
        -
          +---------+---------+---------+---------+---------+---------+------FRACTION
       0.00      0.16      0.32      0.48      0.64      0.80

REGRESS C9 1 'FRACTION'

The regression equation is
SYSTOL = 133 - 15.8 FRACTION

Predictor       Coef          Stdev      t-ratio
Constant     133.496          4.038        33.06
FRACTION     -15.752          9.013        -1.75

s = 12.77      R-sq = 7.6%    R-sq(adj) = 5.1%

Analysis of Variance

SOURCE        DF          SS          MS
Regression     1       498.1       498.1
Error         37      6033.4       163.1
Total         38      6531.4

Unusual Observations
Obs FRACTION     SYSTOL        Fit  Stdev.Fit   Residual   St.Resid
  1    0.048     170.00     132.75       3.67      37.25      3.05R
 39    0.741     152.00     121.83       3.79      30.17      2.47R

R denotes an obs. with a large st. resid.
```

Taking Weight into Consideration

At this point in the analysis it does not appear that there is any significant relationship between fraction of life and blood pressure. But blood pressure is known to depend on how much the person weighs. What happens if we also take weight into consideration?

Exhibit 10.7 shows the result of fitting a multiple regression model involving both fraction of life and weight. The relationship can be written as

$$\text{(fitted SYSTOLIC)} = b_0 + b_1\text{(WEIGHT)} + b_2\text{(FRACTION)}$$

Here, the equation is

$$\text{(fitted SYSTOLIC)} = 60.9 + 1.22\text{(WEIGHT)} - 26.8\text{(FRACTION)}$$

The coefficient for fraction of life since migration is again negative, -26.8. This indicates that blood pressure tends to be lower for those individuals who have lived longer in their new modern low-altitude environment. The t ratio for the coefficient is now -3.71, which is statistically significant. This tends to rule out the possibility of chance variation producing this result and lends support to the theory about the long-term effects of a change in environment on blood pressure. Of course, it does not prove the theory, since those results could have been caused by the effect of a change in altitude, or some other factor, or with some very small probability, by chance alone.

The equation indicates that a person who has lived half a life in the new environment will have, on the average, a systolic blood pressure which is about $26.77 \times .5 = 13.4$ lower than someone of the same weight who has just moved into modern society. Thus it appears that fraction of life since migration is both statistically significant (t ratio $= -3.71$) and practically significant (13.4 change in blood pressure in 1/2 of life).

The coefficient for weight is positive ($+1.22$), as we might expect. This indicates that blood pressure tends to increase as weight increases. This coefficient is statistically significant. It has a t ratio of 5.21 for testing the null hypothesis that there is no association between blood pressure and weight. Weight also is practically significant: The equation indicates that a change in weight of 10 kilograms (about 22 pounds) would lead to a $10 \times 1.22 = 12.2$ change in systolic blood pressure.

Weight as a Suppressor Variable

These data demonstrate what some scientists call a "suppressor variable." Here weight was a suppressor variable for fraction of life since migration, as we found no relationship between blood pressure and fraction of life

Exhibit 10.7 Multiple Regression Model for PERU Data

```
REGRESS C9 2 'WEIGHT' 'FRACTION'

The regression equation is
SYSTOL = 60.9 + 1.22 WEIGHT - 26.8 FRACTION

Predictor        Coef        Stdev     t-ratio
Constant        60.90        14.28        4.26
WEIGHT         1.2169       0.2337        5.21
FRACTION      -26.767        7.218       -3.71

s = 9.777        R-sq = 47.3%   R-sq(adj) = 44.4%

Analysis of Variance

SOURCE         DF          SS          MS
Regression      2      3090.1      1545.0
Error          36      3441.4        95.6
Total          38      6531.4

SOURCE         DF      SEQ SS
WEIGHT          1      1775.4
FRACTION        1      1314.7

Obs    WEIGHT     SYSTOL       Fit  Stdev.Fit  Residual  St.Resid
  1      71.0     170.00    146.02       3.80     23.98     2.66R
  8      53.0     108.00    101.49       5.15      6.51     0.78 X
 39      87.0     152.00    146.93       5.63      5.07     0.63 X

R denotes an obs. with a large st. resid.
X denotes an obs. whose X value gives it large influence.
```

until we took weight into consideration. The explanation seems to be that the migrants' blood pressure would have decreased over time as they stayed in their new environment, except that their blood pressure was simultaneously being driven up by their gaining weight. Increased weight was sending it up, while environmental acclimatization was lowering it. The net effect was essentially zero change.

10.6 Fitting Polynomials

So far we have talked about fitting straight lines (Sections 10.2 and 10.3) and about fitting equations with several variables (Sections 10.4 and 10.5). Now we will show how to fit curved data like those shown in Exhibit 10.8. In this section, we will describe how to use polynomial models.

Exhibit 10.8 Stream Data and Plot

Stream Depth	0.34	0.29	0.28	0.42	0.29	0.41	0.76	0.73	0.46	0.40
Flow Rate	0.636	0.319	0.734	1.327	0.487	0.924	7.350	5.890	1.979	1.124

```
PLOT 'FLOW' 'DEPTH'
         -
FLOW     -                                                  *
         -
         -
     6.0+                                                 *
         -
         -
         -
         -
     4.0+
         -
         -
         -
         -
     2.0+                             *
         -
         -                      * *
         -         *      *       *
         -           2
     0.0+
         -
       ------+---------+---------+---------+---------+---------+DEPTH
           0.30      0.40      0.50      0.60      0.70      0.80
```

In Section 10.8 we show how transformations give another, often preferable, method of analysis.

Polynomials are equations that involve powers of the x variable. For example,

$$y = B_0 + B_1x + B_2x^2$$

This is a statistical model for a second-degree polynomial, or quadratic equation.

An Example

The data in Exhibit 10.8 were gathered in conjunction with an environmental impact study to find the relationship between stream depth and flow rate. Flow rate is the total amount of water that flows past a given point in a fixed amount of time. Data were collected on seven different streams.

The data in Exhibit 10.8 are all from the same site on one stream. We are interested in estimating the flow rate from the depth of the stream. Obviously, a straight line will not give a very good fit—but a second-degree polynomial might.

All we need to do, to fit a second-degree polynomial, is to compute x^2 and put it in another column. Then we simply regress y on the two variables x and x^2, just as we did in multiple regression. Exhibit 10.9 contains the appropriate commands.

The form of the output is essentially the same as for a straight line. The regression equation is

$$\text{(FLOW)} = 1.68 - 10.9\text{(DEPTH)} + 23.5\text{(DEPTH)}^2$$

The standard deviation about the regression line is .2794.

We can do a t-test of the null hypothesis that, in the population, the coefficient of the $(\text{depth})^2$ term is zero. The data give $t = (23.535-0)/4.274 = 5.51$. This is statistically significant, giving us evidence that the population coefficient for $(\text{DEPTH})^2$ is not zero.

One might ask, "Just what is the 'population' in this case?" To answer this question, we need to imagine the amount of water in this stream fluctuating up and down over a long period of time, but without changing the basic flow pattern of the stream. Then all the measurements of flow rate and depth that might be made over this long period of time is our population. We pretend (but don't really believe) that our set of measurements is a random sample from this population. All we really can hope is that our measurements act just about the same as a random sample. If all of our measurements were made in the spring, or early in the morning, or just after a storm, or all in the same week, we could be fairly confident they would not act like a random sample. It is hoped the experimenter took such factors into consideration when the data were collected. If not, our inferences may be seriously incorrect.

Exercise

10–21 In Exercise 10–10, we fitted the equation (volume) $= B_0 + B_1$ (diameter) to the Tree data. In Exercise 10–20, we added height as a second explanatory variable and fitted the equation, (volume) $= B_0 + B_1$ (diameter) $+ B_2$ (height). In this exercise, use just diameter to predict volume, but now fit the quadratic equation, (volume) $= B_0 + B_1$ (diameter) $+ B_2$ $(\text{diameter})^2$.

(a) How well does this quadratic equation fit? Compare its fit to that of the straight line we fitted in Exercise 10–10.

Exhibit 10.9 Fitting a Quadratic Polynomial to the Stream Data

```
NAME C1 = 'DEPTH', C2 = 'FLOW'
READ 'DEPTH', 'FLOW'
 .34  .636
 .29  .319
 .28  .734
 .42 1.327
 .29  .487
 .41  .924
 .76 7.350
 .73 5.890
 .46 1.979
 .40 1.124
END
NAME C3 = 'DEPTH-SQ'
LET C3 = 'DEPTH'**2
BRIEF 3
REGRESS 'FLOW' ON 2, 'DEPTH', 'DEPTH-SQ';
  RESIDUALS C21.

The regression equation is
FLOW = 1.68 - 10.9 DEPTH + 23.5 DEPTH-SQ

Predictor         Coef      Stdev.Coef       t-ratio
Constant         1.683           1.059          1.59
DEPTH          -10.861           4.517         -2.40
DEPTH-SQ        23.535           4.274          5.51

s = 0.2794      R-sq = 99.0%   R-sq(adj) = 98.7%

Analysis of Variance

SOURCE          DF             SS             MS
Regression       2         54.105         27.053
Error            7          0.547          0.078
Total            9         54.652

SOURCE          DF         SEQ SS
DEPTH            1         51.739
DEPTH-SQ         1          2.367

Obs      DEPTH        FLOW         Fit   Stdev.Fit   Residual   St.Resid
  1      0.340      0.6360      0.7107      0.1029    -0.0747      -0.29
  2      0.290      0.3190      0.5123      0.1477    -0.1933      -0.82
  3      0.280      0.7340      0.4868      0.1634     0.2472       1.09
  4      0.420      1.3270      1.2727      0.1344     0.0543       0.22
  5      0.290      0.4870      0.5123      0.1477    -0.0253      -0.11
  6      0.410      0.9240      1.1860      0.1281    -0.2620      -1.05
  7      0.760      7.3500      7.0223      0.2138     0.3277       1.82
  8      0.730      5.8900      6.2961      0.1830    -0.4061      -1.92
  9      0.460      1.9790      1.6667      0.1575     0.3123       1.35
 10      0.400      1.1240      1.1040      0.1218     0.0200       0.08
```

(b) Does a quadratic equation seem a reasonable choice from a theoretical standpoint? Consider the geometry of diameter versus volume for a tree.

(c) Compare the quadratic equation in (a) to the equation in Exercise 10–20 which was based on height and diameter. How well does each fit?

(d) It is considerably more difficult to measure the height of a tree than its diameter. Based on this information, what equation would you most likely use in practice to estimate the volume of a tree?

10.7 Interpreting Residuals in Simple and Polynomial Regression

Whenever we fit a model to a set of data, we always should plot the data and the residuals. Recall that a residual is the difference between the observation and the fitted value determined by the regression equation. Thus, residuals tell us how the model missed in fitting the data. For example, in Exhibit 10.9, for the first observation, the stream depth is .340, the observed flow rate is .6360, and the estimated flow rate is .7107. Thus the first residual is (.6360 − .7107) = −.0747, and the second residual is −.1933. These, and all the other residuals, are listed in the column headed Residual.

Exhibits 10.10 and 10.11 show what happens when we fit an inadequate model to the stream-flow data. The data in Exhibit 10.10 are curved but we tried to represent them with a straight line anyway. Here it is easy to see that the equation is not a good fit. For low x values the observed y values are above the straight line, for middle x values the observed y values are below the straight, and for the largest x value the observed y is above the line. This means the residuals will be positive for low values of x, negative for middle values of x, and positive again for the high value of x. This pattern shows up nicely in the plot of residuals versus x shown in Exhibit 10.11.

To make it easy to do plots such as these, REGRESS has a subcommand which stores the residuals. We use this subcommand in Exhibit 10.10. Patterns like the one in Exhibit 10.10 sometimes show up more clearly in plots of the residuals than in plots of the original data. The plots of the data and residuals help us see when the model we have fit to the data is seriously wrong. A strong pattern in the residual plot indicates that we probably have a poor model.

What happens when we fit a second-degree polynomial to the stream data? Exhibit 10.12 contains a residual plot for such a fit. Is there any

Exhibit 10.10 Stream Data with a Straight Line Regression Equation

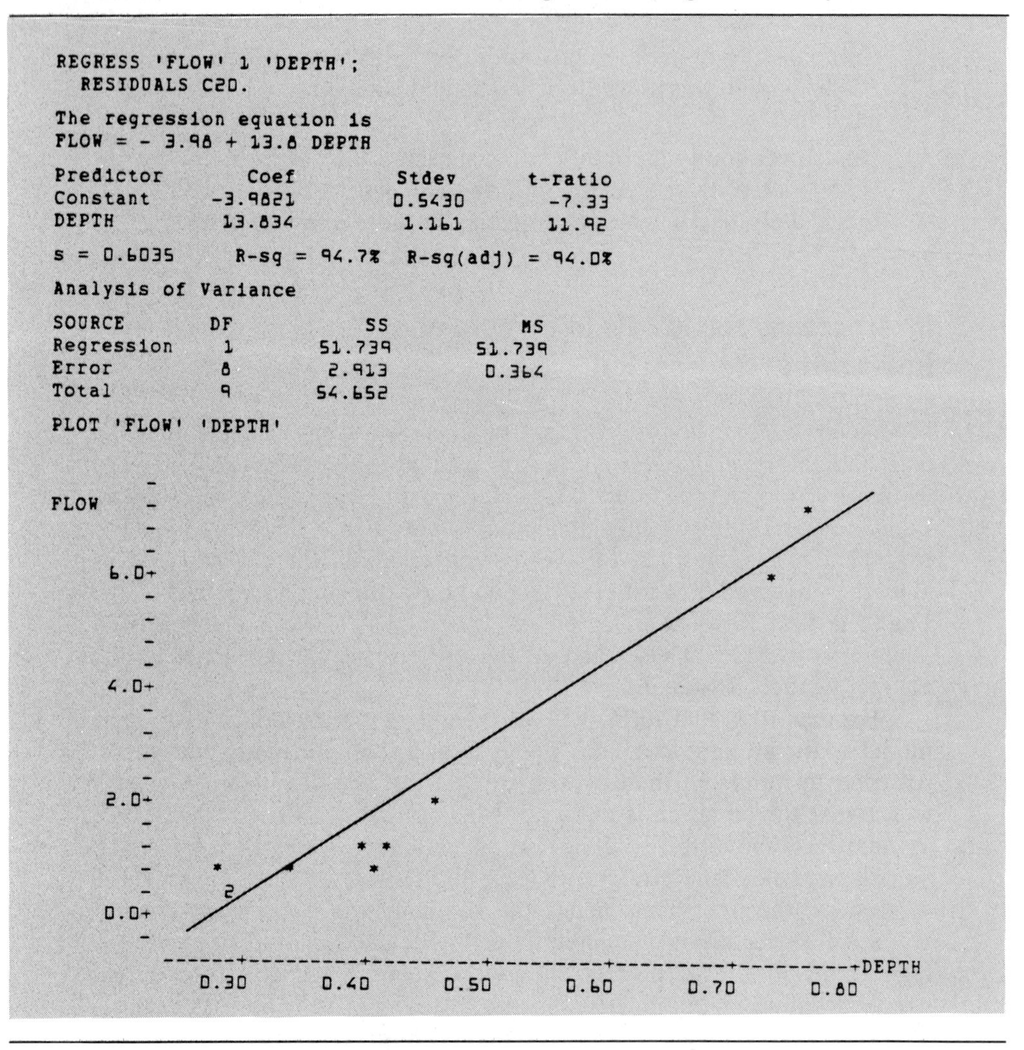

```
REGRESS 'FLOW' 1 'DEPTH';
  RESIDUALS C20.

The regression equation is
FLOW = - 3.98 + 13.8 DEPTH

Predictor       Coef          Stdev       t-ratio
Constant     -3.9821         0.5430         -7.33
DEPTH         13.834          1.161         11.92

s = 0.6035     R-sq = 94.7%  R-sq(adj) = 94.0%

Analysis of Variance

SOURCE        DF          SS            MS
Regression     1      51.739        51.739
Error          8       2.913         0.364
Total          9      54.652

PLOT 'FLOW' 'DEPTH'
```


pattern here? For example, do low x values tend to have low residuals? How about the residuals for middle and high values? Are there any patterns that would indicate our model is not adequate? In this plot, we do not see any. It appears that the model we fit was okay, at least as far as this plot is concerned. There are other types of plots we should do, but we will leave these to Section 10.9.

Exhibit 10.11 Plot of Residuals from Straight Line Fit in Exhibit 10.10

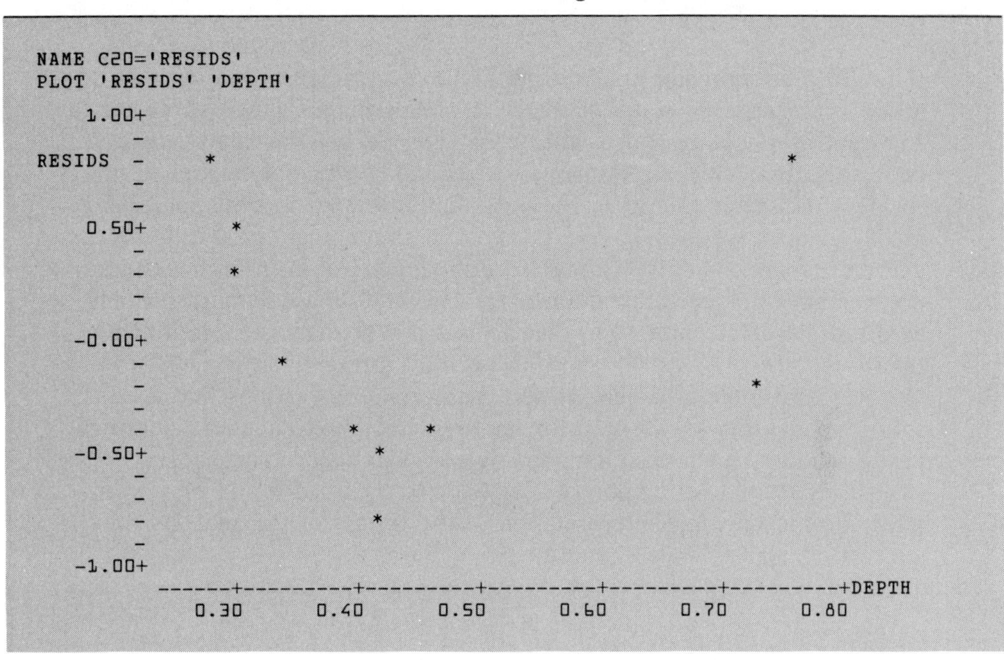

Exhibit 10.12 Residuals from Second-degree Polynomial Fit to Stream Data

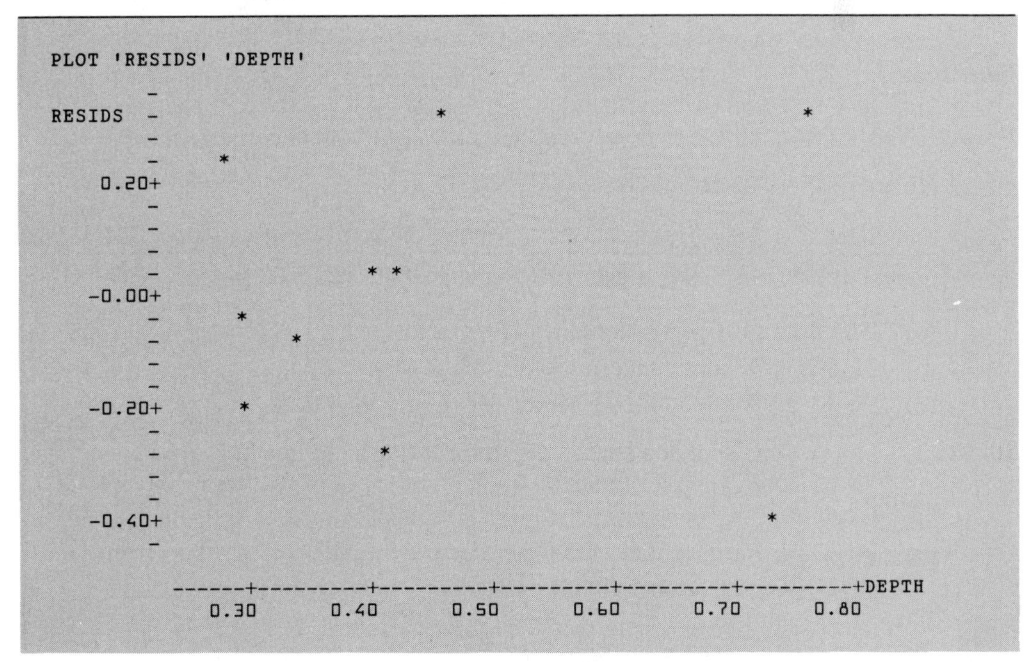

Summary

If we fit a straight line to data when the basic relationship between x and y is a curve, then a plot of the residuals versus x will be curved. The plot of y versus x also will be curved, but the curvature may not be as apparent. We might then fit a second-degree polynomial. If the residual plot after this fit is no longer curved (and no other problems are indicated), we may have a good fit. If the residual plot still shows curvature, we could try fitting a third-degree, or even higher-degree polynomial. As a general rule, however, it is better to use a transformation, as discussed in Section 10.8, than to use polynomials of degree higher than two.

Residual plots also help us spot outliers—observations that are far from the majority of the data or far from the fitted equation. Outliers always should be checked for possible errors or special causes. In some cases they should be temporarily set aside to see whether they have any effect on the practical interpretation of the results of the analysis.

REGRESS C on K in C, ..., C

RESIDUALS into C

The residuals are stored in the indicated column.

Exercises

Exercises 10–22 through 10–25 involve stream flow data from two different sites.

10–22 Here are the data from site 3.

Flow Rate	.820	.500	.433	.215	.120	.172	.106	.094	.129	.240
Stream Depth	.96	.92	.90	.85	.84	.84	.82	.80	.83	.86

(a) Plot flow rate versus stream depth.

(b) Find a 95% confidence interval for the slope, B.

(c) Find a 95% confidence interval for the intercept, A.

10–23 (a) Plot residuals versus stream depth for the straight line fit in Exercise 10–22. Does the plot indicate any lack of fit, or any other problem? Explain.

(b) Fit a quadratic model (second-order polynomial) to the data from site 3. Does it fit any better? Explain.

10–24 Here are data from site 4.

Flow Rate	.352	.320	.219	.179	.160	.113	.043	.095	.278
Stream Depth	.71	.72	.64	.64	.67	.61	.56	.73	.72

(a) Plot flow rate versus stream depth.
(b) Fit a straight line to explain flow rate based on stream depth. Where does this line fall on the plot?
(c) Plot the residuals against stream depth. Is there any indication that another model should be used? Anything else indicated?
(d) Fit a quadratic model. Does it fit any better? Explain.

10–25 Compare the analyses for sites 3 and 4 on the following points:
(a) How well you can predict flow rate from depth;
(b) whether or not a quadratic model fits better than a straight line;
(c) whether a similar relationship between flow rate and depth seems to hold for both sites.

10–26 A simple pendulum experiment from physics consists of releasing a pendulum of a given length (L), allowing it to swing back and forth for 50 cycles, and recording the time it takes to swing these 50 cycles. Data from five trials of this experiment are given below.

	Trial				
	1	*2*	*3*	*4*	*5*
Length	175.2	151.5	126.4	101.7	77.0
Time for 50 Cycles	132.5	123.4	112.8	101.2	88.2

(a) Let T be the average time per cycle. Compute T for each trial by dividing the time by 50. Plot T versus L and fit a straight line to explain T based on L. How well can you estimate time per cycle from pendulum length using a straight line?
(b) Plot the residuals versus L. Are there any indications that a higher-degree polynomial should be used? Fit a better model if one seems needed.

10–27 In Exercise 10–16 we simulated data from a regression equation. Repeat those simulations and each time plot the residuals versus x. This should give you some idea of how a residual plot looks when a correct model is fitted.

10–28 A statistician named Frank Anscombe constructed the data listed below to make an important point. The following steps should dramatically illustrate his point. (Note: y_1, y_2, and y_3 all use the same x values.)

(a) Use REGRESS to fit a straight line to each set of data. Compare the regression output from the different data sets. Do they have anything in common? Based on the regression output, would you tend to think the data sets were pretty much alike?

(b) For each data set make a plot of y versus x.

Now think back to the regression output in part (a). Can you guess what important point Anscombe was trying to make?

x_1, x_2, x_3	y_1	y_2	y_3		x_4	y_4
10.0	8.04	9.14	7.46	:	8.0	6.58
8.0	6.95	8.14	6.77	:	8.0	5.76
13.0	7.58	8.74	12.74	:	8.0	7.71
9.0	8.81	8.77	7.11	:	8.0	8.84
11.0	8.33	9.26	7.81	:	8.0	8.47
14.0	9.96	8.10	8.84	:	8.0	7.04
6.0	7.24	6.13	6.08	:	8.0	5.25
4.0	4.26	3.10	5.39	:	19.0	12.50
12.0	10.84	9.13	8.15	:	8.0	5.56
7.0	4.82	7.26	6.42	:	8.0	7.91
5.0	5.68	4.74	5.73	:	8.0	6.89

Data are in the saved worksheet called FA.

10.8 Using Transformations

Rather than fit a polynomial to curved data, it often is preferable to try to transform them to see if a simpler model can be found. The basic idea was explained in Section 3.4 (p. 72). If our data seem to fall along some curve, we may transform one or both variables using a square root, logarithm, or negative reciprocal. If we are lucky, one of the resulting plots will be approximately a straight line. We then can use REGRESS to find an approximating equation.

We can use the stream data from Exhibit 10.8 as an example. Plotting FLOW versus DEPTH did not give us a straight line. The curvature in the plot of FLOW versus DEPTH is similar to that in Panel (b) of Exhibit 3.9. To achieve a straight line, we need to pull in the upper end of the FLOW scale. The indicated transforms are thus $\sqrt{\text{FLOW}}$, log(FLOW), and $-1/\text{FLOW}$. These are plotted versus DEPTH in Exhibit 10.13. Also shown is log(FLOW) versus log(DEPTH).

Several of these plots seem to be reasonably close to a straight line, although none overwhelm us with their straightness. Perhaps $\sqrt{\text{FLOW}}$

Exhibit 10.13 Transformations of Stream Data

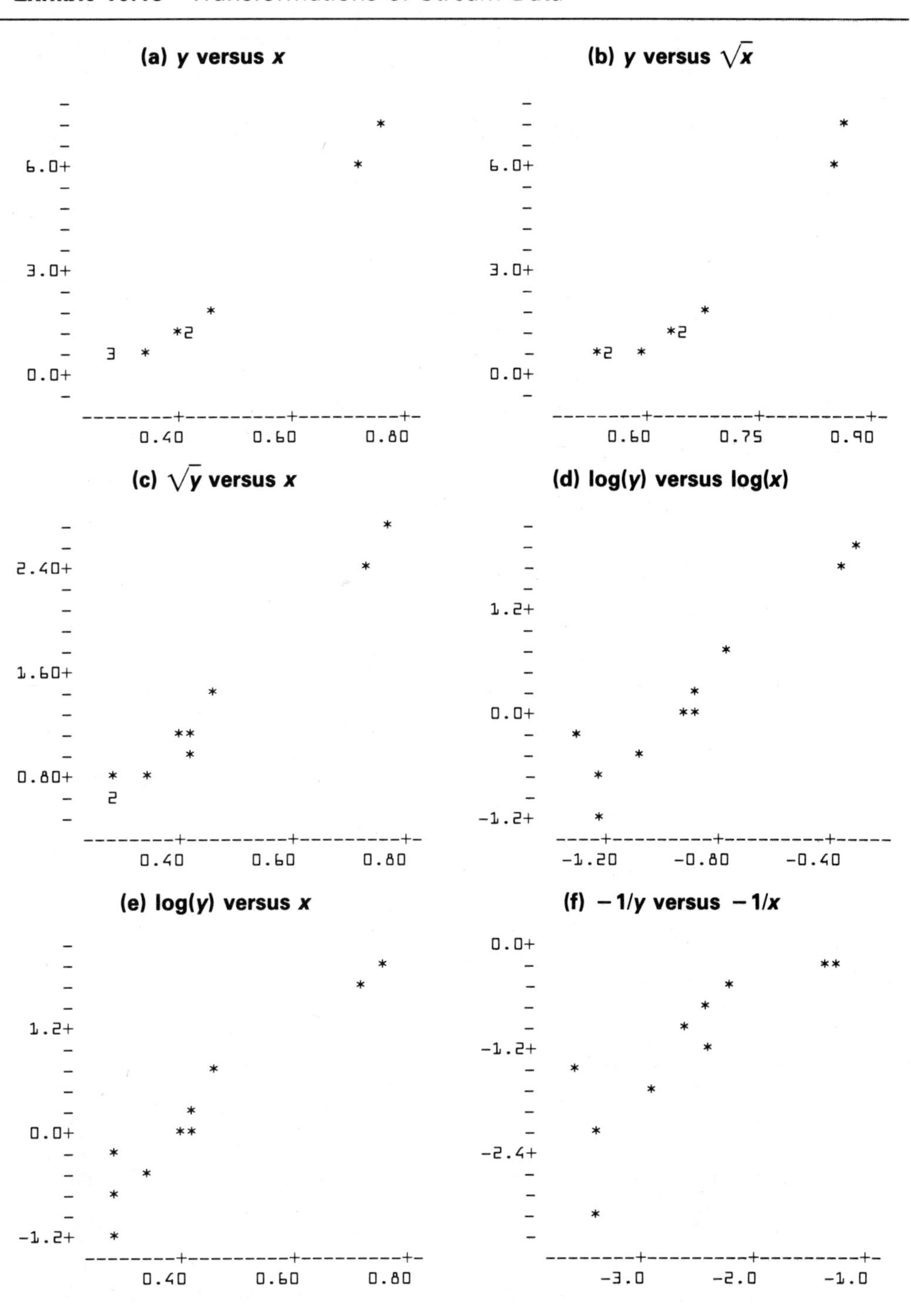

versus DEPTH is close enough to use. If we fit a straight line to this plot the resulting equation would be

$$\sqrt{\text{FLOW}} = a + b(\text{DEPTH})$$

to fit this model, we could use

```
NAME C11 = 'SQRTFLOW'
LET 'SQRTFLOW' = SQRT('FLOW')
REGRESS 'SQRTFLOW' ON 1, 'DEPTH'
```

The output from REGRESS is a little tricky to interpret. Minitab tells us that the best equation is

```
SQRTFLOW = -.558 + (4.16)(DEPTH)
```

If the DEPTH were .7, we would estimate SQRTFLOW as

```
SQRTFLOW = -.558 + 4.16(.7) = 2.35
```

We would thus estimate FLOW, at DEPTH = .7, as

```
FLOW = (2.35)² = 5.52
```

Another plot that looks fairly good is log(FLOW) versus log(DEPTH). Fitting a straight line to this plot would give the equation

$$\log(\text{FLOW}) = a + b \log(\text{DEPTH})$$

Either of these equations might do well enough for estimation, particularly within the range of depths for which we have data. In addition, they might give us some idea of a good theoretical model for the relationship between stream flow rate and depth. Using a polynomial might be equally good for estimation over the range of the data, but it probably would not work very well outside the range of the data, nor would it be likely to give us much theoretical insight.

Effect of Transformations on Assumptions

The conditions necessary for inference in regression were listed in Section 10.3. Whenever we use a transformation, there will be some effect on the validity of the conditions. If we have an exactly straight line, and take the logarithm of x or y, we will not have an exactly straight line any more. If y has exactly the same variance for all values of x, then $\log(y)$, $\sqrt{y}$, and $-1/y$ will not. If y is exactly normally distributed, then $\log(y)$, $\sqrt{y}$, and $-1/y$ will not be.

But, on the other hand, if some of these conditions were not met before we transformed the data, they may be afterward. Sometimes a transformation will help with one of these conditions, but cause problems with another. Surprisingly often, a transformation that helps with one condition also helps with others.

To summarize, our advice is to use transformations as you would use any other statistical technique—not blindly, but with a healthy skepticism that says, "I know none of these assumptions are met exactly, and I know that some are more important than others. I will check them all, particularly the most important ones, and always use the results of my analysis as guidance, not as gospel."

Exercises

10–29 In Exercise 10–26, you fit a line to some pendulum data. A residual plot indicates a curve in the data—a curve which may not have been apparent in a data plot. You then probably fit a quadratic. This gives a residual plot with no apparent pattern (of course, with just five observations, it's difficult to do a very precise analysis). It is known from physics that the correct relationship is $T = (2\pi/\sqrt{q})\sqrt{L}$, where $q = 981$ cm/sec^2 (at the latitude where the experiment was done). Thus if we fit a straight line of T versus $\sqrt{L}$, we should find $A = 0$ and $B = 2\pi/\sqrt{981} = .201$.

(a) Fit the model $T = A + B\sqrt{L}$. How well does it fit compared to a quadratic model? If you didn't know any physics, could you decide between these two models based on these data?

(b) Do your estimates of A and B seem close to the theoretical values?

(c) Now test $H_0{:}A = 0$ and $H_0{:}B = .201$ (use $\alpha = .05$). Are the results of these tests consistent with your answers in (b)? Explain.

10–30 An experiment was run in which a tumor was induced in a laboratory animal. The size of the tumor was recorded as it grew.

Number of Days After Induction	*Size of Tumor (cc)*
14	1.25
16	1.90
19	4.75
21	5.45
23	7.53
26	14.5
28	16.7
30	21.0
33	27.1
35	30.3
37	40.5
41	51.4

Investigate the relationship between time and tumor size. Is the relationship linear? Can it be "linearized" by an appropriate transformation?

10.9 Some More on Regression

REGRESS C on K predictors C, ..., C

This is the Minitab command for regression analyses. REGRESS has many subcommands. Here we will describe the more elementary ones. See the Minitab Manual or type HELP REGRESS to learn about the others.

PREDICT for E, ..., E

This is described on page 240.

RESIDUALS into C

This is described on pages 249–52.

COEF into C

Stores the coefficients b_0, b_1, ..., b_k down the column.

MSE into K

Stores the mean square error, which is the same as s^2, in K.

NOCONSTANT

Fits a model without a constant term, that is, the model

$$y = b_1x_1 + b_2x_2 + \cdots + b_kx_k$$

REGRESS C on K C, ..., C [store st. resid. in C [fits in C]]

This allows you to store the standardized residuals and fits by listing just two storage columns on the REGRESS line. Standardized residuals also are printed on the output, and are calculated by (residual)/(standard deviation of residual). The standard deviation is different for each residual and given by the formula (standard deviation of residual) = $\sqrt{\text{MSE} - (\text{standard deviation of fit})^2}$

BRIEF output at level K

Controls the amount of output from all REGRESS commands that follow. K is an integer from 1 to 3; the larger the integer, the more output.

K=1 The regression line, table of coefficients, line with S, R^2 and R^2-adjusted, and the first part of the analysis of variance table (regression, error, and total sums of squares) are printed.

K=2 In addition, the second part of the analysis of variance table (provided there are two or more predictors) and these "unusual" observations in the table of data, fits, stdev fit, etc. are printed. There are two reasons why an observation is unusual: Its predictors, x_1, ..., x_k are unusual (i.e., far away from most other predictors); its standardized residual is large (specifically, over 2). In both cases you should check the observation to make sure it is correct. This is the usual output.

K=3 In addition, the full table of data, fits, stdev fit, etc. is printed.

11 Chi-Square Tests and Contingency Tables

Many times we count the number of occurrences of an event. For example, we count how many fatal accidents there are on a holiday weekend or how many people in different age groups are opposed to a military draft. We then may be interested in whether or not our counts are in agreement with some theory. In this chapter we show how Minitab may be used to test agreement in such cases.

11.1 Chi-Square Goodness-of-Fit Test

The Missouri Department of Conservation wanted to learn about the migration habits of Canadian geese. They banded over 18,000 geese from one flock and classified each goose into one of four groups according to its age and sex (see Table 11.1). Information on the bands asked all hunters who shot one of these geese to send the band to a central office and tell where they shot the goose. Table 11.1 gives a breakdown of the 112 bands that were returned from geese shot outside the normal migration route for this flock.

An important question was, "Are geese from each of the four groups equally likely to stray?" Suppose this were true, that is, suppose our null hypothesis were:

H_0: The four groups are equally likely to stray.

Table 11.1 Canadian Geese Bands Returned from Areas Outside Normal Migration Routes

Group	*Number Banded*	*Proportion Banded*	*Observed Number Returned*	*Expected Number Returned*
Adult Males	4144	.2264	17	25.36
Adult Females	3597	.1965	21	22.01
Yearling Males	5034	.2750	38	30.80
Yearling Females	5531	.3021	36	33.84
Totals	18,306	1.0000	112	112.01

We then would expect that most of the bands returned would be from yearling females, since that is the group which had the most bands to begin with. It had 30.21% of the original bands, so we might expect about 30.21% of the 112 returned bands to be from yearling females. This gives $.3021 \times 112 = 33.84$ bands expected for yearling females. The last column in Table 11.1 gives the expected number for each of the four groups.

We see that the Observed and Expected numbers returned do not agree exactly. Even if our hypothesis were exactly true and all four groups were equally likely to stray, we would not expect perfect agreement. Some amount of chance variation always will be present. But, on the other hand, if our hypothesis were true, we should not expect "too much" difference between Observed and Expected counts. How much difference is "too much"?

The chi-square test gives us a way of checking whether the difference between the Observed and Expected numbers is greater than could reasonably be due to chance alone. The chi-square statistic is

$$\chi^2 = \Sigma[(\text{observed} - \text{expected})^2/\text{expected}]$$

If the Observed and Expected numbers are very different, the value of χ^2 will be large. Minitab does not have a special command to calculate this statistic, but we can use the following commands:

```
SET C1
17 21 38 36
END
SET C2
25.36 22.01 30.80 33.84
END
LET K1 = SUM((C1-C2)**2/C2)
PRINT K1
```

When these commands were run, Minitab printed 4.62 for the value of the chi-square statistic. Is this value unusually large? To decide we compare it to the value from a table of the chi-square distribution. The appropriate degrees of freedom is the number of groups minus one. Here it is $(4 - 1) = 3$. The corresponding table value, using $\alpha = .10$, is 6.25. Since 4.62 is less than 6.25, the chi-square statistic is not unusually large, and so we cannot reject H_0. Thus, these data give no evidence that one group of geese is any more likely to stray (or be shot) than another.

Exercises

11–1 Table 11.1 contains the expected number of returned bands. Suppose these had not been calculated for you. Give Minitab commands to calculate them from the number banded and the observed number returned.

11–2 Suppose we were only interested in comparing yearling geese with adults. There were 4144 + 3597 = 7741 adults and 5034 + 5531 = 10,565 yearlings banded. Do an appropriate chi-square test to see if yearlings stray any more or less than adults.

11–3 The following table gives accidental deaths from falls, by month, for the year 1979.

Month	*Number of Deaths from Falls*
Jan	1150
Feb	1034
Mar	1080
Apr	1126
May	1142
Jun	1100
Jul	1112
Aug	1099
Sep	1114
Oct	1079
Nov	999
Dec	1181
Total	13,216

(a) Do accidental deaths due to falls seem to occur equally often in all 12 months? Do a chi-square goodness-of-fit test. What is an appropriate null hypothesis? Calculate the expected number of falls under this null hypothesis. Complete the test.

(b) Can you give some reasons for the result in part (a)? What patterns do you see in the data?

11–4 Several years ago, an article in the *Washington Post* told about a high-school boy named Edward who made 17,950 coin flips and "got 464 more heads than tails and so discovered that the United States Mint produces tail-heavy pennies."

(a) Is this result statistically significant? Calculate the observed and expected number of heads and tails and do a chi-square goodness-of-fit test.

(b) The statistician W. J. Youden called Edward and asked how he had done this experiment. Edward explained that he had tossed five pennies at a time and his younger brother had recorded the results as Edward called them out. Here's what was recorded:

Number of Heads in Five Tosses	*Number of Times Found*
0	100
1	524
2	1080
3	1126
4	655
5	105
Total tosses	3590

The number of heads in five tosses should theoretically follow a binomial distribution. Use the PDF command (see p. 137) to compute the theoretical probabilities of each of these six possible outcomes if $p = .5$. Then test the null hypothesis that Edward's data came from a binomial distribution with $p = .5$ (i.e., that the observed number of times found agrees with the binomial probabilities). Try to interpret any discrepancies that seem significant.

11–5 Suppose we count the number of days in a week on which at least .01 inch of precipitation falls. Can we model this by a binomial distribution? Below we give some data collected in State College, Pennsylvania, during the years 1950–1969. All the observations are from the same month, February. This gives us (4 weeks in Feb) $\times$ (20 years) = (80 weeks in all).

Number of Precipitation Days in a Week	0	1	2	3	4	5	6	7
Number of Weeks in Which This Occurred	3	12	17	25	14	5	4	0

(a) Use these data to estimate p = the probability of precipitation on a given day.

(b) Test to see if a binomial distribution with $n = 7$ and p as estimated in part (a) fits the data. You can use the PDF command to help get the expected counts for each group.

(c) Does your answer to part (b) agree with what you'd expect of rainfall data? Explain.

11.2 Contingency Tables

A researcher did a study to investigate the relationship between being an artist and believing in extrasensory perception (ESP). He asked 114 artists and 344 nonartists to classify themselves into one of three categories: (1) believe in ESP, (2) believe more-or-less, (3) do not believe. His results are given in a contingency table, Table 11.2.

One question that is frequently asked of such data is, "Is there any association between being an artist and belief in ESP?" Another version of the same question is, "Do artists and nonartists seem to have about the same degree of belief in ESP?" The corresponding null hypothesis can be phrased as:

H_0: The proportions in the three categories for belief in ESP are the same for artists as for nonartists

If we pretend this null hypothesis is true, we can do the following calculations: Of the 458 people in the study, 196/458 or 42.79% believe in ESP, 224/458 or 48.91% believe more-or-less, and 38/458 or 8.30% do not believe. There are 114 artists. Therefore, 42.79% of the 114 artists, or 48.79 people, should believe in ESP. Similarly, 48.91% of 114, or 55.76 artists, should believe more-or-less, and 8.30% of 114, or 9.46 artists, should not believe. The same proportions should hold for the non-artists. Thus 42.79% of 344 = 147.21 should believe, 48.91% of 344 = 168.26 should believe more-or-less, and 8.30% of 344 = 28.55 should not believe. These are the Expected counts if our null hypothesis is true.

Table 11.2 Contingency Table from a Study of ESP

	Believe In ESP	*Believe More-or-Less*	*Do Not Believe*	*Total*
Artists	67	41	6	114
Non-Artists	129	183	32	344
Total	196	224	38	458

Are the Observed and Expected counts close? To determine if the differences are more than could reasonably be due to chance alone, we use a chi-square test. The test statistic is similar to the one we used in the preceding section.

$$\chi^2 = \Sigma\ [(\text{observed} - \text{expected})^2/\text{expected}]$$

For the ESP data,

$$\begin{aligned}\chi^2 &= (67-48.79)^2/48.79 + (41-55.76)^2/55.76 + (6-9.46)^2/9.46 \\ &\quad + (129 - 147.21)^2/147.21 + (183-168.26)^2/168.26 \\ &\quad + (32-28.55)^2/28.55 \\ &= 15.94\end{aligned}$$

The number of degrees of freedom is computed as follows: Let r = the number of rows in the contingency table. Here there are two: artists and nonartists. Let c = the number of columns. There are three: believe, believe more-or-less, and do not believe. The number of degrees of freedom is $(r - 1) \times (c - 1)$. Here this is $(2 - 1) \times (3 - 1) = 2$.

Exhibit 11.1 shows how to do these calculations using Minitab's

Exhibit 11.1 Output from CHISQUARE for the ESP Data

```
READ C1-C3
  67          41          6
 129         183         32
END
CHISQUARE C1-C3
Expected counts are printed below observed counts
              C1          C2          C3          All
     1        67          41           6          114
            48.8        55.8         9.5

     2       129         183          32          344
           147.2       168.2        28.5

   All       196         224          38          458
ChiSq  =    6.80 +      3.90 +      1.26 +
            2.25 +      1.29 +      0.42 =      15.94
df = 2
```

CHISQUARE command. Notice we enter each row of the contingency table into a separate row of the worksheet. To finish the test, we compare the calculated chi-square value to a table of the chi-square distribution with 2 degrees of freedom. For α = .005, the table gives 7.88. Our observed value is much larger than this. Therefore, we have strong evidence that the distribution of ESP beliefs among artists is different from that among nonartists.

Notice that the output also gives the cell-by-cell contributions to the overall χ^2 value. In this case the largest contribution, 6.80, comes from the "artists who believe in ESP" cell. The Observed count is 67, but the Expected count if the null hypothesis were true is only 48.8. Many more artists than expected believe in ESP.

CHISQUARE analysis of the table in C, ..., C

Does a chi-square test for association on the table of counts given in the specified columns.

Exercises

11–6 Let us look again at the geese data in Table 11.1. The 112 bands that were returned can be cross-classified as follows:

	Male	*Female*
Adult	17	21
Yearling	38	36

(a) Compute the percentage of males among the adult geese. Compute the percentage of males among the yearlings. Do these percentages seem very different to you? Explain briefly.

(b) Is there any appreciable association between the two variables age and sex? Do a chi-square test to see if there is statistically significant evidence of association.

11–7 Two researchers at Penn State studied the relationship between infant mortality and environmental conditions in Dauphin County, Pennsylvania. This county has one large city, Harrisburg. The researchers recorded the season in which the baby was born and whether or not the baby died before one year of age. Infant deaths and births data for Dauphin County, 1970, are given on the facing page.

	Season of Birth			
	Jan Feb Mar	*Apr May Jun*	*Jul Aug Sep*	*Oct Nov Dec*
Died Before One Year	14	10	35	7
Lived One Year	848	877	958	990

(a) Is an infant more likely to die if it is born in one season than another? Which season has the highest risk? The lowest risk?

(b) Newspapers reported severe air pollution, covering the entire East during the end of July. Air pollution, especially during the end of pregnancy and in the first few days after birth, is suspected of increasing the risk of an infant's death. Is this theory consistent with the data and analysis of part (a)?

11–8 The data in Exercise 11–7 (above) are for all of Dauphin County. However, environmental conditions as well as socio-economic conditions are different for the city of Harrisburg and the surrounding countryside. Here we give the data separately for these two regions. Do the analysis in Exercise 11–7 separately for each region. What are your conclusions now?

Data for Harrisburg

	Season of Birth			
	Jan Feb Mar	*Apr May Jun*	*Jul Aug Sep*	*Oct Nov Dec*
Died Before One Year	6	6	25	3
Lived One Year	306	334	347	369

Data for Dauphin County, Excluding Harrisburg

	Season of Birth			
	Jan Feb Mar	*Apr May Jun*	*Jul Aug Sep*	*Oct Nov Dec*
Died Before One Year	8	4	10	4
Lived One Year	542	543	611	621

11–9 A survey on car defects was done in Dane County, Wisconsin, based on a random sample of people who had just purchased used cars. Each owner was sent an invitation to bring the car in for a free safety inspection. Car owners who did not respond to the invitation were sent post card reminders. Eventually about 56% of the owners had their cars inspected, giving a total of 8842 cars. Here we will look at four types of defects: brakes, suspension, tires, and lights. There is one contingency table for each type of defect. Each car was classified as to whether it was purchased from a car dealer or from a private owner.

	Brakes	
	Defective	*Not Defective*
Dealer	931	2723
Private	1690	3498

	Suspension	
	Defective	*Not Defective*
Dealer	1120	2534
Private	1711	3477

	Tires	
	Defective	*Not Defective*
Dealer	1147	2507
Private	1765	3423

	Lights	
	Defective	*Not Defective*
Dealer	602	3052
Private	1098	4090

(a) Is there a relationship between a car having defective brakes and whether it was purchased from a dealer or a private owner? Do an appropriate test. What percent of dealer sold cars have defective brakes? What percent of privately sold cars? (Calculate these percents by hand.) How do the percents compare?

(b) Repeat part (a) for defects in the suspension system.

(c) Repeat part (a) for defective tires.

(d) Repeat part (a) for defective lights.

(e) Compare the results from parts (a)–(d). Is there a common pattern?

(f) Try to think of some possible reasons for the results in parts (a)–(e).

11–10 Here we give more data from the Wisconsin car survey described in Exercise 11–9. The age of each car was also recorded. These ages were then grouped into three categories: cars under three years old, cars between three and six years old, and cars over six years old.

	Under Three Years	*Three–Six Years*	*Over Six Years*
Dealer	570	1792	1292
Private	513	1859	2816

Is there a relationship between the age of a car and whether it was purchased from a dealer or private owner? Do an appropriate test. Describe the relationship. Calculate (by hand) row percents to use in your description.

11–11 Mark Twain has been credited in numerous places with the authorship of ten letters published in 1861 in the *New Orleans Daily Crescent*. The letters were signed "Quintus Curtius Snodgrass." Did Twain really write these letters? In a 1963 paper, Claude S. Brinegar used statistics to compare the Snodgrass letters to works known to be written by Mark Twain. We present some of his very interesting analyses in this exercise.

There are many statistical tests of authorship. The one Brinegar used is quite simple in concept—just compare the distributions of word lengths for the two authors. If these distributions are very different, we will have some evidence that the authors are probably different people. In using this type of test, we must assume that the distribution of word lengths is about the same in all works written by the same author. Parts (a), (b), and (c) below attempt to provide some check on this assumption.

The ten Snodgrass letters were first divided into three groups. Then the number of two-letter words was recorded in each group, the number of three-letter words in each group, the number of four-letter words, and so on. One-letter words were omitted. There are only two such words, "I" and "a," and the use of "I" tends to characterize content (work written in first person or not) more than an unconscious style. Data for Mark Twain were obtained from letters he wrote to friends at about the same time and from two other works written at a later time. These will be used to see if the word-length distribution remained about the same throughout Twain's writings.

(a) Compare the three groups of Snodgrass letters to see if they are consistent in word-length distribution. First compare them graphically. To make the numbers comparable, change the frequencies into relative frequencies (proportions). Then plot each column versus word length. Put all three plots on the same axes by using MPLOT. Do the distributions look similar? Next do a chi-square test of the null hypothesis that all three sets of letters have the same word-length distribution. Is there any evidence that they differ?

(b) Next, compare the three groups of Mark Twain letters. Do both a plot and a chi-square test, as in part (a). Is there any evidence that these three writings differ in word-length distribution?

(c) Now compare Mark Twain's writings over a large span of years. The samples from *Roughing It* and *Following the Equator* were taken for this purpose. First combine the three columns of Mark Twain letters into one column of "early works." Compare the "early," "middle," and "late" samples for Mark Twain. Do both a plot and chi-square test as in part (a). Is there any evidence that the word-length distribution changed over the years?

(d) Finally, now that we've checked the authors for consistency, let's compare the Twain and Snodgrass works. As in part (c), combine the three columns of Twain's early works into one column. Also combine the three columns of Snodgrass letters into one column. Finally, compare Twain's letters with the Snodgrass letters. Do both a plot and a chi-square test. Do you think Mark Twain wrote the Snodgrass letters?

(e) In what ways do the two authors differ, as far as word-length distribution is concerned? Examine both the plot and chi-square output from part (d) to find out. Does the chi-square output tell you anything the plot doesn't, or vice-versa?

Word Counts for Quintus Curtius Snodgrass Letters

Word Length	*First Three Letters*	*Second Three Letters*	*Last Four Letters*
2	997	831	857
3	1026	828	898
4	856	669	777
5	565	420	446
6	366	326	300
7	318	293	285
8	258	183	197
9	186	150	129
10	96	94	86
11	63	49	40
12	42	30	29
13 and over	25	25	11
Totals	4798	3998	4055

Word Counts for Known Mark Twain Writings

Word Length	*Two Letters from 1858 and 1861*	*Four Letters from 1863*	*Letter from 1867*	*Sample from* Roughing It, *1872*	*Sample from* Following the Equator, *1897*
2	349	1146	496	532	466
3	456	1394	673	741	653
4	374	1177	565	591	517
5	212	661	381	357	343
6	127	442	249	258	207
7	107	367	185	215	152
8	84	231	125	150	103

Word Counts for Known Mark Twain Writings (continued)

Word Length	*Two Letters from 1858 and 1861*	*Four Letters from 1863*	*Letter from 1867*	*Sample from* Roughing It, *1872*	*Sample from* Following the Equator, *1897*
9	45	181	94	83	92
10	27	109	51	55	45
11	13	50	23	30	18
12	8	24	8	10	12
13 and over	9	12	8	9	9
Totals	1811	5794	2858	3031	2617

Data are in the saved worksheet called TWAIN.

11.3 Making the Table and Computing Chi-Square

In the preceding section the contingency table was already prepared; all we had to do was compute the value of chi-square. In this section we show how Minitab can be used to make the table and then do the chi-square test all in one operation. The procedure is a simple extension of the one discussed in Chapter 4 for making tables.

Consider again the Restaurant data. On page 78 we used the instruction

```
TABLE 'OWNER' BY 'SIZE'
```

This made a table that classified the restaurants by type of ownership and size. The same instruction will do a chi-square analysis, if we add the CHISQUARE subcommand, as shown in Exhibit 11.2.

The 2 on the CHISQUARE subcommand says to print two values in each cell of the table. The first is the Observed count, which we saw in Exhibit 4.1 (p. 78); the second is the Expected count—the count we would expect if the null hypothesis of no association were true.

In Exhibit 11.2, the calculated value of chi-square is 67.9. This is quite large for a table with $(3 - 1) \times (3 - 1) = 4$ degrees of freedom. For example, if we used $\alpha = .005$, the chi-square table gives 14.86. We therefore have strong evidence that the size of a restaurant is related to the type of ownership.

To see how size and ownership are related, we can compare the Observed and Expected counts in Exhibit 11.2. For sole proprietorships, there are more small restaurants than expected and fewer large ones than expected. For corporations, the reverse is true—there are fewer

Exhibit 11.2 Chi-square Output from TABLE Using the Restaurant Data

```
RETRIEVE 'RESTRNT'
TABLE 'OWNER' BY 'SIZE';
  CHISQUARE 2.

ROWS: OWNER      COLUMNS: SIZE

            1          2          3        ALL

 1         83         18          2        103
        54.85      26.05      22.10     103.00

 2         16          6          4         26
        13.85       6.57       5.58      26.00

 3         40         42         50        132
        70.30      33.38      28.32     132.00

ALL       139         66         56        261
       139.00      66.00      56.00     261.00

CHI-SQUARE =      67.917,     WITH D.F. =      4

 CELL CONTENTS --
                        COUNT
                        EXP FREQ
```

small restaurants than expected and more large ones. The overall conclusion is not surprising: Corporate owned restaurants tend to be larger than ones owned by a single person.

TABLE C, ..., C

CHISQUARE [K]

This is the subcommand for doing a chi-square test. It tells Minitab to compute and print the value of chi-square below the table of counts.

If K = 2 is specified, both the Observed and Expected counts are printed in each cell. If K is omitted, or set equal to 1, only the Observed counts are printed.

More than two variables may be listed on the TABLE command. A separate chi-square test is then done for each two-way table printed. The LAYOUT subcommand may not be used when the CHISQUARE subcommand is used.

Exercises

11–12 In the Restaurant survey (p. 321), is there a relationship between the type of food sold and the size of the restaurant? Do an appropriate chi-square test.

11–13 In the Restaurant survey (p. 321), is there a relationship between the type of ownership and the overall outlook of the owner? Do an appropriate test. Describe this relationship using appropriate row and/or column percents. (Recall TABLE has subcommands to calculate these.)

11–14 Using the Pulse data (p. 318), test to see if there is a relationship between
- (a) sex and activity;
- (b) sex and smokers;
- (c) smokers and activity.

11–15 Using the Cartoon data (p. 303), see if there is a relationship between education and whether or not a person took the delayed test. Note, you will have to create a variable for the second classification. To do this, you can use the delayed cartoon score and the CODE command (described on p. 339).
- (a) Do an appropriate test.
- (b) Calculate appropriate percents and describe the relationship. Can you give some possible reasons for this relationship?

11.4 Tables with Small Expected Counts

A survey was done in an introductory statistics class. Each student was asked to give his or her year in college and political preference. This gave the contingency table shown as Table 11.3.

Exhibit 11.3 uses the CHISQUARE command. Notice the output contains the message

```
5 cells with expected frequencies less than 5
```

This indicates that the chi-square analysis done on this table may not be quite appropriate. As with many statistical tests, the chi-square test

Table 11.3 Data from a Survey of Political Preference

	Freshman	*Sophomore*	*Junior*	*Senior*
Democrat	4	16	16	6
Republican	1	7	7	3
Other	3	4	12	5

is an approximate test. In this case the approximation becomes better and better as the expected cell frequencies increase. Consequently, if too many cells have expected frequencies that are too small, a chi-square analysis may not be appropriate.

A good rule of thumb is that not more than 20% of the cells should have expected cell frequencies less than 5, and no cell should have an expected frequency less than 1. The table in Exhibit 11.3 has 12 cells. So, no more than $.2 \times 12 = 2.4$ cells should have expected frequencies less than 5. But, as the message says, and as we can see ourselves from the table, there are 5 such cells.

There are several procedures we can follow with tables having too many small expected frequencies. We can try to combine cells so that we reduce the number of rows and/or columns. For example, if we combine columns 1 and 2 into one level called "lower classmen," and

Exhibit 11.3 CHISQUARE Output for Political Preference Data

```
READ C1-C4
 4        16      16       6
 1         7       7       3
 3         4      12       5
END
NAME C1='FRESHMAN' C2='SOPHOMRE' C3='JUNIOR' C4='SENIOR'
CHISQUARE C1-C4

Expected counts are printed below observed counts

       FRESHMAN SOPHOMRE   JUNIOR   SENIOR      All
 1            4       16       16        6       42
            4.0     13.5     17.5      7.0

 2            1        7        7        3       18
            1.7      5.8      7.5      3.0

 3            3        4       12        5       24
            2.3      7.7     10.0      4.0

All           8       27       35       14       84

ChiSq =    0.00 +   0.46 +   0.13 +   0.14 +
           0.30 +   0.25 +   0.03 +   0.00 +
           0.22 +   1.79 +   0.40 +   0.25 =    3.98
df = 6
5 cells with expected counts less than 5.0
```

combine columns 3 and 4 into one level called "upper classmen," we get the following 3×2 table:

	Lower Classmen	*Upper Classmen*
Democrat	20	22
Republican	8	10
Other	7	17

Now, if no more than one cell in this smaller table has expected frequency less than 5, we can comfortably do the chi-square analysis. The smaller table was read into the worksheet and a second analysis done. In this case, no cell had expected frequency less than 5, so the chi-square approximation is probably good enough. Of course, we have changed our analysis slightly. We are now testing a different null hypothesis, that there is no relationship between a student's political preference and whether he is an upper or lower classman. Before we were testing the null hypothesis that there is no relationship between a student's political preference and whether he is a freshman, sophomore, junior, or senior. We do not have quite enough data here to analyze the larger 3×4 table. So, if we want to get something out of the data, we can try looking at a table with fewer levels.

Another way to handle the problem is to remove one or more levels of a classification. In the political preference table, we could omit the first column, corresponding to freshmen. This leads to the following table:

	Sophomore	*Junior*	*Senior*
Democrat	16	16	6
Republican	7	7	3
Other	4	12	5

This table provides an analysis relevant to students above the freshman level. Here, however, the omission doesn't help much. There are still too many cells with small expected frequencies.

12 Nonparametric Statistics

There are several reasons why it may be more appropriate to use nonparametric procedures instead of the normal theory (parametric) procedures described in Chapters 7, 8, 9, and 10. These reasons include the following:

1. *Populations not normal.* To be strictly valid, the procedures in Chapters 7, 8, 9, and 10 require that we have random samples from normal populations. If the samples do not come from normal populations, these procedures may give misleading results. For example, if we construct a 95% t confidence interval from nonnormal data, the real confidence may be only 91%, not 95%, as we had planned. Another, and usually more serious, problem is loss in efficiency. Loosely speaking, a more efficient procedure makes better use of the data and enables us to get a better estimate or test with a smaller sample size. Some nonparametric procedures are more efficient than normal theory procedures if we are sampling from a nonnormal population.
2. *Concern about occasional "outliers."* Normal theory methods are quite sensitive to even a few extreme, or outlying, observations. As a simple example, consider the five numbers: 3, 7, 9, 10, 11. Normal theory uses their mean while many of the methods in this chapter use the median. The median of these numbers is 9 and the mean is 8. But what would happen if one of the observations had been measured or recorded incorrectly? Suppose the 11 had been recorded as a 61 and we failed to notice this. Then the median would remain unchanged but the mean would more than double, to 18. The median, and the other procedures described in this chapter, are more *resistant* than normal theory methods to distortion by a few gross errors.

Discrete Data

A distribution is discrete if it takes on just a few distinct values (e.g., the binomial distribution) or if it takes on a list of values (e.g., the Poisson distribution). A continuous distribution, on the other hand, takes on all values on the real line (e.g., the normal distribution and the t distribution) or on a portion of the real line (e.g., the F distribution takes on all values greater than or equal to zero and the uniform distribution on the interval 2 to 2.5 takes on all values between 2 and 2.5).

Strictly speaking, the procedures discussed in this chapter need a continuous population. However, if you use them on discrete data you will not go too far wrong. In most cases you will be "conservative;" that is, if you do a test, the true p-value will be less than the one you calculate and if you construct a confidence interval, the true confidence will be greater than the confidence you calculate.

12.1 Sign Procedures

Exhibit 12.1 summarizes some data from the restaurant survey, which was discussed in Chapter 4. These data are for sales by supper clubs owned by one person. Notice that the stem-and-leaf display is not symmetric but skewed toward high values: 16 observations are on the top three lines while the remaining 13 observations are spread over the next eight lines.

Sign Tests

Suppose we want to test a hypothesis about these values. We might not want to assume we have a sample from a normal distribution. A sign test does not make any assumptions about the shape of the population in our hypothesis: It could be very skewed; it could have "fat tails" like the t distribution; it could be bimodal (i.e., have two peaks).

As an example, let us suppose that the median sales for all supper clubs in Michigan, owned by one person, is \$115,000. Are supper clubs in Wisconsin doing better? Let η be the (unknown) median sales of the population of all supper clubs owned by one person in Wisconsin. Then H_0: η = 115,000 and H_1: η > 115,000.

Minitab's STEST command will do the calculations for a sign test. It first counts how many values are above the hypothesized median value of η = 115,000 and how many are below. Here there are 17 above and 12 below. If η were 115,000 we would expect, on the average, about half our observations to be above 115,000 and half below. In the restaurant example we have 17 above and only 12 below. Is this surprising? The

Exhibit 12.1 Sales for the 29 Supper Clubs Owned by One Person in the Restaurant Survey. (Note: SALES were recorded in thousands. Thus, 184 represents $184,000.)

```
DESCRIBE 'SALES'

                N      MEAN    MEDIAN    TRMEAN     STDEV    SEMEAN
SALES          29     184.3     144.0     179.1     130.5      24.2

              MIN       MAX        Q1        Q3
SALES         0.0     507.0      89.5     265.0

STEM-AND-LEAF 'SALES'

Stem-and-leaf of SALES       N  =   29
Leaf Unit = 10

    1    0  0
    8    0  5577889
  (8)    1  01112444
   13    1  58
   11    2  0124
    7    2  9
    6    3  022
    3    3  3
    3    4  0
    2    4
    2    5  00

SORT 'SALES' into 'SALES'
PRINT 'SALES'

SALES
     0     50     56     72     75     80     80     99    101    110    110
   110    120    140    144    145    150    180    201    210    220    240
   290    309    320    325    400    500    507
```

output from STEST in Exhibit 12.2 tells us that it is not. The p-value is .229. That is, over 20% of the time we will get 17 or more above the median even if $\eta = 115{,}000$.

Ties. Sometimes there are several values equal to the hypothesized median value. This would be the case if our hypothesis were $\eta = 110{,}000$. Exhibit 12.3 shows there were nine values below 110,000, three equal to 110,000 and 17 above 110,000. In such cases, it is conventional to set aside the ties and apply the sign test to the remaining data. Here we would have $n = 26$ values left, nine below, and 17 above.

How Minitab Does a Sign Test. The sign test is based on the binomial distribution. Suppose $\eta = 115{,}000$. If we pick one supper club in Wisconsin

Exhibit 12.2 Sign Test of $H_0: \eta = \$115,000$ Versus $H_1: \eta > \$115,000$ on SALES

```
STEST 115 'SALES';
 ALT = 1.

SIGN TEST OF MEDIAN = 115 VERSUS G.T. 115

                N    BELOW    EQUAL    ABOVE    P-VALUE    MEDIAN
SALES          29       12        0       17      0.229     144.0
```

Exhibit 12.3 Sign Test of $H_0: \eta = \$110,000$ Versus $H_1: \eta > \$110,000$ on SALES

```
STEST  110 'SALES';
  ALT=1.

SIGN TEST OF MEDIAN = 110 VERSUS G.T. 110

                N    BELOW    EQUAL    ABOVE    P-VALUE    MEDIAN
SALES          29        9        3       17      0.084     144.0
```

at random, there is a 50–50 chance that its sales are above \$115,000. If we pick a second club, again there is a 50–50 chance that its sales are above \$115,000, and so on, for all 29 observations. On each trial (sampling a supper club at random), the probability of a success (sales over \$115,000) is 1/2. Let X be the total number of successes. Then X has a binomial distribution with $p = 1/2$ and $n = 29$. Suppose we test H_0: $\eta = 115{,}000$ versus H_1: $\eta > 115{,}000$. In our sample $X = 17$. The probability of getting this many, or even more, observations over 115,000 is $P(X \geq 17)$. The command

```
CDF 16;
   BINOMIAL 29 .5.
```

gives $P(X \leq 16) = .7709$, so $P(X \geq 17) = .2291$, which is the p-value given by STEST in Exhibit 12.2

Suppose we test H_0: $\eta = 250{,}000$ versus H_1: $\eta < 250{,}000$. Now we let X be the number of observations below 250,000. Look at the ordered data in Exhibit 12.1. Since there are no ties, X has a binomial distribution with $n = 29$ and $p = 1/2$. There are 22 sales below 250, $P(X \geq 22) =$ .0041, and thus the p-value is .0041.

Now suppose we do a two-sided sign test. For example, suppose we test H_0: $\eta = 150{,}000$ versus H_1: $\eta \neq 150{,}000$. Here there are 16 observations below 150,000, one equal, and 12 above. We discard the tie and find the probability corresponding to the larger of the two remaining counts (if they are equal, either will do). Here the number below is the larger. So we calculate $P(X \geq 16) = .2858$, for $n = 28$ and $p = .5$. The p-value of the two-sided test is twice this value, or $2(.2858) = .5716$. Thus, over 57% of the time we can expect to get a split this extreme or more so, that is, with 16 or more on one side or the other.

STEST of median = K for C, ..., C

Does a separate sign test on each column. The subcommand ALT allows you to specify a one-sided alternative hypothesis as follows:

ALT = −1 for H_1: $\eta < K$
ALT = 1 for H_1: $\eta > K$

If you do not use ALT a two-sided test is done.

Sign Confidence Intervals

In Exhibit 12.4, we used the SINTERVAL command to get a 95% confidence interval for the median sales of all supper clubs owned by one person. First notice that SINT prints out two confidence intervals, and that neither has exactly 95% confidence. This is one characteristic of nonparametric methods—rarely can we get the exact confidence we want. In the restaurant example, we can get a 93.9% confidence interval and a 97.6% confidence interval, but nothing in between.

Let's look at the 93.9% interval. The 10 under POSITION tells us that this interval goes from the tenth smallest observation to the tenth largest observation. Exhibit 12.1 has the observations in order. If we count ten observations from the bottom, we get 110,000 and if we count ten observations down from the top, we get 210,000. This gives the interval ($110,000, $210,000).

How Minitab Calculates a Sign Confidence Interval. Suppose we have a sample of 29 observations. Suppose we form the confidence interval which goes from the tenth smallest to the tenth largest observations. Let X be the number of observations which are less than η. If $X = 0,1, \ldots,$ or 9, then the tenth smallest observation must be above η; if $X = 20,21, \ldots,$ or 29, then the tenth largest observation must be below η. In both cases, η is not in the confidence interval. On the other hand, if $X =$

Exhibit 12.4 Sign Confidence Interval for Median Sales of All Supper Clubs Owned by One Person

```
SINTERVAL  95  C1

SIGN CONFIDENCE INTERVAL FOR MEDIAN

                                 ACHIEVED
                N   MEDIAN   CONFIDENCE   CONFIDENCE INTERVAL    POSITION
SALES          29     144         0.939   (      110,      210)        10
                                  0.976   (      101,      220)         9
```

10,11, ..., or 19, then η is in the confidence interval. Therefore P (η is in the confidence interval) $= P(X = 10,11, ..., 19) = 1 - [P(X = 0,1, ..., 9) + P(X = 20,21, ..., 29)]$. For a binomial with $n = 29$ and $p = 1/2$, $P(X = 0,1, ..., 9) = P(X = 20,21, ..., 29) = .0307$. Therefore, P(η is in the confidence interval) $= 1 - 2(.0307) = .9386$. In general, the interval that goes from the dth smallest to the dth largest observation has confidence given by the formula $1 - 2P(X < d)$. The second interval in Exhibit 12.4 goes from the ninth smallest to the ninth largest observation. Therefore, it has confidence $1 - 2P(X < 9) = 1 - 2P(X \leq 8) = 1 - 2(.0121) = .9758$.

The problem of ties, which we discussed for the sign test, does not arise in constructing a confidence interval. Therefore, we always use all the data for the interval.

Paired Data

One natural use of sign tests and confidence intervals is for paired data. (See Chapter 5, p. 92, for a discussion of paired data.) With paired data, the null hypothesis is often that the median difference is zero. To test this, we first compute the differences, then use STEST. Exhibit 12.5 gives an example. We are interested in the change from the second to the fourth day. Therefore, we first calculated those changes and put them in C3. Then we tested H_0: $\eta = 0$ versus H_1: $\eta \neq 0$, where η is the median of the population of all changes. We also found a 90% confidence interval for η.

SINTERVAL confidence = K for C, ..., C

Calculates a pair of sign confidence intervals separately for each column. The first interval of the pair corresponds to the achievable confidence just below K and the second to that just above K. (Rarely can you achieve confidence exactly equal to K.) If d is the POSITION printed on the output for a given interval, then the interval goes from the dth smallest to the dth largest observation.

Exhibit 12.5 Example of Sign Procedures for Paired Data (Data Are from the Cholesterol Study in Table 5.1, page 93)

```
RETRIEVE 'CHOLEST'
LET C3 = '4-DAY' - '2-DAY'
STEM-AND-LEAF C3

Stem-and-leaf of C3         N  =   28
Leaf Unit = 10

    1   -1  0
    3   -0  98
    5   -0  66
    9   -0  5554
   14   -0  33322
   14   -0  11100
    9    0  000011
    3    0  23
    1    0  4

STEST 0 C3

SIGN TEST OF MEDIAN = 0 VERSUS N.E. 0

                 N   BELOW   EQUAL   ABOVE   P-VALUE    MEDIAN
C3              28      19       0       9    0.0872    -19.00

SINT 90 C3

SIGN CONFIDENCE INTERVAL FOR MEDIAN

                               ACHIEVED
                 N   MEDIAN   CONFIDENCE   CONFIDENCE INTERVAL   POSITION
C3              28   -19.00       0.8151   (   -30.00   -8.000)        11
                                  0.9128   (   -38.00   -2.000)        10
```

Exercises

12–1 Consider the Cartoon experiment (p. 303). The national median of all OTIS scores is 100. Do the OTIS scores for the participants in this study differ significantly from the national median? Do a sign test.

12–2 Suppose we wanted to test H_0: η = \$115,000 versus H_1: $\eta \neq$ \$115,000 using the SALES data in Exhibit 12.1. Use the output in Exhibit 12.2 to find the corresponding p-value.

12–3 The data below, collected in a chemistry class, are the results of a titration to determine the acidity of a solution.

.123	.109	.110	.109	.112	.109	.110	.110
.110	.112	.110	.101	.110	.110	.110	.110
.106	.115	.111	.110	.107	.111	.110	.113
.109	.108	.109	.111	.104	.114	.110	.110
.110	.113	.114	.110	.110	.110	.110	.110
.090	.109	.111	.098	.109	.109	.109	.109
.111	.109	.108	.110	.112	.111	.110	.111
.111	.107	.111	.112	.105	.109	.109	.110
.110	.109	.110	.104	.111	.110	.111	.109
.110	.111	.112	.123	.110	.109	.110	.109
.110	.109	.110	.110	.111	.111	.109	.107
.120	.133	.107	.103	.111	.110	.122	.109
.108	.109	.109	.114	.107	.104	.110	.114
.107	.101	.111	.109	.110	.111	.110	.126
.110	.109	.114	.110	.110	.110	.110	.110
.111	.107	.110	.107				

These data are in C1 of the worksheet ACID.

(a) The instructor knew the correct value for this solution was .110. Do a two-sided sign test of the null hypothesis H_0: η = .110. (This is a check to see if the class is "biased"—i.e., to see if it tends to be systematically too high or too low.)

(b) Make a histogram of the data. Do you think the population is symmetric? If it is, then $\eta = \mu$.

(c) A distribution is called "heavy tailed" if there is a higher probability of very extreme values than in a normal distribution. Does the histogram of part (b) give any indication that the distribution of titration results from this class is heavy tailed? (It may help to compare your histogram with those of normal distributions.)

12–4 (a) Suppose we test H_0: η = \$220,000 versus H_1: $\eta <$ \$220,000 on the SALES data. Use Exhibit 12.1 and the CDF command to find the corresponding p-value "by hand."

(b) Use STEST to check your answer to part (a).

12–5 The data from a second titration experiment are given below.

.109	.111	.110	.110	.105
.110	.111	.110	.110	.111
.109	.111	.109	.112	.109
.109	.111	.110	.112	.112
.109	.110	.110	.109	.113
.108	.105	.110	.109	.109
.110	.110	.110	.104	.109
.110	.111			

These data are in C2 of the worksheet ACID.

(a) Make a histogram. Do you think $\mu = \eta$? It will be if the population is symmetric. Exercise 12–3 gives us some more information about the shape of the distribution for titration data. What did you conclude there?

(b) Find (as close as possible) a 95% sign confidence interval for η.

(c) Find a 95% confidence interval for η using normal theory methods and compare it to the sign confidence interval in part (a).

(d) Repeat parts (a) and (b) using 90% confidence intervals.

12–6 The following is a sample (which we got by simulation) from a normal distribution:

62, 60, 65, 70, 60, 67, 61, 66, 64, 64, 62, 63

(a) Calculate an approximately 95% sign confidence interval for η (the closest confidence will be 96.1).

(b) Find a 96.1% t confidence interval using the TINTERVAL command. Compare this with the sign interval of part (a).

(c) Now suppose a mistake was made in recording the last observation and 36 was recorded instead of 63. Repeat parts (a) and (b).

(d) Repeat part (c), only now suppose the last observation was mistakenly recorded as 630.

12–7 We can use simulation to get a feel for how well sign confidence intervals work in various cases. If your data came from a normal population, a t confidence interval is the best way to estimate $\mu = \eta$. But how much worse might a sign confidence interval be? The following commands simulate ten samples, each of 18 observations, from a normal distribution with $\mu = 50$ and $\sigma = 8$, and then get a 90% sign confidence interval and a 90% t confidence interval for each sample. Compare the two confidence intervals for each sample. Do both cover μ? Which is narrower?

```
RANDOM 18 C1-C10;
  NORMAL 50 8.
SINT 90 C1-C10
TINT 90 C1-C10
```

12–8 (a) Repeat Exercise 12–7, except simulate data from a Cauchy distribution. Use:

```
RANDOM 18 C1-C6;
  CAUCHY 0 1.
```

This gives an example of data from a very "heavy tailed" distribution.

(b) Repeat Exercise 12–7, except simulate data from a uniform distribution. Use:

```
RANDOM 18 C1-C6;
  UNIFORM 0 1.
```

A uniform distribution is an example of a distribution that has very "skinny tails."

12–9 Consider the SALES data in Exhibit 12.1. Use the CDF command to find the achieved confidence for the sign confidence interval ($110,000, $220,000).

12–10 The job of President of the United States is a very demanding, high-pressure job. This might cause premature deaths of Presidents. On the other hand, only vigorous people are going to run for President, so Presidents might tend to live longer than other people. We list below the "modern" (since Lincoln) Presidents who died by September 1975, the number of years they lived after inauguration, and the life expectancy of a man the same age as the President was on his first inauguration.

Longevity of U.S. Presidents

	Life Expectancy After First Inauguration	*Actual Years Lived After First Inauguration*
Andrew Johnson	17.2	10.3
Ulysses S. Grant	22.8	16.4
Rutherford B. Hayes	18.0	15.9
James A. Garfield	21.2	.5
Chester A. Arthur	20.1	5.2
Grover Cleveland	22.1	23.3
Benjamin Harrison	17.2	12.0
William McKinley	18.2	4.5
Theodore Roosevelt	26.1	17.3
William H. Taft	20.3	21.2

Longevity of U.S. Presidents

	Life Expectancy After First Inauguration	*Actual Years Lived After First Inauguration*
Woodrow Wilson	17.1	10.9
Warren G. Harding	18.1	2.4
Calvin Coolidge	21.4	9.4
Herbert C. Hoover	19.0	35.6
Franklin D. Roosevelt	21.7	12.1
Harry S. Truman	15.3	27.7
Dwight D. Eisenhower	14.7	16.2
John F. Kennedy	28.5	2.8
Lyndon B. Johnson	19.3	9.2

These data are in the worksheet PRES.

(a) Do a paired sign test of the null hypothesis that being President has no effect on length of life.

(b) Find an approximately 95% sign confidence interval for the median difference between the expected and attained life span of Presidents.

(c) If our main interest is the effect of stress on length of life, then perhaps we should not include the Presidents who were assassinated (Garfield, McKinley, and Kennedy). Carry out the analysis of (a) and (b) without these three Presidents.

(d) We have, perhaps, stretched the use of statistics rather far here, as we often must in real problems. Comment on this statement, paying particular attention to the assumptions needed for a sign test.

12.2 Wilcoxon Signed Rank Procedures

With sign procedures, the population can have any shape; with Student *t* procedures, the population should be approximately normal. Wilcoxon procedures are in between: the population should be approximately symmetric but it need not be normal or have any other specific shape. Now recall that for any symmetric population the mean and median are equal. The Wilcoxon test, which tests for the location of the center of the population, can then be viewed as a test for either the population mean μ or the population median η. Similarly, a Wilcoxon confidence interval can be viewed as either an interval for μ or for η.

Wilcoxon Test

Exhibit 12.6 contains a very simple example that shows how to calculate the Wilcoxon test statistic. Here the test is done for the null hypothesis

Exhibit 12.6 Example of Calculating the Wilcoxon Statistic W for H_0: η = 10

X	*20*	*18*	*23*	*5*	*14*	*8*	*18*	*22*
X–10	10	8	13	−5	4	−2	8	12
|X–10|	10	8	13	5	4	2	8	12
Ranks	6	4.5	8	3	2	1	4.5	7

W = 6 + 4.5 + 8 + 2 + 4.5 + 7 = 32

H_0: η = 10. The first line contains the eight observations. First we subtract the null hypothesis value of η from each observation. Then we take absolute values. Next we assign ranks. The smallest absolute value is 2 and gets rank 1. The second smallest absolute value is 4 and gets rank 2, and so on. Notice that the fourth and fifth smallest are both 8s. Sometimes several absolute values are tied. Then we average all the ranks associated with these absolute values and give this average rank (often called the midrank) to each one. For example, the two 8s have ranks 4 and 5, so we give each of them the rank (4 + 5)/2 = 4.5. Finally we compute *W* as the sum of all the ranks corresponding to the observations that were above 10 (i.e., had positive deviations). If *W* is large, then many observations are above 10 and we suspect η is also above 10. If *W* is small, then many observations are below 10 and we suspect η is also below 10. Here *W* is 32.

Suppose we test H_0: η = 10 versus H_1: $\eta > 10$, using α = .10. Minitab instructions are shown in Exhibit 12.7. The output tells us that the *p*-value for this test is .029. Thus there is little chance of *W* being as large as 32 by chance alone. So we conclude our data are inconsistent with the hypothesis that they are a random sample from a population which has μ = 10. (We will explain ESTIMATED CENTER when we discuss estimation of η.) Now suppose we test H_0: η = 10 versus the alternative H_1: $\eta \neq 10$. Since this is a two-sided alternative hypothesis we need two probabilities—one for the lower tail and one for the upper tail. The distribution of the test statistic is symmetric so all that one needs to do is to double the one-sided *p*-value giving 2(.029) = .058 as the two-sided *p*-value. This is shown in Exhibit 12.8.

Ties. If any observations are equal to the null hypothesis value of η, we omit these and do the Wilcoxon test using the observations remaining.

How Minitab Does the Wilcoxon Test. If the null hypothesis is true then *W* has mean = $n(n + 1)/4$ and variance = $n(n + 1)(2n + 1)/24$, where

Exhibit 12.7 Minitab Instructions to Do Wilcoxon Test

```
SET C1
20 18 23 5 14 8 18 22
END
WTEST 10  C1;
  ALT 1.

TEST OF CENTER = 10.00 VERSUS CENTER G.T. 10.00

                    N FOR   WILCOXON             ESTIMATED
             N      TEST   STATISTIC   P-VALUE      CENTER
C1           8         8        32.0     0.029       16.50
```

Exhibit 12.8 Minitab Instructions to do Two-sided Wilcoxon Test

```
SET C1
20 18 23 5 14 8 18 22
END
WTEST 10  C1

SIGNED RANK TEST OF CENTER = 10.00 VERSUS CENTER N.E. 10.00

                  N FOR   WILCOXON             ESTIMATED
             N     TEST  STATISTIC   P-VALUE      CENTER
C1           8        8       32.0     0.059       16.50
```

n is the number of observations in our sample. The example in Exhibit 12.7 has $n = 8$. So, the mean of $W = 8(9)/4 = 18$ and the variance $= 8(9)(17)/24 = 51$. The standard deviation then is $\sqrt{51} = 7.14$.

The exact distribution of W is given in special tables (see *Nonparametric Statistical Methods,* M. Hollander and D. Wolfe, Wiley, 1973). If we don't have such tables, we can use the fact that W has approximately a normal distribution. Minitab uses the normal approximation (with a continuity correction). This is what is given in Exhibit 12.7.

How we calculate the p-value for a test depends on the alternative hypothesis. Suppose we do the test in Exhibit 12.7. Here H_1: $\eta > 10$. The p-value is then $P(W \geq$ the value of W in our sample$) = P(W \geq 32)$. Notice that W is almost two standard deviations above its mean, so its p-value should be fairly small. Exhibit 12.7 gives this value as .029.

Suppose we test H_0: $\eta = 10$ versus H_1: $\eta \neq 10$ using the data in Exhibit 12.6. Since this is a two-sided test we need two probabilities—one for small values of W and one for large values of W. In our sample, $W = 32$. Since this value is above the mean of W, the probability for large values of W is $P(W \geq 32)$. The probability for small values then is given by $P(W \leq R - 32)$, where R is the sum of all the ranks. This is given by the formula $R = n(n+1)/2$. For $n = 8$, $R = 8(9)/2 = 36$. Thus $P(W \leq R - 32) = P(W \leq 4)$. Since the distribution of W is symmetric, $P(W \geq 32) = P(W \leq 4)$. Thus the p-value $= P(W \leq 4) + P(W \geq 32) = 2P(W \geq 32) = 2(.029) = .058$. Exhibit 12.8 does this test using Minitab. (Note: The p-value on this output is .059, not .058. The discrepancy is due to round-off error.)

Suppose we test H_0: $\eta = 20$ versus H_1: $\eta < 20$ using the data in Exhibit 12.6. Here the p-value is $P(W \leq$ the value of W in our sample). Notice there is one observation equal to 20. We set aside this observation when we do the test. This gives $n = 7$ and $W = 6$. The p-value is then $P(W \leq 6)$.

Confidence Intervals and Estimation of η

There is a second way to calculate the value of W and this is the method that we will use to find a point estimate of η and get a confidence interval. Exhibit 12.9 contains the calculations for testing H_0: $\eta = 10$ using the data in Exhibit 12.6. We first write the observations across the top and down the left side of the table. Each entry in the table is the average (often called the Walsh average) of the observations in the corresponding row and column. The entries in the bottom part of the table are not used since they are just repeats of the entries in the top part of the table. Therefore, we did not write them in. Underneath this table are the ordered Walsh averages. We can compute the Wilcoxon statistic W from the ordered Walsh averages: W = (number of Walsh averages above 10) + 1/2(number of Walsh averages equal to 10) = 32 + 1/2(0) = 32, which is the same value we got in Exhibit 12.6.

Estimation of η. Since we are working with a population that we believe to be approximately symmetric, there are several ways to estimate η. We might use the sample mean or the sample median. Or we could use the median of all the Walsh averages, a method that involves both averaging and finding a median. For the data in Exhibit 12.9, there are 36 Walsh averages, so the median of the Walsh averages is the average of the eighteenth and nineteenth ordered averages. These are circled in Exhibit 12.9. The average of these two values is $(16 + 17)/2 = 16.5$. This is the estimate of η that WTEST and WINTERVAL print.

Exhibit 12.9 Walsh Averages for Use in Wilcoxon Procedure

		Data							
		20	*18*	*23*	*5*	*14*	*8*	*18*	*22*
Data	20	20	19	21.5	12.5	17	14	19	21
	18		18	20.5	11.5	16	13	18	20
	23			23	14	18.5	15.5	20.5	22.5
	5				5	9.5	6.5	11.5	13.5
	14					14	11	16	18
	8						8	13	15
	18							18	20
	22								22

Walsh Averages in Order from Smallest to Largest

5	6.5	8	9.5	11	11.5	11.5	12.5	13	13
13.5	14	14	14	15	15.5	16	⑯	⑰	18
18	18	18	18.5	19	19	20	20	20	20.5
20.5	21	21.5	22	22.5	23				

The Walsh averages can also be used to form a confidence interval for η. The method is very similar to the method we used for sign confidence intervals. The statistic W is analogous to the statistic X = the number of observations less than η. If our interval goes from the dth smallest to the dth largest Walsh average, then it has confidence given by the formula $1-2P(W < d)$. Unfortunately, W does not have a simple distribution, so we must either use special tables (see *Nonparametric Statistical Methods,* M. Hollander and D. Wolfe, Wiley, 1973) or the fact that W is approximately normal. (Minitab uses the normal approximation with a continuity correction.) Exhibit 12.10 contains output from the command WINTERVAL for 95% confidence. As with most nonparametric procedures, we cannot exactly achieve the confidence we request. Minitab gives the closest it can get. Here it is 95.8% and the confidence interval is (9.500, 21.500). (If we look at Exhibit 12.9, we see that in this case d was 4 since 9.500 is the fourth smallest and 21.500 is the fourth largest Walsh average.)

WTEST median = K for C, ..., C

Does a Wilcoxon signed rank test of H_0: $\eta = K$, separately for each column. The value of η is estimated using signed-rank procedures.

One-sided alternatives may be specified by the subcommand ALT as follows:

ALT = −1 H_1: $\eta < K$
ALT = 1 H_1: $\eta > K$

If you do not use ALT, a two-sided test is done.

WINTERVAL[confidence = K]for C, ..., C

Calculates a confidence interval for the median separately for each column. The achievable confidence closest to K is used.

Exhibit 12.10 Confidence Interval Associated with Wilcoxon Signed Rank Test

```
SET C1
20 18 23 5 14 8 18 22
END
WINTERVAL .95  C1

                      ESTIMATED     ACHIEVED            WILCOXON
                  N      CENTER   CONFIDENCE   CONFIDENCE INTERVAL
C1                8       16.50         95.8   (    9.500,    21.500)
```

Exercises

12–11 Do Exercise 12–1 using a Wilcoxon test in place of the sign test. Do you think a Wilcoxon test is appropriate for these data?

12–12 Do Exercise 12–3 using a Wilcoxon test. Do you think a Wilcoxon test is appropriate for these data?

12–13 (a) Do Exercise 12–5 using a Wilcoxon confidence interval.
(b) Compare this interval to the sign interval and t interval of Exercise 12–5.

12–14 Do Exercise 12–6 using Wilcoxon confidence intervals in place of sign confidence intervals.

12–15 Do Exercise 12–7 using Wilcoxon confidence intervals in addition to sign confidence intervals. (Thus use SINT, TINT, and WINT for each sample.) Compare all three intervals for each sample.

12–16 (a) Find the Wilcoxon point estimate of η by hand using the sample

5, 3, 5, 7, 6, 6

(b) Use WINT to check your results.

12.3 Two-Sample Rank Procedures

In this section we describe a nonparametric procedure for comparing two populations. We assume that (1) we have a random sample from each population, (2) the samples were taken independently of each other, and (3) the populations have approximately the same shape (this means the variances must be approximately equal). We use η_1 to represent the median of the first population and η_2 for the second population.

Our null hypothesis is that the medians of the two populations are equal. Our alternative is that one population is shifted from the other, that is, has a different median. Since we assume the two populations have the same shape, this procedure is analogous to the pooled t procedures discussed in Section 8.2.

This two-sample rank test was introduced by Wilcoxon and is often called the Wilcoxon rank sum test. It was further developed by Mann and Whitney. To avoid confusion with the Wilcoxon test in Section 12.2, we chose the name MANN-WHITNEY for the Minitab command.

An Example

The data in Table 12.1 are from a study on Parkinson's disease which, among other things, affects a person's ability to speak. Eight of the people in this study had received one of the most common operations to treat the disease. This operation seemed to improve the patients' condition overall, but how did it affect their ability to speak? Each patient was given several tests. The results of one of these tests are shown in Exhibit 12.11.

First we will show how to do a two-sample rank test by hand, then how to do it with Minitab. We begin by combining the two samples. Then we rank the combined sample, giving rank 1 to the smallest observation, rank 2 to the second smallest, etc. This procedure is illustrated in Exhibit 12.11. Whenever we have two or more observations that are tied, we assign the average rank (or midrank) to each. For example, the

Table 12.1 Speaking Ability for Patients in Study of Parkinson's Disease (The higher the score, the more problems when speaking.)

Patients Who Had Operation	*Patients Who Did Not Have Operation*
2.6	1.2
2.0	1.8
1.7	1.8
2.7	2.3
2.5	1.3
2.6	3.0
2.5	2.2
3.0	1.3
	1.5
	1.6
	1.3
	1.5
	2.7
	2.0

three observations that equal 1.3 are tied for ranks 2, 3, and 4. We therefore give each the average rank of $(2 + 3 + 4)/3 = 3$. Next we sum all the ranks corresponding to the observations in the first sample. This sum is usually denoted by W. Here $W = 126.5$.

The value of W reflects the relative locations of the two samples. If the values in the first sample tend to be larger than those in the second sample, W will be large. If the values in the first sample tend to be smaller, then W will be small. Minitab's MANN-WHITNEY command does all this work for us. See Exhibit 12.12.

The output gives the value of $W = 126.5$ and tells us that the attained significance of the two-sided test is 0.0203. This means the chance of observing two samples as separated as these, when in fact the two populations have the same median, is only 0.0203. We therefore have statistical evidence that the two populations differ.

Let's look for a moment at the practical significance of these results. The test says there is statistically significant evidence that the two populations differ. If we look at the data in Table 12.1, we see the direction of the difference—the patients who had the operation had more severe speech problems than those who did not have the operation. This, of course, does not prove that the operation causes difficulties in speaking. Perhaps the operation was only done on patients who were already in poor condition. The output also contains a confidence interval and a

Exhibit 12.11 Two-Sample Rank Test, Done by Hand

Samples		*Ranks*	
First	*Second*	*First*	*Second*
	1.2		1
	1.3		3
	1.3		3
	1.3		3
	1.5		5.5
	1.5		5.5
	1.6		7
1.7		8	
	1.8		9.5
	1.8		9.5
2.0		11.5	
	2.0		11.5
	2.2		13
	2.3		14
2.5		15.5	
2.5		15.5	
2.6		17.5	
2.6		17.5	
2.7		19.5	
	2.7		19.5
3.0		21.5	
	3.0		21.5

W = sum of ranks in first sample = 126.5

point estimate. Both of these are calculated using procedures developed from the two-sample rank test.

Exhibit 12.13 contains the calculations that are used when this technique is done by hand. The first sample is written down the left side of the table and the second sample across the top. Each entry inside the table is X minus Y, where X is the observation in the corresponding row and Y is the observation in the corresponding column. This gives a total of $8 \times 14 = 112$ differences. We can calculate the Mann-Whitney statistic, W, using the formula W = (number of positive differences) + (1/2) (number of zeros) + $n_1(n_1 + 1)/2$, where n_1 is the number of observations in the first sample. In the example, $W = 89 + (1/2)(3) + 8(9)/2 = 126.5$, the value we got in Exhibit 12.11.

Exhibit 12.12 Mann–Whitney Test and Confidence Interval

```
SET C1
2.6 2.0 1.7 2.7 2.5 2.6 2.5 3.0
END
SET C2
1.2 1.8 1.8 2.3 1.3 3.0 2.2 1.3 1.5 1.6 1.3 1.5 2.7 2.0
END
MANNWHITNEY C1 C2

Mann-Whitney Confidence Interval and Test

C1          N =   8     MEDIAN =        2.5500
C2          N =  14     MEDIAN =        1.7000
POINT ESTIMATE FOR ETA1-ETA2 IS         0.7000
95.6  PCT C.I. FOR ETA1-ETA2 IS (       0.20,      1.20)
W =      126.5
TEST OF ETA1 = ETA2 VS. ETA1 N.E. ETA2 IS SIGNIFICANT AT 0.0203
```

We estimate the difference in the population medians ($\eta_1 - \eta_2$) by the median of these differences. Since there are 112 differences, the median is the average of the 56th and 57th observations. These are circled in Exhibit 12.13. The average of these two values is $(.7 + .7)/2 = .7$. The confidence interval is also based on these differences. It goes from the dth smallest difference to the dth largest difference, where the value of d depends on the confidence you specify. Minitab does not print the value of d, but just gives the confidence interval. Here the confidence interval, given in Exhibit 12.12, goes from .2 to 1.2. Notice again that we cannot achieve the exact confidence we specified. The closest value to 95% that we can achieve is 95.6%. Thus we can be 95.6% confident that if these data were two independent random samples, the median of population 1 is between .2 and 1.2 units higher than the median of population 2.

How Minitab Does the Calculations. If there is no difference between the two populations, then the mean of W is $n_1(n_1 + n_2 + 1)/2$, where n_1 = number of observations in the first sample and n_2 = number of observations in the second sample. Here this average is $8(8 + 14 + 1)/2 = 92$. If W is much larger than 92, then many numbers in the first sample must have been large. In that case, we might suspect that the first population has a larger median. If W is much smaller than 92, then many numbers in the first sample must have been small, so we might suspect that the first population has a smaller median than the second.

Exhibit 12.13 Calculations for the Estimate and Confidence Interval Associated with the Mann–Whitney Test

	Second Sample													
	1.2	1.8	1.8	2.3	1.3	3.0	2.2	1.3	1.5	1.6	1.3	1.5	2.7	2.0
2.6	1.4	.8	.8	.3	1.3	−.4	.4	1.3	1.1	1.0	1.3	1.1	−.1	.6
2.0	.8	.2	.2	−.3	.7	−1.0	−.2	.7	.5	.4	.7	.5	.7	0.0
1.7	.5	−.1	−.1	−.6	.4	−1.3	−.5	.4	.2	.1	.4	.2	−1.0	−.3
2.7	1.5	.9	.9	.4	1.4	−.3	.5	1.4	1.2	1.1	1.4	1.2	0.0	.7
2.5	1.3	.7	.7	.2	1.2	−.5	.3	1.2	1.0	.9	1.2	1.0	−.2	.5
2.6	1.4	.8	.8	.3	1.3	−.4	.4	1.3	1.1	1.0	1.3	1.1	−.1	.6
2.5	1.3	.7	.7	.2	1.2	−.5	.3	1.2	1.0	.9	1.2	1.0	−.2	.5
3.0	1.8	1.2	1.2	.7	1.7	0.0	.8	1.7	1.5	1.4	1.7	1.5	.3	1.0

The differences in order

−1.3	−1.0	−1.0	−.7	.6	−.5	−.5	−.5	−.4	−.4
−.3	−.3	−.3	−.2	−.2	−.2	−.1	−.1	−.1	−.1
0.0	0.0	0.0	.1	.2	.2	.2	.2	.2	.2
.3	.3	.3	.3	.3	.4	.4	.4	.4	.4
.4	.4	.5	.5	.5	.5	.5	.5	.6	.6
.7	.7	.7	.7	.7	.7	.7	.7	.7	.8
.8	.8	.8	.8	.8	.9	.9	.9	.9	1.0
1.0	1.0	1.0	1.0	1.0	1.0	1.0	1.1	1.1	1.1
1.1	1.2	1.2	1.2	1.2	1.2	1.2	1.2	1.2	1.2
1.2	1.3	1.3	1.3	1.3	1.3	1.3	1.3	1.3	1.4
1.4	1.4	1.4	1.4	1.4	1.5	1.5	1.5	1.7	1.7
1.7	1.8								

To do the test, we compare the value of W to a special table of critical values.* If we don't have such a table, we can use the fact that W has approximately a normal distribution, with variance $n_1 n_2 (n_1 + n_2 + 1)/12$. (A slightly more complicated version of this formula is often used when there are ties.†) In our example, the variance is $(8)(14)(8 + 14 + 1)/12 = 214.67$ and, therefore, the standard deviation is $\sqrt{214.67} = 14.65$. If the two populations were the same, W would be near its mean of 92. But, in fact, W is 126.5, over 2.3 standard deviations above 92.

* See *Nonparametric Statistical Methods,* M. Hollander and D. Wolfe, Wiley, 1973.
† Op. cit., p. 69.

MANN-WHITNEY test and c. i. for samples in C and C

Given independent random samples from two populations with population medians η_1 and η_2, this command

1. Performs a two-sample rank test of the null hypothesis H_0: $\eta_1 = \eta_2$ against the two-sided alternative hypothesis H_1: $\eta_1 \neq \eta_2$. The test statistic, W, and the attained significance level of the test are printed out.
2. Finds a 95% confidence interval for $\eta_1 - \eta_2$, using the two-sample rank method. The closest confidence to 95% is used.
3. Prints a point estimate for $\eta_1 - \eta_2$, using two-sample rank methods.

Exercises

12–17 (a) By hand, compute the W statistic of a two-sample rank test, using the following data:

Sample A	10	7	6	12	14
Sample B	8	4	6	11	

Use Minitab's MANN-WHITNEY command to check your answer.

(b) By hand, compute the point estimate for the difference of the two population medians, using two-sample rank procedures. Compare your answer to Minitab's.

12–18 Do migratory birds store, then gradually use up a layer of fat as they migrate? To investigate this question, two samples of migratory song sparrows were caught, one sample on April 5 and one sample on April 6. The amount of stored fat on each bird was subjectively estimated by an expert. The higher the fat class the more fat on the bird.

	Number of Birds in Class	
Fat Class	*Found on April 5*	*Found on April 6*
1	0	3
2	0	1
3	0	11
4	2	6
5	0	1

Fat Class	*Number of Birds in Class* Found on April 5	Found on April 6
6	10	9
7	9	2
8	7	2
9	6	1
10	4	0
11	1	0
12	2	0

(a) Use DOTPLOT to compare these two groups visually.

(b) Do an appropriate test to see if there is any evidence that birds use up a layer of fat as they migrate.

12–19 A study was done at Penn State to see how much one type of air pollution, ozone, damages azalea plants. Eleven varieties of azaleas were included in the study. We will look at data from just two varieties.

During week 1, ten plants of each variety were fumigated with ozone. A short time later each plant was measured for leaf damage. The procedure was repeated four more times, each time using new plants.

Leaf Damage for Variety A

Week 1	*Week 2*	*Week 3*	*Week 4*	*Week 5*
1.58	1.09	.00	2.22	.20
1.62	1.03	.00	2.40	.40
2.04	.00	.07	2.47	.34
1.28	.46	.18	1.85	.00
1.43	.46	.40	2.50	.00
1.93	.85	.20	1.20	.00
2.20	.30	.63	1.33	.10
1.96	.90	.63	2.40	.06
2.23	.00	.56	2.23	.17
1.54	.00	.26	2.57	.25

Leaf Damage for Variety B

Week 1	*Week 2*	*Week 3*	*Week 4*	*Week 5*
1.29	.00	.78	.40	.00
.70	.20	.64	.00	.40
1.93	.00	1.00	.20	.47
.98	.98	.42	.40	.00
.94	.00	.97	.14	.00
1.06	.62	2.43	.44	.00

Leaf Damage for Variety B

Week 1	*Week 2*	*Week 3*	*Week 4*	*Week 5*
.94	.67	.65	1.23	.00
1.65	.00	.00	.35	.00
.70	.00	.30	.17	.00
.35	.00	.00	.20	.00

These data are in C1–C5 (variety A) and C6–C10 (variety B) of the worksheet AZALEA.

(a) Compare the first week's data for the two varieties, using a two-sample rank test. Does there appear to be any difference in the two varieties' susceptibility to ozone damage?

(b) Repeat the test of part (a) for each of the four other weeks. Overall, how do the two varieties compare?

(c) Susceptibility to ozone varies with weather conditions, and weather conditions vary from week to week. Does this seem to show up in the data? One simple way to look for a "week effect" is to calculate the median leaf damage for all 20 plants sprayed during week 1, the median for week 2, etc. Do these five medians seem to be very different?

(d) Another way we could test for a "week effect" is to use a contingency table analysis (discussed in Section 11.2). First, form a contingency table as follows: For each week count the number of damaged plants and the number of undamaged plants (a plant is undamaged if its leaf damage is .00); form a table which has two rows and five columns. Then do a chi-square test for association between damage and week. Do the results agree with what was indicated in part (c)?

APPENDIXES

APPENDIX A
Data Sets Used in This Handbook

APPENDIX B
Additional Features in Minitab

APPENDIX C
A List of Minitab Commands in Releases 5–7

APPENDIX D
Changes in Release 6 and Release 7

APPENDIX E
Using this Book with Minitab Release 8 on a PC or a Macintosh

APPENDIX A
Data Sets Used in This Handbook

Many of the data sets used in this Handbook are available as saved worksheets (to be input with the command RETRIEVE) and as data files (to be input with the command READ). Table A.1 gives a complete list. The same name is used for both the worksheet and the data file (with two exceptions). Thus, you can type either of the following:

```
RETRIEVE 'ACID'
READ 'ACID' C1 C2
```

If you use RETRIEVE to input the saved worksheet, your columns will be given names. You can type the command INFO to find out what these names are. A few computers restrict file names to five letters. On those machines, type just the first five letters of the file name.

In this appendix we also give some data sets that are used in various places throughout this *Handbook*.

Table A.1 Data Sets Available as Saved Worksheets and Data Files

Name	*Description*	*Text Citation*
ACID	Results of a titration to determine the acidity of a solution. ACID is the worksheet name.	p. 283 and p. 284
ACID1 ACID2	ACID1 is the data file containing the data from Exercise 12–3. ACID2 is the data file for Exercise 12–5. Use SET (not READ) to input ACID1 and ACID2.	
ALFALFA	Yields for six varieties of alfalfa grown on four fields.	pp. 117–119
AZALEA	Effects of ozone on azalea plants.	pp. 298–299
CARTOON	Data from a study to evaluate the relative effectiveness of cartoon sketches and realistic photographs when used in an instructional film.	pp. 303–309
CHOLEST	Blood cholesterol levels of 28 heart-attack patients and 30 people who had not had a heart attack. CHOLEST is the worksheet name.	pp. 92–94
CHOLESTE CHOLESTC	CHOLESTE is the data file containing the three variables for the experimental group. CHOLESTC contains the one variable for the control group. Use READ to input both of these.	
CITIES	Monthly temperatures in five U.S. cities.	p. 63
EMPLOY	Number of employees in Wisconsin per month, in three areas: trade, food products, and fabricated metals	pp. 66–68
FA	Data constructed by Frank Anscombe to illustrate why data plots are important in regression.	pp. 253–254
FABRIC	Data from a study of the flammability of cloth.	pp. 193–195
FURNACE	Data on the effectiveness of two devices for improving the efficiency of gas home-heating systems.	pp. 312–315
GRADES	SAT verbal and math scores and freshman grade point averages for 200 students.	pp. 309–315

Name	Description	Text Citation
GA	Data for the first 100 students from the file GRADES.	
GB	Data for the last 100 students from the file GRADES.	
HCC	Data on psychiatric patients at a health care center in Wisconsin.	p. 82
LAKE	Measurements on 71 lakes in northern Wisconsin.	pp. 315–317
MAPLE	Data on maple trees.	p. 235
MEATLOAF	Data from a study to compare different methods of freezing meat loaf.	pp. 200–205
PERU	Study to determine the effects of change in environment on blood pressure of Indians in Peru.	pp. 317–318
PLYWOOD	Study of how temperature, chuck penetration, and log diameter affect the torque applied when cutting layers of wood from a log.	pp. 59–62
POTATO	Data from an experiment to see how temperature and amount of oxygen affect potatoes during storage.	pp. 211, and 214–215
PRES	Longevity of U.S. presidents.	pp. 285–286
PULSE	Pulse rates before and after exercise.	pp. 318–321
RADON	Radiation exposure rates from radon, measured by four different devices.	pp. 107–109
RESTRNT	Survey of restaurants in Wisconsin.	Ch. 4 and pp. 321–328
SCHOOLS	Number of teachers per school who reported sick during a dispute in the public school system in Madison, Wisconsin.	pp. 109–111
TREES	Diameter, height, and volume of black cherry trees.	pp. 328–329
TWAIN	Data from a study to determine the authorship of ten letters thought to be written by Mark Twain.	pp. 269–271

Cartoon

When educators make an instructional film, they have two objectives: Will the people who watch the film learn the material as efficiently as possible? Will they retain what they have learned?

To help answer these questions, an experiment was conducted to evaluate the relative effectiveness of *cartoon* sketches and *realistic* photographs, in both *color* and *black and white* visual materials.

A short instructional slide presentation was developed. The topic chosen for the presentation was the behavior of people in a group situation and, in particular, the various roles or character types that group members often assume. The presentation consisted of a five-minute lecture on tape, accompanied by 18 slides. Each role was identified as an animal. Each animal was shown on two slides: once in a cartoon sketch and once in a realistic picture. All 179 participants saw all of the 18 slides, but a randomly selected half of the participants saw them in black and white while the other half saw them in color.

After they had seen the slides, the participants took a test (*immediate test*) on the material. The 18 slides were presented in a random order, and the participants wrote down the character type represented by that slide. They received two scores: one for the number of cartoon characters they correctly identified and one for the number of realistic characters they correctly identified. Each score could range from 0 to 9, since there were nine characters.

Four weeks later, the participants were given another test (*delayed test*) and their scores were computed again. Some participants did not show up for this delayed test, so their scores were given the missing value code, *.

The primary participants in this study were preprofessional and professional personnel at three hospitals in Pennsylvania involved in an in-service training program. A group of Penn State undergraduate students also were given the test as a comparison. All participants were given the OTIS Quick Scoring Mental Ability Test, which yielded a rough estimate of their natural ability.

Some questions that are of interest here are as follows: Is there a difference between color and black and white visual aids? Between cartoon and realistic? Is there any difference in retention? Does any difference depend on educational level or location? Does adjusting for OTIS scores make any difference?

The data are given below. They have been sorted partially so that various parts may easily be studied separately.

Description of Cartoon Data

Variable	*Description*
1 ID	Identification number
2 COLOR	0 = black and white, 1 = color (no participant saw both)
3 ED	Education: 0 = preprofessional, 1 = professional, 2 = college student
4 LOCATION	Location: 1 = hospital A, 2 = hospital B, 3 = hospital C, 4 = Penn State student
5 OTIS	OTIS Score: from about 70 to about 130

Description of Cartoon Data

Variable	*Description*
6 CARTOON1	Score on cartoon test given immediately after presentation (possible scores are 0, 1, 2, ..., 9)
7 REAL1	Score on realistic test given immediately after presentation (possible scores are 0, 1, 2, ..., 9)
8 CARTOON2	Score on cartoon test given four weeks (delayed) after presentation (possible scores are 0, 1, 2, ..., 9; * is used for a missing observation)
9 REAL2	Score on realistic test given four weeks (delayed) after presentation (possible scores are 0, 1, 2, ..., 9; * is used for a missing observation)

Variable								
1	*2*	*3*	*4*	*5*	*6*	*7*	*8*	*9*
1	0	0	1	107	4	4	*	*
2	0	0	2	106	9	9	6	5
3	0	0	2	94	4	2	3	0
4	0	0	2	121	8	8	6	8
5	0	0	3	86	5	5	*	*
6	0	0	3	99	7	8	7	5
7	0	0	3	114	8	9	5	4
8	0	0	3	100	2	1	*	*
9	0	0	3	85	3	2	*	*
10	0	0	3	115	8	7	8	5
11	0	0	3	101	7	6	*	*
12	0	0	3	84	7	5	*	*
13	0	0	3	94	4	3	*	*
14	0	0	3	87	1	3	2	0
15	0	0	3	104	9	9	5	6
16	0	0	3	104	5	6	*	*
17	0	0	3	97	6	5	*	*
18	0	0	3	91	1	0	*	*
19	0	0	3	83	4	4	*	*
20	0	0	3	93	0	1	*	*
21	0	0	3	92	2	2	*	*
22	0	0	3	91	5	2	3	1
23	0	0	3	88	2	1	*	*
24	0	0	3	90	5	4	4	3
25	0	0	3	103	6	2	*	*
26	0	0	3	93	9	9	8	4
27	0	0	3	106	2	0	6	3
28	1	0	1	98	3	3	*	*
29	1	0	1	103	6	5	2	2

Variable								
1	*2*	*3*	*4*	*5*	*6*	*7*	*8*	*9*
30	1	0	2	109	5	4	1	2
31	1	0	2	107	8	8	*	*
32	1	0	2	108	8	8	7	6
33	1	0	2	107	3	2	*	*
34	1	0	3	87	6	4	2	2
35	1	0	3	113	5	4	4	4
36	1	0	3	80	0	3	1	1
37	1	0	3	91	5	6	*	*
38	1	0	3	102	8	9	5	5
39	1	0	3	83	4	1	2	1
40	1	0	3	108	9	9	*	*
41	1	0	3	86	4	4	*	*
42	1	0	3	96	6	3	*	*
43	1	0	3	101	5	3	*	*
44	1	0	3	97	6	3	4	4
45	1	0	3	88	3	1	2	0
46	1	0	3	104	4	2	2	0
47	1	0	3	87	7	3	*	*
48	1	0	3	86	1	1	*	*
49	1	0	3	90	6	5	4	1
50	1	0	3	102	6	2	*	*
51	1	0	3	105	2	2	*	*
52	1	0	3	115	7	8	*	*
53	1	0	3	88	4	3	*	*
54	1	0	3	111	8	8	*	*
55	1	0	3	95	5	4	*	*
56	1	0	3	104	5	5	*	*
57	0	1	1	79	7	4	6	4
58	0	1	1	82	3	2	*	*
59	0	1	1	123	8	8	7	5
60	0	1	1	106	9	7	8	6
61	0	1	1	125	9	9	4	3
62	0	1	1	98	7	6	*	*
63	0	1	1	95	7	7	4	4
64	0	1	2	129	9	9	7	7
65	0	1	2	90	7	6	3	5
66	0	1	2	111	6	2	3	1
67	0	1	2	99	4	5	3	1
68	0	1	2	116	9	7	7	7
69	0	1	2	106	8	7	6	4
70	0	1	2	107	8	5	*	*
71	0	1	2	100	7	6	2	1
72	0	1	2	124	8	9	3	5
73	0	1	3	98	6	7	1	1
74	0	1	3	124	9	6	6	5

Variable								
1	*2*	*3*	*4*	*5*	*6*	*7*	*8*	*9*
75	0	1	3	84	1	4	*	*
76	0	1	3	91	8	3	*	*
77	0	1	3	118	6	6	3	4
78	0	1	3	102	6	4	*	*
79	0	1	3	95	7	4	*	*
80	0	1	3	90	4	3	*	*
81	0	1	3	86	1	0	*	*
82	0	1	3	104	6	4	*	*
83	1	1	1	111	9	9	6	3
84	1	1	1	105	1	0	*	*
85	1	1	1	110	1	0	0	0
86	1	1	1	80	0	0	0	0
87	1	1	1	78	4	1	1	1
88	1	1	2	120	9	9	*	*
89	1	1	2	110	9	6	6	5
90	1	1	2	107	8	6	*	*
91	1	1	2	125	7	8	*	*
92	1	1	2	117	9	9	*	*
93	1	1	2	126	8	8	5	5
94	1	1	2	98	4	5	*	*
95	1	1	2	111	8	6	*	*
96	1	1	2	110	8	7	*	*
97	1	1	2	120	9	7	*	*
98	1	1	2	114	8	7	6	4
99	1	1	2	117	6	7	*	*
100	1	1	3	105	7	6	*	*
101	1	1	3	97	6	6	*	*
102	1	1	3	86	1	1	*	*
103	1	1	3	111	7	5	*	*
104	1	1	3	93	1	0	*	*
105	1	1	3	115	8	7	*	*
106	1	1	3	102	2	3	5	2
107	1	1	3	111	7	3	4	4
108	1	1	3	82	1	1	*	*
109	1	1	3	117	8	5	4	3
110	0	2	4	132	9	9	*	*
111	0	2	4	113	7	8	*	*
112	0	2	4	130	9	7	1	4
113	0	2	4	122	9	9	6	4
114	0	2	4	133	9	9	*	*
115	0	2	4	103	7	5	3	0
116	0	2	4	118	9	9	*	*
117	0	2	4	119	9	9	7	8
118	0	2	4	97	8	8	6	4
119	0	2	4	123	9	9	7	4

Variable								
1	*2*	*3*	*4*	*5*	*6*	*7*	*8*	*9*
120	0	2	4	113	8	7	6	6
121	0	2	4	110	8	7	3	5
122	0	2	4	119	8	7	6	6
123	0	2	4	116	5	7	*	*
124	0	2	4	113	8	6	5	5
125	0	2	4	128	9	9	*	*
126	0	2	4	113	8	5	4	2
127	0	2	4	110	5	7	*	*
128	0	2	4	114	7	6	5	5
129	0	2	4	132	9	8	4	6
130	0	2	4	110	7	8	2	5
131	0	2	4	122	7	7	4	2
132	0	2	4	123	9	9	6	7
133	0	2	4	131	9	9	7	7
134	0	2	4	131	9	9	8	8
135	0	2	4	121	9	8	7	8
136	0	2	4	125	9	8	*	*
137	0	2	4	101	6	6	4	6
138	0	2	4	120	8	9	6	7
139	0	2	4	99	9	6	*	*
140	0	2	4	128	8	9	8	7
141	0	2	4	129	8	6	5	2
142	0	2	4	125	8	6	7	4
143	0	2	4	107	8	8	8	5
144	0	2	4	102	8	7	6	4
145	0	2	4	125	9	8	*	*
146	1	2	4	129	8	8	*	*
147	1	2	4	122	3	0	2	3
148	1	2	4	124	7	6	6	7
149	1	2	4	115	8	8	*	*
150	1	2	4	117	8	6	5	2
151	1	2	4	132	7	6	5	7
152	1	2	4	109	8	5	5	5
153	1	2	4	107	9	5	9	2
154	1	2	4	116	8	7	6	5
155	1	2	4	118	8	5	6	5
156	1	2	4	124	9	9	6	7
157	1	2	4	102	9	5	5	2
158	1	2	4	110	9	7	7	7
159	1	2	4	119	7	5	2	4
160	1	2	4	99	3	2	4	0
161	1	2	4	102	7	8	5	6
162	1	2	4	115	7	7	*	*
163	1	2	4	105	8	6	3	0
164	1	2	4	104	7	6	*	*

Variable								
1	*2*	*3*	*4*	*5*	*6*	*7*	*8*	*9*
165	1	2	4	112	7	7	*	*
166	1	2	4	117	9	9	6	5
167	1	2	4	108	9	9	*	*
168	1	2	4	135	8	8	8	8
169	1	2	4	133	8	8	7	7
170	1	2	4	105	6	4	5	3
171	1	2	4	124	7	7	9	8
172	1	2	4	112	9	9	9	8
173	1	2	4	128	9	9	9	9
174	1	2	4	96	8	8	7	6
175	1	2	4	110	8	8	4	5
176	1	2	4	108	8	8	6	8
177	1	2	4	125	7	6	8	8
178	1	2	4	111	4	3	4	1
179	1	2	4	103	4	3	2	1

Grades (Also GA, GB)

Scholastic Aptitude Tests (SAT) often are used as a criterion for admission to college, as predictors of college performance, or as indicators for placement in courses. The data below are a sample of SAT scores and freshman-year grade-point averages (GPA) from a northeastern university. (The university wishes to remain anonymous.) The sample of 200 students was broken down randomly into two samples of size 100 for ease of use in this *Handbook*. These two samples are saved as GA and GB. The combined sample is saved as GRADES.

Description of Grades Data

Variable	*Description*
VERB	Score on verbal aptitude test
MATH	Score on mathematical aptitude test
GPA	Grade-point average (0 to 4, with 4 the best grade)

Sample A		
VERB	*MATH*	*GPA*
623	509	2.6
454	471	2.3
643	700	2.4
585	719	3.0
719	710	3.1

Sample B		
VERB	*MATH*	*GPA*
545	643	3.0
558	602	2.3
544	665	2.0
646	573	2.0
655	719	3.8

Sample A			*Sample B*		
VERB	*MATH*	*GPA*	*VERB*	*MATH*	*GPA*
693	643	2.9	585	602	3.4
571	665	3.1	634	515	2.9
646	719	3.3	759	734	2.8
613	693	2.3	532	653	2.5
655	701	3.3	653	668	2.8
662	614	2.6	682	764	2.9
585	557	3.3	641	605	2.1
580	611	2.0	547	602	1.4
648	701	3.0	634	602	2.4
405	611	1.9	609	695	3.0
506	681	2.7	620	773	3.1
669	653	2.0	634	710	3.0
558	500	3.3	585	556	3.4
577	635	2.0	558	656	2.0
487	584	2.3	689	614	2.8
682	629	3.3	780	692	1.3
565	624	2.8	448	645	2.0
552	665	1.7	523	614	2.1
567	724	2.4	571	674	1.6
745	746	3.4	680	490	2.0
610	653	2.8	550	782	2.5
493	605	2.4	544	575	1.4
571	566	1.9	580	677	2.1
682	724	2.5	626	724	2.0
600	677	2.3	617	621	2.0
740	729	3.4	578	609	0.3
593	611	2.8	430	710	2.4
488	683	1.9	662	624	2.6
526	777	3.0	494	561	2.5
630	605	3.7	520	618	2.3
586	653	2.3	760	710	1.1
610	674	2.9	604	700	3.0
695	634	3.3	523	643	2.3
539	601	2.1	484	620	2.0
490	701	1.2	584	567	2.7
509	547	3.3	613	626	3.3
667	753	2.0	696	620	2.0
597	652	3.1	649	621	2.6
662	664	2.6	649	665	3.6
566	664	2.4	578	635	2.9
597	602	2.4	585	710	3.1
604	557	2.3	610	634	2.6
519	529	3.0	641	656	3.1
643	715	2.9	465	683	2.4
606	593	3.4	667	611	2.3

Sample A			Sample B		
VERB	*MATH*	*GPA*	*VERB*	*MATH*	*GPA*
500	661	2.3	578	584	2.1
460	692	1.4	564	575	1.8
717	672	2.8	578	665	3.0
592	441	2.4	539	586	2.5
752	729	3.4	495	748	2.8
695	681	2.5	537	638	2.3
610	777	3.6	558	557	2.3
620	638	2.6	564	593	2.9
682	701	3.6	648	611	3.3
524	700	2.9	673	748	2.6
552	692	2.6	666	621	2.9
703	710	3.8	571	729	3.1
584	738	3.0	487	686	2.1
550	638	2.5	659	575	2.3
659	672	3.5	649	746	3.0
585	605	2.0	675	629	3.0
578	614	3.0	552	662	2.7
533	630	2.0	636	592	2.7
532	586	1.8	580	624	2.8
708	701	2.3	643	583	2.3
537	681	2.1	688	643	2.6
635	647	3.0	620	555	2.7
591	614	3.3	523	737	3.0
552	669	3.0	727	602	3.7
557	674	3.2	502	528	2.3
599	664	2.3	686	800	3.4
540	658	3.3	547	649	3.0
752	737	3.3	481	575	2.0
726	800	3.9	600	621	3.1
630	668	2.1	604	719	2.4
558	567	2.6	573	526	2.5
646	771	2.4	558	576	3.2
643	719	3.3	586	677	2.0
606	755	3.1	597	737	3.6
682	652	3.6	545	692	2.4
565	672	2.9	547	724	3.3
578	629	2.4	601	682	2.9
488	611	1.8	659	649	2.4
361	602	2.4	544	629	2.4
560	639	2.9	507	624	2.3
630	647	3.5	641	764	2.5
666	705	3.4	585	576	2.6
719	668	2.3	630	624	3.4
669	701	2.9	613	677	2.3
571	647	1.8	710	647	3.0

Sample A		
VERB	*MATH*	*GPA*
520	583	2.8
571	593	2.3
539	601	2.5
580	630	2.4
629	695	2.9

Sample B		
VERB	*MATH*	*GPA*
509	538	3.0
480	526	2.4
487	672	2.9
526	796	1.8
532	710	2.1

Furnace

Wisconsin Power and Light studied the effectiveness of two devices for improving the efficiency of gas home-heating systems. The electric vent damper (EVD) reduces heat loss through the chimney when the furnace is in its off cycle by closing off the vent. It is controlled electrically. The thermally activated vent damper (TVD) is the same as the EVD except it is controlled by the thermal properties of a set of bimetal fins set in the vent. Ninety test houses were used, 40 with TVDs and 50 with EVDs. For each house, energy consumption was measured for a period of several weeks with the vent damper active and for a period with the damper not active. This should help show how effective the vent damper is in each house.

Both overall weather conditions and the size of a house can greatly affect energy consumption. A simple formula was used to try to adjust for this. Average energy consumed by the house during one period was recorded as (consumption)/[(weather)(house area)], where consumption is total energy consumption for the period, measured in BTU's, weather is measured in number of degree days, and house area is measured in square feet. In addition, various characteristics of the house, chimney, and furnace were recorded for each house. A few observations were missing and recorded as *, Minitab's missing data code.

Description of Furnace Data

Variable		*Description*
1	TYPE	Type of furnace: 1 = forced air, 2 = gravity, 3 = forced water
2	CH.AREA	Chimney area
3	CH.SHAPE	Chimney shape: 1 = round, 2 = square, 3 = rectangular
4	CH.HT	Chimney height (in feet)
5	CH.LINER	Type of chimney liner: 0 = unlined, 1 = tile, 2 = metal
6	HOUSE	Type of house: 1 = ranch, 2 = two-story, 3 = tri-level, 4 = bi-level, 5 = one and a half stories

Variable		*Description*
7	AGE	House age in years (99 means 99 or more years)
10	DAMPER	Type of damper: 1 = EVD, 2 = TVD
8	BTU.IN	Average energy consumption with vent damper in
9	BTU.OUT	Average energy consumption with vent damper out

Variable									
1	*2*	*3*	*4*	*5*	*6*	*7*	*8*	*9*	*10*
1	28	1	20	2	3	8	7.87	8.25	1
2	144	2	26	0	2	75	9.43	9.66	1
1	80	3	30	1	2	44	7.16	8.33	1
2	100	2	24	0	2	75	8.67	8.82	1
3	168	3	35	1	2	30	12.31	12.06	1
3	28	1	17	2	3	4	9.84	9.67	1
1	64	2	24	1	2	45	16.90	17.51	1
1	64	2	18	1	1	16	10.04	10.79	1
3	96	3	25	1	5	45	12.62	13.59	1
3	108	3	27	1	5	40	7.62	7.99	1
1	64	2	16	1	1	22	11.12	12.64	1
2	63	3	30	1	2	40	13.43	14.42	1
1	42	3	15	1	1	13	9.07	9.25	1
1	117	3	25	0	2	99	6.94	7.79	1
1	64	2	18	1	1	19	10.28	11.29	1
1	28	1	17	2	2	30	9.37	10.26	1
2	64	2	28	0	2	60	7.93	9.46	1
1	64	2	19	1	2	30	13.96	14.77	1
1	28	1	26	2	2	10	6.80	7.21	1
1	80	3	27	0	2	60	4.00	4.29	1
1	28	1	14	2	1	24	8.58	9.81	1
1	28	1	23	2	2	70	8.00	8.41	1
1	64	2	17	1	1	12	5.98	6.78	1
3	*	*	30	*	2	60	15.24	16.30	1
1	64	2	27	0	2	40	8.54	9.01	1
1	64	2	19	1	1	17	11.09	11.41	1
1	50	1	18	1	4	15	11.70	12.37	1
1	50	1	18	1	1	18	12.71	13.28	1
1	50	1	18	1	1	4	6.78	7.24	1
1	28	1	16	2	1	5	9.82	10.55	1
1	80	3	26	0	2	75	12.91	13.89	1
1	50	1	18	1	1	14	10.35	10.72	1
1	28	1	15	2	1	8	9.60	9.22	1
1	100	2	31	0	2	99	9.58	10.61	1
1	28	1	16	2	1	99	9.83	10.04	1
1	50	1	20	1	1	34	9.52	10.20	1

Variable									
1	*2*	*3*	*4*	*5*	*6*	*7*	*8*	*9*	*10*
1	108	3	25	0	2	80	18.26	20.55	1
1	64	2	25	0	2	99	10.64	11.75	1
1	36	2	26	0	2	99	6.62	7.08	1
1	28	1	16	2	4	6	5.20	5.50	1
1	49	2	32	1	2	50	12.28	13.07	2
1	38	1	16	2	1	10	7.23	7.60	2
1	28	1	18	2	3	2	2.97	3.20	2
1	64	2	20	1	2	99	8.81	9.28	2
1	72	3	31	1	2	15	9.27	9.73	2
1	70	3	39	1	2	45	11.29	11.73	2
1	28	1	15	2	1	1	8.29	9.67	2
1	72	3	32	0	2	30	9.96	10.76	2
1	96	3	25	0	2	40	10.30	11.05	2
1	49	2	21	1	2	50	16.06	17.63	2
1	100	2	23	0	5	60	14.24	15.58	2
3	49	2	20	1	1	12	11.43	12.53	2
1	38	1	16	2	1	6	10.28	11.87	2
1	28	1	17	2	1	9	13.60	14.19	2
1	38	1	31	1	2	99	5.94	6.84	2
1	72	3	27	0	2	90	10.36	11.89	2
1	28	1	14	2	1	3	6.85	7.41	2
1	84	3	29	1	2	55	6.72	7.42	2
1	28	1	17	2	1	14	10.21	10.83	2
1	64	2	16	1	1	12	8.61	9.44	2
1	64	2	28	0	2	55	11.62	12.94	2
1	64	2	19	1	1	28	11.21	13.15	2
1	28	1	17	2	1	12	10.95	11.69	2
1	64	2	29	0	2	80	7.62	7.73	2
1	28	1	17	2	1	19	10.40	11.94	2
1	80	3	20	1	1	32	12.92	13.62	2
1	80	3	22	0	1	99	15.12	17.07	2
1	64	2	33	0	2	60	13.47	14.66	2
2	64	2	30	0	2	65	8.47	9.56	2
1	50	1	18	1	4	15	11.70	12.37	2
1	50	1	19	1	2	55	7.73	8.33	2
1	28	1	15	2	4	7	8.37	8.67	2
1	50	1	18	1	4	13	7.29	11.27	2
1	50	1	18	1	1	10	10.49	11.67	2
1	50	1	15	1	2	6	8.69	9.37	2
1	50	1	16	1	1	5	8.26	8.93	2
1	28	1	17	2	1	5	7.69	8.41	2
1	28	1	15	2	1	8	12.19	12.85	2
1	28	1	14	2	4	5	5.56	5.27	2
3	144	2	30	1	1	50	9.76	10.02	2
1	28	1	17	2	1	10	7.15	7.87	2

Variable									
1	*2*	*3*	*4*	*5*	*6*	*7*	*8*	*9*	*10*
1	49	2	18	2	1	14	12.69	11.82	2
1	100	2	27	1	1	40	13.38	14.42	2
1	144	2	22	0	1	70	13.11	13.69	2
1	144	1	30	0	2	85	10.50	10.77	2
2	100	2	24	0	2	70	14.35	15.26	2
2	96	1	17	0	1	40	13.42	14.53	2
1	100	2	20	1	2	99	6.35	6.84	2
1	100	2	20	1	1	14	9.83	10.92	2
1	28	1	28	1	2	55	12.16	13.05	2

Lake

These lakes are all in the Vilas and Oneida counties of northern Wisconsin. Measurements were made in 1959–1963.

Description of Lake Data

Variable	*Description*
AREA	Area of lake in acres
DEPTH	Maximum depth of lake in feet
PH	pH, a measure of acidity (lower pH is more acidic; a pH of 7 is neutral; a higher pH is more alkaline)
WSHED	Watershed area in square miles
HIONS	Concentration of hydrogen ions

Variable				
Area	*Depth*	*PH*	*WSHED*	*HIONS*
55.0	19.0	7.1	.8	.0000000794
26.0	14.0	6.1	.3	.0000007943
1065.0	36.0	7.6	6.3	.0000000251
213.0	71.0	7.6	4.0	.0000000251
1463.0	35.0	8.2	33.0	.0000000063
180.0	24.0	7.1	5.0	.0000000794
433.0	56.0	6.8	2.5	.0000001585
437.0	30.0	7.4	2.0	.0000000398
207.0	34.0	7.4	1.7	.0000000398
98.0	17.0	7.0	1.5	.0000001000
33.0	32.0	6.6	.2	.0000002512
30.0	10.0	6.2	1.0	.0000006310
176.0	17.0	7.3	1.0	.0000000501
55.0	43.0	6.0	.5	.0000010000
96.0	14.0	7.8	1.0	.0000000158

Variable				
Area	*Depth*	*PH*	*WSHED*	*HIONS*
23.0	18.0	6.5	.3	.0000003162
282.0	24.0	7.4	2.0	.0000000398
124.0	17.0	7.0	1.5	.0000001000
22.0	14.0	6.9	.2	.0000001259
223.0	17.0	5.7	1.5	.0000019953
107.0	26.0	6.8	.8	.0000001585
112.0	33.0	7.2	1.2	.0000000631
161.0	25.0	6.4	1.0	.0000003981
301.0	24.0	8.6	88.0	.0000000025
59.0	7.0	7.3	2.0	.0000000501
88.0	13.0	6.0	.7	.0000010000
97.0	50.0	8.6	2.0	.0000000025
126.0	37.0	6.9	1.0	.0000001259
356.0	17.0	7.0	2.0	.0000001000
148.0	50.0	6.8	.8	.0000001585
397.0	26.0	6.9	2.0	.0000001259
89.0	9.0	5.8	1.0	.0000015849
237.0	9.0	6.6	13.0	.0000002512
29.0	33.0	6.2	.2	.0000006310
238.0	19.0	6.1	1.8	.0000007943
189.0	15.0	6.5	2.5	.0000003162
599.0	43.0	8.6	3.0	.0000000025
149.0	11.0	6.9	65.0	.0000001259
34.0	15.0	5.8	2.0	.0000015849
533.0	32.0	7.8	12.0	.0000000158
47.0	65.0	7.1	.3	.0000000794
170.0	11.0	7.5	6.4	.0000000316
113.0	58.0	7.0	1.0	.0000001000
352.0	16.0	8.8	8.0	.0000000016
187.0	36.0	6.4	4.0	.0000003981
48.0	13.0	6.2	.8	.0000006310
76.0	9.0	5.9	.7	.0000012589
52.0	7.0	6.7	.4	.0000001995
175.0	25.0	7.1	.9	.0000000794
191.0	45.0	6.7	3.0	.0000001995
1285.0	60.0	6.4	80.0	.0000003981
124.0	40.0	7.5	.7	.0000000316
53.0	23.0	6.6	.7	.0000002512
125.0	89.0	6.8	1.0	.0000001585
3585.0	39.0	7.4	10.0	.0000000398
211.0	55.0	7.2	.5	.0000000631
372.0	28.0	7.3	1.4	.0000000501
33.0	21.0	6.0	1.0	.0000010000
172.0	33.0	7.2	2.0	.0000000631
716.0	42.0	7.0	8.0	.0000001000
130.0	7.0	7.2	2.0	.0000000631

		Variable		
Area	*Depth*	*PH*	*WSHED*	*HIONS*
610.0	39.0	7.0	3.0	.0000001000
223.0	70.0	7.0	1.5	.0000001000
1352.0	45.0	6.9	17.0	.0000001259
35.0	14.0	6.2	.4	.0000006310
132.0	27.0	7.0	1.0	.0000001000
95.0	33.0	6.1	.9	.0000007943
77.0	23.0	7.6	.6	.0000000251
185.0	31.0	7.1	12.0	.0000000794
97.0	71.0	6.8	1.0	.0000001585
28.0	38.0	7.1	.4	.0000000794

Peru

A study was conducted by some anthropologists to determine the long-term effects of a change in environment on blood pressure. In this study they measured the blood pressure of a number of Indians who had migrated from a very primitive environment, high in the Andes mountains of Peru, into the mainstream of Peruvian society, at a much lower altitude.

A previous study in Africa had suggested that migration from a primitive society to a modern one might increase blood pressure at first, but that the blood pressure would tend to decrease back to normal over time.

The anthropologists also measured the height, weight, and a number of other characteristics of the subjects. A portion of their data is given below. All these data are for males over 21 who were born at a high altitude and whose parents were born at a high altitude. The skin-fold measurements were taken as a general measure of obesity. Systolic and diastolic blood pressure usually are studied separately. Systolic is often a more sensitive indicator.

Description of Peru Data

Variable		*Description*
1	AGE	Age in years
2	YEARS	Years since migration
3	WEIGHT	Weight in kilograms (1 kg = 2.2 lb.)
4	HEIGHT	Height in millimeters (1 mm = 0.039 in.)
5	CHIN	Chin skin fold in millimeters
6	FOREARM	Forearm skin fold in millimeters
7	CALF	Calf skin fold in millimeters
8	PULSE	Pulse rates in beats per minute
9	SYSTOL	Systolic blood pressure
10	DIASTOL	Diastolic blood pressure

Variable									
1	*2*	*3*	*4*	*5*	*6*	*7*	*8*	*9*	*10*
21	1	71.0	1629	8.0	7.0	12.7	88	170	76
22	6	56.5	1569	3.3	5.0	8.0	64	120	60
24	5	56.0	1561	3.3	1.3	4.3	68	125	75
24	1	61.0	1619	3.7	3.0	4.3	52	148	120
25	1	65.0	1566	9.0	12.7	20.7	72	140	78
27	19	62.0	1639	3.0	3.3	5.7	72	106	72
28	5	53.0	1494	7.3	4.7	8.0	64	120	76
28	25	53.0	1568	3.7	4.3	0.0	80	108	62
31	6	65.0	1540	10.3	9.0	10.0	76	124	70
32	13	57.0	1530	5.7	4.0	6.0	60	134	64
33	13	66.5	1622	6.0	5.7	8.3	68	116	76
33	10	59.1	1486	6.7	5.3	10.3	72	114	74
34	15	64.0	1578	3.3	5.3	7.0	88	130	80
35	18	69.5	1645	9.3	5.0	7.0	60	118	68
35	2	64.0	1648	3.0	3.7	6.7	60	138	78
36	12	56.5	1521	3.3	5.0	11.7	72	134	86
36	15	57.0	1547	3.0	3.0	6.0	84	120	70
37	16	55.0	1505	4.3	5.0	7.0	64	120	76
37	17	57.0	1473	6.0	5.3	11.7	72	114	80
38	10	58.0	1538	8.7	6.0	13.0	64	124	64
38	18	59.5	1513	5.3	4.0	7.7	80	114	66
38	11	61.0	1653	4.0	3.3	4.0	76	136	78
38	11	57.0	1566	3.0	3.0	3.0	60	126	72
39	21	57.5	1580	4.0	3.0	5.0	64	124	62
39	24	74.0	1647	7.3	6.3	15.7	64	128	84
39	14	72.0	1620	6.3	7.7	13.3	68	134	92
41	25	62.5	1637	6.0	5.3	8.0	76	112	80
41	32	68.0	1528	10.0	5.0	11.3	60	128	82
41	5	63.4	1647	5.3	4.3	13.7	76	134	92
42	12	68.0	1605	11.0	7.0	10.7	88	128	90
43	25	69.0	1625	5.0	3.0	6.0	72	140	72
43	26	73.0	1615	12.0	4.0	5.7	68	138	74
43	10	64.0	1640	5.7	3.0	7.0	60	118	66
44	19	65.0	1610	8.0	6.7	7.7	74	110	70
44	18	71.0	1572	3.0	4.7	4.3	72	142	84
45	10	60.2	1534	3.0	3.0	3.3	56	134	70
47	1	55.0	1536	3.0	3.0	4.0	64	116	54
50	43	70.0	1630	4.0	6.0	11.7	72	132	90
54	40	87.0	1542	11.3	11.7	11.3	92	152	88

Pulse

Students in an introductory statistics course participated in a simple experiment. The students took their own pulse rate (which is easiest to do by holding the thumb and forefinger of one hand on the pair of arteries on the side of the neck). They then were asked to flip a coin. If their coin came up heads, they were to run in place for one minute. Then everyone took their own pulse again. The pulse rates and some other data are given below.

Description of Pulse Data

Variable		*Description*
1	PULSE1	First pulse rate
2	PULSE2	Second pulse rate
3	RAN	1 = ran in place, 2 = did not run in place
4	SMOKES	1 = smokes regularly, 2 = does not smoke regularly
5	SEX	1 = male, 2 = female
6	HEIGHT	Height in inches
7	WEIGHT	Weight in pounds
8	ACTIVITY	Usual level of physical activity: 1 = slight, 2 = moderate, 3 = a lot

Variable							
1	*2*	*3*	*4*	*5*	*6*	*7*	*8*
64	88	1	2	1	66	140	2
58	70	1	2	1	72	145	2
62	76	1	1	1	73.5	160	3
66	78	1	1	1	73	190	1
64	80	1	2	1	69	155	2
74	84	1	2	1	73	165	1
84	84	1	2	1	72	150	3
68	72	1	2	1	74	190	2
62	75	1	2	1	72	195	2
76	118	1	2	1	71	138	2
90	94	1	1	1	74	160	1
80	96	1	2	1	72	155	2
92	84	1	1	1	70	153	3
68	76	1	2	1	67	145	2
60	76	1	2	1	71	170	3
62	58	1	2	1	72	175	3
66	82	1	1	1	69	175	2
70	72	1	1	1	73	170	3
68	76	1	1	1	74	180	2
72	80	1	2	1	66	135	3
70	106	1	2	1	71	170	2
74	76	1	2	1	70	157	2
66	102	1	2	1	70	130	2
70	94	1	1	1	75	185	2

Variable							
1	*2*	*3*	*4*	*5*	*6*	*7*	*8*
96	140	1	2	2	61	140	2
62	100	1	2	2	66	120	2
78	104	1	1	2	68	130	2
82	100	1	2	2	68	138	2
100	115	1	1	2	63	121	2
68	112	1	2	2	70	125	2
96	116	1	2	2	68	116	2
78	118	1	2	2	69	145	2
88	110	1	1	2	69	150	2
62	98	1	1	2	62.75	112	2
80	128	1	2	2	68	125	2
62	62	2	2	1	74	190	1
60	62	2	2	1	71	155	2
72	74	2	1	1	69	170	2
62	66	2	2	1	70	155	2
76	76	2	2	1	72	215	2
68	66	2	1	1	67	150	2
54	56	2	1	1	69	145	2
74	70	2	2	1	73	155	3
74	74	2	2	1	73	155	2
68	68	2	2	1	71	150	3
72	74	2	1	1	68	155	3
68	64	2	2	1	69.5	150	3
82	84	2	1	1	73	180	2
64	62	2	2	1	75	160	3
58	58	2	2	1	66	135	3
54	50	2	2	1	69	160	2
70	62	2	1	1	66	130	2
62	68	2	1	1	73	155	2
48	54	2	1	1	68	150	0
76	76	2	2	1	74	148	3
88	84	2	2	1	73.5	155	2
70	70	2	2	1	70	150	2
90	88	2	1	1	67	140	2
78	76	2	2	1	72	180	3
70	66	2	1	1	75	190	2
90	90	2	2	1	68	145	1
92	94	2	1	1	69	150	2
60	70	2	1	1	71.5	164	2
72	70	2	2	1	71	140	2
68	68	2	2	1	72	142	3
84	84	2	2	1	69	136	2
74	76	2	2	1	67	123	2
68	66	2	2	1	68	155	2
84	84	2	2	2	66	130	2

Variable							
1	*2*	*3*	*4*	*5*	*6*	*7*	*8*
61	70	2	2	2	65.5	120	2
64	60	2	2	2	66	130	3
94	92	2	1	2	62	131	2
60	66	2	2	2	62	120	2
72	70	2	2	2	63	118	2
58	56	2	2	2	67	125	2
88	74	2	1	2	65	135	2
66	72	2	2	2	66	125	2
84	80	2	2	2	65	118	1
62	66	2	2	2	65	122	3
66	76	2	2	2	65	115	2
80	74	2	2	2	64	102	2
78	78	2	2	2	67	115	2
68	68	2	2	2	69	150	2
72	68	2	2	2	68	110	2
82	80	2	2	2	63	116	1
76	76	2	1	2	62	108	3
87	84	2	2	2	63	95	3
90	92	2	1	2	64	125	1
78	80	2	2	2	68	133	1
68	68	2	2	2	62	110	2
86	84	2	2	2	67	150	3
76	76	2	2	2	61.75	108	2

Restaurant (Saved worksheet is called RESTRNT)

The survey is described in Chapter 4.

Description of Restaurant Data

Variable		*Description*
1	ID	Identification number
2	OUTLOOK	Values 1, 2, 3, 4, 5, 6, 7, denoting from very unfavorable to very favorable
3	SALES	Gross 1979 sales in $1000s
4	NEWCAP	New capital invested in 1979, in $1000s
5	VALUE	Estimated market value of the business, in $1000s
6	COSTGOOD	Cost of goods sold as a percentage of sales
7	WAGES	Wages as a percentage of sales
8	ADS	Advertising as a percentage of sales
9	TYPEFOOD	1 = fast food, 2 = supper club, 3 = other
10	SEATS	Number of seats in dining area
11	OWNER	1 = sole proprietorship, 2 = partnership, 3 = corporation
12	FT.EMPL	Number of full-time employees

Variable	*Description*
13 PT.EMPL	Number of part-time employees
14 SIZE	Size of restaurant: 1 = 1 to 9.5 full-time equivalent employees, 2 = 10 to 20 full-time equivalent employees, 3 = over 20 full-time equivalent employees, where full-time equivalent employees equals (number of full time) + (1/2)(number of part time)

	Variable												
ID	*1*	*2*	*3*	*4*	*5*	*6*	*7*	*8*	*9*	*10*	*11*	*12*	*13*
1	2	480	0	600	35	25	2	2	200	3	8	30	3
2	4	507	22	375	59	20	5	2	150	1	6	25	2
3	5	210	25	275	40	24	3	1	46	1	0	17	1
4	5	246	*	80	43	30	1	1	28	3	2	13	1
5	2	148	*	85	45	35	1	3	44	1	*	*	*
6	3	50	*	135	40	30	10	2	50	3	2	*	*
7	2	72	0	125	85	10	5	2	50	1	0	5	1
8	3	99	7	150	43	25	1	2	130	1	1	8	1
9	4	160	5	85	*	*	*	*	*	2	2	10	1
10	4	243	7	150	38	15	2	2	50	2	2	19	2
11	4	200	3	225	42	22	2	1	64	1	3	12	1
12	4	1000	20	1500	20	20	10	1	240	3	30	40	3
13	4	350	*	*	31	35	*	1	111	3	10	19	2
14	3	550	0	410	50	26	2	2	125	3	6	16	2
15	3	500	10	1000	50	40	10	2	120	1	4	28	2
16	4	1100	8	900	*	*	*	*	*	3	13	47	3
17	3	416	0	400	40	21	4	1	92	3	7	15	2
18	2	650	*	*	63	32	5	1	90	3	20	25	3
19	5	292	0	425	42	13	1	2	150	3	1	16	1
20	3	400	10	350	30	25	5	2	90	1	15	10	2
21	3	42	0	15	64	35	1	3	15	2	0	0	1
22	2	100	15	185	50	15	1	2	80	3	0	7	1
23	4	75	0	160	*	*	*	2	76	1	0	10	1
24	3	180	0	180	50	20	2	2	65	1	1	14	1
25	4	201	0	250	70	27	3	2	178	1	0	20	2
26	6	273	60	300	32	28	10	3	110	3	7	13	2
27	4	150	0	150	*	*	*	2	60	*	51	80	3
28	5	60	4	100	*	*	*	1	0	1	0	0	1
29	4	1200	50	800	35	32	3	3	150	3	35	45	3
30	3	247	4	*	38	*	2	1	60	3	4	8	1
31	2	290	3	200	39	29	1	1	85	3	16	14	3
32	2	58	2	75	45	28	5	2	25	3	2	2	1
33	4	400	0	100	40	35	1	3	85	3	18	15	3
34	3	75	*	26	40	40	5	3	20	2	4	4	1
35	5	*	*	*	32	40	4	2	200	*	*	*	*

	Variable												
ID	*1*	*2*	*3*	*4*	*5*	*6*	*7*	*8*	*9*	*10*	*11*	*12*	*13*
36	4	144	0	25	45	25	0	3	0	1	6	3	1
37	4	65	0	25	48	20	1	1	0	1	2	2	1
38	3	*	*	*	*	*	*	1	210	*	*	*	*
39	4	465	0	75	38	28	7	1	111	3	6	32	3
40	5	*	*	*	50	40	10	3	0	3	10	5	2
41	5	510	3	750	35	29	4	1	152	3	30	25	3
42	3	440	0	*	38	20	5	1	62	3	9	16	2
43	4	608	30	395	31	28	2	3	165	3	30	12	3
44	3	200	3	350	43	21	4	1	68	2	3	18	2
45	1	90	5	40	60	30	10	1	60	1	3	3	1
46	5	45	3	40	40	20	3	3	0	1	0	7	1
47	6	36	1	*	65	25	5	1	0	2	3	0	1
48	4	249	6	275	65	30	5	1	52	1	8	10	2
49	5	200	5	60	35	20	2	3	24	3	4	20	2
50	2	80	0	150	60	30	5	2	70	1	1	4	1
51	1	500	5	350	40	30	3	1	72	3	20	6	3
52	2	125	10	140	50	20	5	2	68	3	0	8	1
53	2	101	0	140	54	13	1	2	58	1	2	3	1
54	4	110	2	160	60	20	1	2	46	1	0	6	1
55	3	1200	0	2500	37	29	3	1	200	3	80	45	3
56	1	*	*	*	33	25	25	3	120	3	25	5	3
57	4	4700	20	1500	50	20	4	1	200	3	15	50	3
58	3	48	0	45	45	25	2	3	10	3	0	12	1
59	2	150	20	150	45	25	5	3	0	3	10	20	2
60	1	185	40	*	40	40	4	1	62	3	2	18	2
61	4	157	2	250	*	*	*	1	99	1	3	8	1
62	6	621	9	0	36	23	3	1	120	3	5	45	3
63	2	257	10	365	40	22	0	1	100	1	14	3	2
64	2	137	0	75	55	30	3	1	0	3	1	3	1
65	1	190	0	400	60	40	0	1	125	1	2	11	1
66	1	*	*	*	*	*	*	*	0	*	*	*	*
67	2	320	6	350	50	20	3	2	96	1	10	10	2
68	5	650	*	*	50	30	1	2	140	3	20	15	3
69	5	610	61	*	38	19	5	1	100	3	10	30	3
70	2	385	4	150	36	29	4	3	48	3	20	28	3
71	5	360	75	325	29	23	3	1	120	3	4	15	2
72	1	276	*	200	65	30	5	1	0	3	20	3	3
73	6	600	20	500	38	22	2	2	125	3	28	5	3
74	2	330	0	100	45	25	2	1	0	1	2	14	1
75	3	215	10	125	45	30	2	1	15	1	11	0	2
76	3	425	15	1750	39	27	12	3	250	1	2	70	3
77	4	250	10	10	40	40	10	2	80	3	10	4	2
78	2	120	0	80	*	*	*	3	30	1	1	2	1
79	3	60	30	45	60	30	10	1	16	2	0	4	1
80	3	141	6	80	85	10	5	1	34	2	0	4	1
81	5	800	50	500	50	25	5	3	120	2	35	13	3

	Variable												
ID	*1*	*2*	*3*	*4*	*5*	*6*	*7*	*8*	*9*	*10*	*11*	*12*	*13*
82	3	207	4	200	48	20	1	1	0	*	*	*	*
83	5	1016	16	1000	40	36	1	3	200	3	20	40	3
84	3	60	0	40	50	30	20	3	80	3	2	4	1
85	3	309	10	500	52	18	2	2	80	1	6	14	2
86	2	960	20	400	54	22	2	3	0	3	7	40	3
87	3	150	*	650	70	20	10	2	220	3	3	18	2
88	4	56	5	125	40	20	10	2	44	1	0	2	1
89	6	250	5	100	33	30	2	1	55	3	10	15	2
90	4	275	10	295	50	25	5	2	85	3	10	3	2
91	5	150	50	300	30	30	0	1	77	3	4	13	2
92	4	325	2	175	45	25	5	3	125	3	20	6	3
93	5	110	5	235	50	30	20	2	65	2	2	10	1
94	3	250	5	230	50	30	4	2	90	3	4	12	2
95	3	550	0	500	48	22	2	2	100	3	13	6	2
96	1	100	3	200	35	25	0	1	50	3	0	6	1
97	3	32	1	42	35	10	2	1	30	1	3	0	1
98	7	366	10	300	42	25	1	2	150	3	12	40	3
99	5	70	3	150	50	7	2	2	50	3	3	2	1
100	3	531	2	450	46	30	1	3	72	3	1	39	3
101	4	225	0	300	50	40	0	1	43	1	10	4	2
102	3	108	5	110	*	*	*	*	*	2	0	8	1
103	3	100	*	*	86	14	0	3	0	1	3	2	1
104	4	40	4	75	30	1	1	3	20	1	*	2	*
105	4	750	0	1000	40	22	5	2	140	2	15	25	3
106	2	312	*	250	40	34	2	3	110	3	6	20	2
107	2	50	5	75	40	20	5	3	56	1	0	5	1
108	5	163	3	115	50	28	5	2	75	3	5	4	1
109	1	75	1	55	*	*	*	3	32	1	0	0	1
110	6	550	6	600	48	24	1	2	76	2	3	30	2
111	1	3450	8	100	*	*	*	2	80	3	3	9	1
112	3	50	4	305	45	9	4	2	60	1	4	6	1
113	5	80	0	50	43	18	2	3	0	1	0	6	1
114	3	435	10	250	30	14	5	3	36	1	1	11	1
115	1	70	2	75	30	35	0	1	0	1	1	2	1
116	1	78	0	125	90	0	10	3	62	1	5	0	1
117	5	210	20	225	80	18	2	1	28	1	0	5	1
118	4	280	*	300	40	16	8	1	50	3	2	18	2
119	4	192	*	300	35	85	5	2	82	3	8	4	2
120	4	116	15	135	50	*	*	1	28	1	2	7	1
121	5	245	0	450	36	24	2	2	100	3	7	20	2
122	1	110	*	160	*	*	*	2	80	1	0	0	1
123	3	229	*	150	35	25	5	1	72	3	7	8	2
124	4	275	*	1100	60	40	0	1	96	3	4	7	1
125	1	100	*	75	50	20	2	3	46	1	4	4	1
126	4	647	10	350	50	30	5	1	90	3	5	12	2

	Variable												
ID	*1*	*2*	*3*	*4*	*5*	*6*	*7*	*8*	*9*	*10*	*11*	*12*	*13*
127	4	300	1	100	*	*	*	*	*	3	*	*	*
128	7	54	0	*	35	20	15	3	70	*	0	0	1
129	5	400	*	300	*	*	*	3	78	3	3	25	2
130	2	120	2	100	84	15	1	2	55	1	1	5	1
131	5	*	*	*	30	15	9	1	40	3	10	15	2
132	1	179	6	70	43	23	2	1	27	1	1	15	1
133	4	300	3	175	35	30	1	1	30	3	7	8	2
134	3	500	125	300	45	78	5	1	125	3	10	22	3
135	2	150	12	210	45	15	2	2	60	1	0	10	1
136	3	135	2	90	60	18	2	1	42	1	3	4	1
137	5	400	4	250	42	35	1	1	36	3	13	3	2
138	1	480	*	450	57	38	0	2	200	2	30	8	3
139	3	530	40	200	40	30	8	2	180	3	25	12	3
140	5	*	*	*	*	*	*	3	20	1	0	2	1
141	5	600	12	90	35	30	5	1	30	3	36	4	3
142	4	*	4	150	42	0	2	3	50	1	7	3	1
143	2	125	1	*	55	20	2	1	35	3	3	8	1
144	1	382	0	190	*	*	*	1	51	3	4	0	1
145	5	*	*	400	30	13	4	1	100	3	2	7	1
146	3	200	10	200	30	30	5	3	50	3	4	12	2
147	5	800	21	750	38	25	3	3	144	3	20	21	3
148	4	144	*	200	40	20	1	2	50	1	2	5	1
149	4	130	1	150	0	40	1	2	60	3	3	9	1
150	2	1010	50	*	50	25	3	2	127	3	25	35	3
151	5	60	5	150	*	*	*	3	25	1	0	2	1
152	4	292	20	100	49	30	8	2	75	2	6	24	2
153	3	100	56	*	45	25	5	2	75	3	6	14	2
154	3	98	0	70	70	28	2	1	32	1	0	7	1
155	2	250	6	250	50	25	1	2	90	3	7	7	2
156	4	172	1	200	35	14	3	1	0	1	0	20	2
157	3	145	12	155	*	*	*	2	74	1	1	9	1
158	4	*	*	*	*	*	*	2	0	3	*	0	*
159	1	*	*	*	*	*	*	*	*	*	*	*	*
160	4	*	*	*	*	*	*	1	0	3	3	1	1
161	3	37	1	20	45	10	1	3	12	1	1	1	1
162	3	*	*	*	*	*	3	1	82	2	5	22	2
163	4	77	0	150	*	*	*	1	35	2	4	1	1
164	3	400	*	400	*	*	*	3	44	3	6	8	2
165	4	1000	20	*	40	34	2	2	*	2	*	*	*
166	2	250	15	750	40	20	5	2	95	3	26	13	3
167	4	50	*	90	*	*	*	3	24	1	0	0	1
168	1	120	6	*	80	15	5	1	70	1	1	6	1
169	4	750	78	0	30	32	3	3	94	2	40	6	3
170	5	190	8	75	42	31	1	3	60	3	6	3	1
171	1	140	5	180	40	5	5	2	87	1	3	3	1

	Variable												
ID	*1*	*2*	*3*	*4*	*5*	*6*	*7*	*8*	*9*	*10*	*11*	*12*	*13*
172	4	80	52	60	36	23	1	2	100	1	0	8	1
173	5	55	0	50	50	20	10	3	40	3	0	6	1
174	5	690	0	250	45	21	3	3	196	3	8	35	3
175	4	200	1	175	49	19	2	1	100	1	0	18	1
176	4	28	2	55	33	0	1	3	34	3	0	0	1
177	5	40	1	0	*	*	*	3	24	3	1	3	1
178	4	2	0	2	30	5	0	1	10	3	0	2	1
179	2	217	0	750	51	29	4	2	95	3	3	25	2
180	4	250	0	300	40	30	10	1	20	3	10	10	2
181	3	990	*	1500	40	29	5	1	175	3	12	43	3
182	4	2	2	*	90	10	0	1	0	2	0	4	1
183	3	50	20	325	50	30	20	2	75	3	2	3	1
184	7	290	150	450	51	59	3	2	110	1	5	14	2
185	1	75	10	140	*	*	*	1	0	1	0	0	1
186	1	*	*	*	*	*	*	*	*	1	0	0	1
187	5	400	10	300	20	25	5	2	85	3	10	10	2
188	3	30	0	60	49	30	1	1	0	2	2	4	1
189	2	70	0	32	40	30	1	3	65	1	2	3	1
190	2	250	0	0	37	14	4	1	16	3	3	13	1
191	5	1600	20	1000	34	32	4	1	52	3	20	55	3
192	2	290	0	125	60	35	5	1	70	3	6	2	1
193	4	203	2	40	39	31	4	1	0	1	8	4	2
194	5	*	*	*	60	30	0	1	16	1	0	3	1
195	2	100	5	300	60	0	1	2	50	2	0	0	1
196	3	551	0	1500	*	*	*	1	100	3	10	25	3
197	3	220	10	70	42	40	4	1	85	3	6	7	1
198	3	225	10	550	50	15	5	1	200	3	1	15	1
199	3	140	14	175	33	10	5	1	0	1	0	4	1
200	1	154	0	20	45	28	0	*	80	3	6	5	1
201	4	39	0	65	42	42	0	1	75	1	0	30	2
202	4	565	0	500	*	*	*	2	85	3	15	35	3
203	3	0	2	0	43	15	2	2	75	1	1	4	1
204	2	1096	73	2000	34	29	1	1	142	3	42	30	3
205	4	35	1	0	90	10	0	1	60	1	4	0	1
206	5	53	20	125	40	20	5	1	30	3	5	7	1
207	2	390	8	450	*	*	*	*	*	3	30	16	3
208	3	*	*	*	32	34	2	1	104	3	2	16	2
209	4	*	*	80	*	*	*	3	45	1	1	9	1
210	5	500	25	450	45	35	1	2	132	3	18	20	3
211	4	180	5	300	58	40	2	3	30	3	8	3	1
212	6	89	4	120	45	15	2	1	0	1	1	2	1
213	1	77	10	175	35	10	1	3	0	1	0	5	1
214	1	460	3	75	40	23	6	1	94	3	2	35	2
215	3	440	35	1000	38	39	3	3	110	3	40	30	3
216	5	56	8	125	40	33	2	3	0	1	12	0	2

	Variable												
ID	*1*	*2*	*3*	*4*	*5*	*6*	*7*	*8*	*9*	*10*	*11*	*12*	*13*
217	4	15	23	30	52	46	2	1	0	1	0	0	1
218	5	150	*	150	*	*	*	*	*	2	0	7	1
219	5	8064	300	12000	37	31	3	1	550	3	250	60	3
220	3	200	*	*	20	20	*	1	20	3	1	8	1
221	3	30	*	350	50	20	5	1	80	3	0	4	1
222	5	71	0	185	40	8	0	3	0	1	0	3	1
223	4	11	0	0	99	0	0	1	4	1	0	0	1
224	1	267	2	125	40	25	5	1	44	3	2	13	1
225	7	325	10	400	46	25	3	2	70	1	3	9	1
226	1	155	0	85	35	35	5	3	70	1	10	3	2
227	4	1000	100	5000	35	20	7	1	180	3	30	20	3
228	3	85	30	45	45	25	2	1	54	1	1	6	1
229	3	250	50	1000	35	35	0	3	150	3	25	30	3
230	2	30	5	40	30	30	3	3	55	1	1	4	1
231	2	20	1	0	45	20	2	3	0	3	0	3	1
232	*	*	*	*	40	10	5	1	40	*	*	*	*
233	2	125	5	125	50	30	10	1	65	1	0	6	1
234	4	720	13	650	37	24	6	1	150	3	6	25	2
235	4	*	*	*	40	30	5	3	150	3	25	100	3
236	1	240	3	225	50	20	2	2	30	1	5	6	1
237	4	10	*	10	50	38	2	3	35	2	3	10	1
238	5	240	0	125	45	25	1	1	0	1	7	3	1
239	6	59	*	*	*	*	*	3	0	2	2	4	1
240	3	1080	20	1000	32	30	5	3	170	3	40	50	3
241	1	225	1	150	34	22	4	3	120	3	18	5	3
242	3	*	*	*	*	*	*	*	*	3	4	20	2
243	2	70	7	225	25	35	10	3	43	1	4	2	1
244	1	430	35	500	42	26	2	2	0	3	6	30	3
245	4	198	*	130	45	21	1	1	62	1	5	8	1
246	5	65	12	150	35	30	2	1	35	1	10	4	2
247	2	69	3	18	44	26	1	3	43	1	0	9	1
248	4	230	0	0	35	36	3	3	150	3	20	12	3
249	4	250	25	850	40	15	1	1	40	3	0	*	*
250	5	140	80	140	*	*	*	*	*	1	2	12	1
251	5	180	5	150	40	25	5	1	130	3	2	6	1
252	3	60	7	*	40	10	2	1	18	1	1	0	1
253	1	80	0	150	51	18	0	1	0	1	1	2	1
254	1	42	0	75	65	25	10	1	36	1	0	3	1
255	4	8	0	14	25	15	0	3	0	1	*	2	*
256	3	210	*	350	*	*	*	2	100	1	0	16	1
257	3	95	0	70	45	25	2	3	42	1	4	1	1
258	4	55	50	89	65	30	5	3	32	3	0	2	1
259	5	121	1	160	55	30	5	1	30	1	3	2	1
260	3	75	10	80	40	26	5	3	26	1	2	2	1
261	1	*	*	*	45	21	4	3	205	3	8	32	3

	Variable												
ID	*1*	*2*	*3*	*4*	*5*	*6*	*7*	*8*	*9*	*10*	*11*	*12*	*13*
262	6	250	10	300	38	29	4	1	50	3	5	20	2
263	4	220	10	350	55	25	10	2	70	1	0	15	1
264	4	120	10	80	35	60	2	1	80	1	2	22	2
265	2	25	1	40	40	10	6	1	0	*	*	*	*
266	1	500	0	175	75	20	5	2	200	3	32	1	3
267	3	*	*	475	45	20	10	2	80	1	2	15	1
268	3	200	10	70	45	20	25	1	70	3	0	10	1
269	4	250	3	5	35	50	0	3	15	2	1	3	1
270	5	215	1	100	36	33	2	3	98	3	5	17	2
271	4	*	*	*	*	*	*	2	36	2	7	6	2
272	3	733	35	500	53	21	0	1	0	1	6	40	3
273	1	*	*	*	*	*	*	3	0	1	0	0	1
274	1	200	1	210	50	20	5	2	70	*	*	*	*
275	5	305	0	450	58	27	2	2	85	3	3	25	2
276	1	110	5	175	*	*	*	2	99	1	0	7	1
277	2	*	*	100	*	*	*	3	45	1	3	6	1
278	3	100	20	250	24	30	10	3	100	3	0	7	1
279	4	355	*	95	40	20	5	1	130	3	8	12	2

Trees

People in forestry need to be able to estimate the amount of timber in a given area of a forest. Therefore, they need a quick and easy way to determine the volume of any given tree. Of course, it is difficult to measure the volume of a tree directly. But it is not too difficult to measure the height, and even easier to measure the diameter. Thus, the forester would like to develop an equation or table that makes it easy to estimate the volume of a tree from its diameter and/or height. A sample of trees of various diameters and heights were cut, and the diameter, height, and volume recorded. Below are the results of one such sample. This sample is for black cherry trees in Allegheny National Forest, Pennsylvania. (Of course, different varieties of trees and different locations will yield different results. So, separate tables are prepared for each species and each location.)

Description of Tree Data

Variable	*Description*
DIAMETER	Diameter in inches at 4.5 feet above ground level
HEIGHT	Height of tree in feet
VOLUME	Volume of tree in cubic feet

	Variables	
DIAMETER	*HEIGHT*	*VOLUME*
8.3	70	10.3
8.6	65	10.3
8.8	63	10.2
10.5	72	16.4
10.7	81	18.8
10.8	83	19.7
11.0	66	15.6
11.0	75	18.2
11.1	80	22.6
11.2	75	19.9
11.3	79	24.2
11.4	76	21.0
11.4	76	21.4
11.7	69	21.3
12.0	75	19.1
12.9	74	22.2
12.9	85	33.8
13.3	86	27.4
13.7	71	25.7
13.8	64	24.9
14.0	78	34.5
14.2	80	31.7
14.5	74	36.3
16.0	72	38.3
16.3	77	42.6
17.3	81	55.4
17.5	82	55.7
17.9	80	58.3
18.0	80	51.5
18.0	80	51.0
20.6	87	77.0

APPENDIX B
Additional Features in Minitab

This appendix gives additional information on some commands that already have been described and introduces a few new commands.

B.1 Input and Output of Data in Computer Files

READ and SET

On pages 11–13, we showed how READ and SET can be used when data are typed as part of a Minitab program. READ and SET can also input data that previously have been stored in a computer file. As an example, suppose the data in Exhibit B.1 had been stored in a file named PLANT. The following commands input the data and assign names.

```
READ 'PLANT' INTO C10, C1-C5
NAME C1='MON' C2='TUES' C3='WED' C4='THURS' &
     C5='FRI' C10='LINE'
```

Exhibit B.1 Example to Illustrate Input and Output Commands

Production	*Yield for Each Day*				
Line	*Monday*	*Tuesday*	*Wednesday*	*Thursday*	*Friday*
1	25	29	30	26	22
2	32	36	35	33	28
3	21	24	23	21	20
4	17	18	16	17	17

READ data from 'FILENAME' into C, ..., C

This form of READ inputs data from a computer file. The name of the file must be enclosed in single quotes (apostrophes). The computer file that READ inputs could have been created by your computer's editor, by Minitab's WRITE command, or by another program that outputs standard data files. The file must be in a form suitable for READ (see p. 12). *Note:* You should not give a column of your worksheet the same name as a file you plan to use.

SET data from 'FILENAME' into C

This form of SET inputs data from a computer file. The file name must be enclosed in single quotes. The computer file that SET inputs could have been created by your computer's editor or by Minitab's WRITE command. The file must be in a form suitable for a SET command (see p. 13). *Note:* You should not give a column of your worksheet the same name as a file you plan to use.

WRITE to a File

In some cases you may want to put data from the worksheet into a computer file for later use with Minitab or some other program. This can be done with WRITE. For example,

```
WRITE 'YIELD' C1-C5
```

puts the data that are in C1–C5 into a computer file named YIELD. If we used this instruction after the commands on page 330, the file YIELD would look like this:

25	29	30	26	22
32	36	36	33	28
21	24	24	21	20
17	18	18	17	17

Note: The column headers that occur with PRINT are not used with WRITE. You can use READ to input this file. You also can use your computer's editor to print or modify the file.

WRITE data into 'FILENAME' from C, ..., C

WRITE creates a computer file from columns of the worksheet. The columns must all be the same length. The file name must be enclosed in single quotes.

Usually one line is put in the data file for each row of the worksheet. The numbers on a line are separated by blanks. If you WRITE many columns, they may not all fit on one line. Minitab then uses two or more lines per row and the continuation symbol, & (see p. 10). If you WRITE just one column, as many numbers as will fit are put on each line.

If you WRITE one column, you can use SET to input the resulting file. If you WRITE more than one column, you can use READ to input the resulting file. You may use your computer's editor or other computer programs to read, print, or modify files created by WRITE.

B.2 Entering Patterned Data with SET and INSERT

The SET command has a feature that makes it easy to input certain kinds of patterned data. For example, you can put the integers from 1 to 12 into C2 by typing

```
SET C2
  1:12
END
```

In general, any list of consecutive integers can be abbreviated with a colon. For example,

```
SET C1
37:40, 6:4, -3:2
END
```

puts 37, 38, 39, 40, 6, 5, 4, −3, −2, −1, 0, 1, 2, into C1.

A slash allows you to abbreviate other types of patterns. For example,

```
SET C2
10:20/2
END
```

puts 10, 12, 14, 16, 18, 20 into C2 and

```
SET C3
2:1.4/.1
END
```

puts 2, 1.9, 1.8, 1.7, 1.6, 1.5, 1.4 into C3. In general, you put the distance between consecutive numbers after the slash.

Parentheses can be used to repeat a list of numbers. For example, suppose C1 contains the blood pressure for 27 people. Suppose the first 15 are women and the last 12 are men. You could enter the sex of each person, coded 1 for men and 2 for women, by typing

```
SET C2
15(2)  12(1)
END
```

The number in front of the open parenthesis is a repeat factor. The 15 says to repeat the number inside the parentheses 15 times. *Note:* There must not be any blanks (or other symbols) between the repeat factor and the parenthesis. Thus, the 15 must be next to the open parenthesis. You may, however, put blanks and commas inside the parentheses.

In general, any list of numbers may be repeated. Just enclose the list in parentheses and put the repeat factor in front. For example,

```
SET C1
2(0:2, 10)
END
```

puts 0, 1, 2, 10, 0, 1, 2, 10 into C1.

If you put the repeat factor after the parentheses, each number is repeated individually. For example,

```
SET C1
(0,1)4
END
```

puts 0, 0, 0, 0, 1, 1, 1, 1, into C1. Again, do not leave any blanks between the right parenthesis and the repeat factor.

You can use a repeat factor both before and after the parentheses. For example,

```
SET C1
2(0,1)3
END
```

Here, each number inside the parentheses is repeated three times to give the list 0, 0, 0, 1, 1, 1. Then this list is repeated twice. Thus, C1 will contain 0, 0, 0, 1, 1, 1, 0, 0, 0, 1, 1, 1.

If you use INSERT to enter data into just one column, you can use all the patterned data features of SET.

B.3 Calculations with LET

The LET command makes it easy to perform relatively complicated calculations. In most data analysis, however, you will need just a few of the simpler forms of this command.

Unlike most other Minitab commands, no extra text may be used on a LET line (unless you use the line terminator, #; see p. 343).

Arithmetic. The simplest form of LET is

```
LET C = arithmetic expression
```

or

```
LET K = arithmetic expression
```

The arithmetic expression uses the symbols + for add, − for subtract, * for multiply, / for divide, and ** for raise to a power (exponentiation). Parentheses may be used for grouping. (This was discussed in Section

1.3.) In addition, you can use column statistics, such as MEAN and MAX in LET. (This was described in Section 2.5.)

You can access an individual number in a column by using a subscript. For example, suppose C2 contains the five numbers 13, 12, 10, 14, 12. Then

```
LET K2 = C2(2)
```

takes the second number, 12, from C2 and puts it into K2. You can also store a number in one spot in a column. For example,

```
LET C2(3) = 17
```

puts 17 into the third row of C2. In Section 1.3, we showed how this form of LET could be used to correct numbers in the worksheet.

Any expression which evaluates to an integer may be used as a subscript. For example, (COUNT (C2) − 1) is 5 − 1, or 4, so

```
LET C2(COUNT(C2)-1) = 15
```

puts the value 15 into row 4 of C2.

Subscripted columns can be part of a more complicated expression. For example,

```
LET C7 = (C2-MIN(C2))*C2(1)/C2(COUNT(C2)-3)
```

subtracts 10 from each number in C2, then multiplies by 13, and divides by 12.

Functions. LET can use the functions described below. In all cases the input is either a single column or a constant.

ABSOLUTE	Computes the absolute value
SIGNS	Returns −1, 0, or 1 for negative, zero, or positive values, respectively
SQRT	Computes the square root
ROUND	Rounds to the nearest integer
LOGE	Computes the logarithm to base e
LOGTEN	Computes the logarithm to base 10
EXPONENTIAL	Computes the value of e^x
ANTILOG	Computes the value of 10^x
SIN	Computes the sine of the angle given in radians
COS, TAN	Cosine or tangent, angle in radians
ASIN	Inverse sine or arcsine; the answer is in radians
ACOS, ATAN	Inverse cosine and inverse tangent

RANKS	Assigns 1 to the smallest number in the column, 2 to the second smallest, 3 to the third smallest, etc; the average rank is assigned when there are ties
SORT	Orders the numbers in a column, putting the smallest on top
NSCORES	Calculates normal scores (see p. 179)
PARSUMS	Calculates partial sums; for example, if C1 contains 2, 4, 1, 9, then LET C2 = PARSUMS(C1) puts 2, 2 + 4 = 6, 2 + 4 + 1 = 7, 2 + 4 + 1 + 9 = 16 into C2
PARPRODUCTS	Calculates partial products
LAG	Moves all numbers down one row, puts * (the missing data code, see p. 342) in the first row, and omits the number in the last row; for example, if C1 contains 2, 4, 1, 9, then LET C2 = LAG(C1) puts *, 2, 4, 1 into C2

Here are some examples using these functions. Suppose C1 contains the five numbers 4, −3.6, 0, −8, 4.2. Then after

```
LET C2 = ABSO(C1)
LET C3 = SIGNS(C1)
LET C4 = ROUND(C1)
LET C5 = SORT(C1)
LET K1 = SQRT(9)
```

the worksheet is

C1	*C2*	*C3*	*C4*	*C5*	*K1*
4	4	1	4	−8	3
−3.6	3.6	−1	−4	−3.6	
0	0	0	0	0	
−8	8	−1	−8	4	
4.2	4.2	1	4	4.2	

B.4 The COPY Command

The COPY command has two functions: It allows you to copy data from one place in the worksheet to another and it allows you to select a subset of data. For example, it can be used to select the data for just the women in a study.

COPY C, ..., C into C, ..., C
COPY K, ..., K into K, ..., K
COPY C into K, ..., K
COPY K, ..., K into C

The copy command allows you to make a second copy of data and to move data from columns to constants, and vice versa.

Selecting Subsets of Data

COPY has two subcommands, USE and OMIT. We will explain these with a simple example. Suppose we've entered sex (coded 1 = male, 2 = female), height, and weight for seven people into C1–C3, which we have named HT, WT, SEX. Then the following commands put the data for women into C12–C13.

```
COPY C2 C3 INTO C12 C13;
  USE 'SEX' = 2.
NAME C12 = 'HT.F', C13 = 'WT.F'
```

All rows where SEX = 2 are copied into C12 and C13. Now the worksheet contains:

C1	*C2*	*C3*	*C12*	*C13*
SEX	*HT*	*WT*	*HT.F*	*WT.F*
2	66	130	66	130
1	70	155	64	125
2	64	125	65	115
2	65	115	63	108
2	63	108		
1	66	145		
1	69	160		

In general, you may specify any list of values or range of values for the variable on USE. For example,

```
COPY C2 C3 INTO C22 C23;
  USE 'HT' = 64, 65, 66.
```

copies all rows that have any of the values 64, 65, or 66 for HT. You can use a colon to indicate a range of values. For example,

```
COPY C2 C3 INTO C22 C23;
  USE 'HT' = 0:64.
```

copies the data for all people who are 64 inches or shorter.

You also can subset by giving row numbers. For example,

```
COPY C2 C3 INTO C22 C23;
  USE 1, 3:5.
```

copies rows 1, 3, 4, 5.

The OMIT subcommand is the opposite of USE. USE says which rows to copy, OMIT says which rows not to copy. For example,

```
COPY C2 C3 INTO C22 C23;
  OMIT SEX = 1.
```

again copies just the women. You also can use row numbers with OMIT. For example, to copy all but rows 2, 6, and 7, type

```
COPY C2 C3 INTO C22 C23;
  OMIT 2, 6, 7.
```

COPY C, ..., C into C, ..., C

USE ROWS
USE C = VALS
OMIT ROWS
OMIT C = VALS

These subcommands copy selected rows. The USE subcommand says which rows to copy; OMIT says which rows not to copy. These rows can be specified by their row numbers or by the values in a column.

ROWS can be any list of row numbers. Consecutive numbers can be abbreviated with a colon. Thus, USE 2, 5:8 is the same subcommand as USE 2, 5, 6, 7, 8.

VALS can be any list of values. An interval can be abbreviated with a colon. Thus, USE C1 = 5:8, 20 says to copy all rows where C1 is a value in the range of 5 through 8, or equal to 20.

B.5 How to STACK and UNSTACK Data

In Section 5.2 (p. 95) we introduced the concept of stacked versus unstacked data. Here we will show how to change from one form to the other.

Suppose we have height and weight for two groups of people, men and women. In Exhibit B.2(a) the data for men are in one pair of columns and the data for women are in a second pair. In Exhibit B.2(b) the data

Exhibit B.2 Illustrations of Stacked and Unstacked Data

(a) Unstacked Data

C1	*C2*	*C3*	*C4*
66	130	70	155
64	125	66	145
65	115	69	160
63	108		

(b) Stacked Data

C5	*C6*	*SEX*
70	155	1
66	145	1
69	160	1
66	130	2
64	125	2
65	115	2
63	108	2

for men and women are all stacked together. The column SEX is used to indicate which rows are for males (SEX = 1) and which rows are for females (SEX = 2).

Sometimes we need to change data from one format to the other. The STACK and UNSTACK commands make this relatively easy. For example, to stack the data in Exhibit B.2(a) so they look like the data in part (b), use the command

```
STACK (C3, C4) (C1, C2) into C5, C6;
   SUBSCRIPTS 'SEX'.
```

Similarly, to unstack the data in Exhibit B.2(b) so they look like the data in part (a), use the command

```
UNSTACK (C5, C6) into (C3, C4) (C1, C2);
   SUBSCRIPTS 'SEX'.
```

STACK also can be used with numbers and stored constants. For example,

```
STACK C1, 61, 64, C3, 66 into C20
```

creates one column which contains 10 numbers, 66, 64, 65, 63, 61, 64, 70, 66, 69, 66. Note that we omitted the parentheses in this command. We can omit the parentheses if everything is stacked into one column.

STACK (E, ..., E), ..., (E, ..., E) store in (C, ..., C)

This command stacks blocks of columns and constants on top of each other. In general, each block must be enclosed in parentheses. However, if each block contains just one argument, the parentheses may be omitted. If you want to create a column of subscripts, use the subcommand

SUBSCRIPTS into C

In this case, all rows in the first block will be given subscripts of 1, all rows in the second block subscripts of 2, and so on.

UNSTACK (C, ..., C) into (E, ..., E), ..., (E, ..., E)

SUBSCRIPTS are in C

This command separates one block of columns into several blocks of columns and/or stored constants. In general, each block must be enclosed in parentheses. However, if each block contains just one argument, the parentheses may be omitted.

For most applications, the subcommand SUBSCRIPTS will be needed. The rows with the smallest subscript are stored in the first block, the rows with the second smallest subscript stored in the second block, and so on. If you do not use SUBSCRIPTS, each row is stored in a separate block. The numbers in the subscript column must be integers between −10000 and 10000. They need not be in order; nor do they need to be consecutive integers.

B.6 Changing Values with CODE

The CODE command changes values in columns. For example,

```
CODE (-2,-1) to 0 in C1 C2, put in C11 C12
```

changes the worksheet as follows:

C1	*C2*
8	16
−2	13
6	−2
0	−2
−1	14

C11	*C12*
8	16
0	13
6	0
0	0
0	14

Every −2 and −1 in C1 and C2 is changed into 0. All other values are kept the same.

You can make many changes at once and you can specify a range with a colon. For example,

```
CODE (30:55)1 (56:80)2 (81:120)3 C1-C5, C11-C15
```

changes all values from 30 up through 55 into 1, values from 56 up through 80 into 2, and values from 81 up through 120 into 3. These changes are made on columns C1–C5 and the results are stored in C11–C15. The original columns, C1–C5, are not changed.

CODE (VALS) to K, ..., (VALS) to K for C ... C put in C ... C

This command changes the specified values. All other values in the columns are left as is. VALS can be any list of values. In addition, an interval can be abbreviated with a colon. For example, CODE (1:1.5, 2)0 changes all values in the range 1 through 1.5, and the value 2 into zeros. VALS must be enclosed in parentheses.

If two instances of VALS overlap, the last one is used. For example, CODE (10:20)1 (20:30)2 changes the value 20 into a 2.

CODE can change values into the missing value code, *. For example, to change all occurrences of −99 to missing values, use

```
CODE (-99) '*' C1-C10, put in C1-C10
```

B.7 How to SORT and RANK Data

The SORT Command

Here is a simple example.

```
SORT C1, carry along C2-C4, put into C11-C14
```

changes the worksheet as follows:

C1	*C2*	*C3*	*C4*
10	0.2	31	131
10	0.1	35	210
12	0.1	37	176
12	0.1	36	190
10	0.4	31	140
12	0.1	30	110
11	0.2	29	180
11	0.1	33	182

C11	*C12*	*C13*	*C14*
10	0.2	31	131
10	0.1	35	210
10	0.4	31	140
11	0.2	29	180
11	0.1	33	182
12	0.1	37	176
12	0.1	36	190
12	0.1	30	110

The first column listed, C1, determined the order. C2–C4 were carried along as the ordering was done. The smallest value in C1 is 10. So, all rows where C1 = 10 were put first. The second smallest value in C1 is 11. So the two rows where C1 = 11 were stored next, and so on.

The RANK Command

The RANK command does not reorder the data. It creates a new column containing the rank of each value in the original column. The smallest value is given rank 1, the second smallest rank 2, the third smallest rank 3, and so on. For example,

```
RANK C1 into C11
```

changes the worksheet as follows:

C1	*C11*
1.4	3
1.1	1
2.0	4.5
3.1	6
2.0	4.5
1.3	2

Ranks for C1 were put into C11. The smallest number in C1 is 1.1, so a 1 was put into C11. The second smallest number in C1 is 1.3 and so a 2 was put into C11, and so on. If there are ties, the average rank is assigned to each value. For example, the fourth and fifth values in C1 are both 2.0. They both were given the rank of (4 + 5)/2 = 4.5.

SORT C, carry along C, ..., C store in C, ..., C

Sorts the first column into increasing order. The next group of columns are just carried along as the sorting is done.

RANK C into C

Ranks are determined as follows: The number 1 is put next to the smallest value in the column, the number 2 next to the second smallest, and so on. If several entries in a column are equal, they are all given the average rank.

B.8 Missing Data

Section 1.5 introduced Minitab's missing data code, *. Here we will give some additional information.

The Missing Data Code on a Command Line

In general, an * may be used on a command or subcommand line in place of a number whenever it makes sense. On command and subcommand lines, the * must be enclosed in single quotes (apostrophes). For example,

```
COPY C2 INTO C12;
  OMIT C2 = '*'.
```

copies all rows that do not have an * in C2. The statement

```
LET C1(5) = '*'
```

changes the value in row 5 of C1 to *.

Another use of * is to code values that already have been entered with some other missing data code. For example, if missing data have been coded as −99, then the CODE command (see p. 339) can be used to change −99 to *:

```
CODE (-99) TO '*' IN C1-C10, PUT BACK IN C1-C10
```

Calculations that Result in Missing Values

Suppose C1 contains the three values 2, 0, 3 and we use

```
LET C2 = 12/C1
```

Then C2 will contain 6, *, 4. The * was put in row two, since 12/0 is undefined and the answer therefore is missing. There are a number of algebraic calculations that are undefined and result in *, for example, $\sqrt{-5}$ and log (-3).

How Minitab Commands Treat Missing Data

The arithmetic operations (+, −, *, /, and **) and functions, such as SQRT and LOGE, set the answer to missing if any input value is missing. The column summaries such as SUM and MEAN, and the row summaries such as RSUM and RMEAN, just omit all missing values from the calculations. For example, the commands

```
LET C4 = C1 + C2
LET C5 = C1 + C2 + C3
```

```
LET C6 = SQRT(C2)
RSUM C1-C3 PUT INTO C7
RMEAN C1-C3 PUT INTO C8
LET K1 = SUM(C3)
LET K2 = MEAN(C3)
```

change the worksheet as follows:

C1	C2	C3
2	4	3
5	*	7
5	9	*

C1	C2	C3	C4	C5	C6	C7	C8	K1	K2
2	4	3	6	9	2	9	3	10	5
5	*	7	*	*	*	12	6		
5	9	*	14	*	3	14	7		

Most other commands simply omit missing values from all calculations the way SUM and RSUM do. For example, suppose C1 and C2 both contain 50 observations but C2 has *s in rows 2 and 34. Then HISTOGRAM C1 C2 would produce one histogram of the 50 values in C1 and one histogram of the 48 nonmissing values in C2. PLOT C1 C2 would plot the 48 pairs in which both values are nonmissing.

The command SORT (see p. 340) involves ordering data. This command treats * as if it were the largest number in the column.

B.9 Some Miscellaneous Features

How to Annotate a Program

There are two ways to annotate a Minitab program: with the NOTE command and with the symbol #. If you type NOTE as the first word on a line, then everything after it is ignored by Minitab. You can type any words or symbols you wish. The symbol # may be typed anywhere on a line. All text after the # is ignored by Minitab. For example,

```
#Now we will analyze the data for men.
LET C4 = C1-MEAN(C1)    #centered data
```

There is one difference between a NOTE command and a line beginning with #. In a stored command file, the NOECHO command (see p. 350) tells Minitab not to print any commands except NOTE commands. Thus in this case, # lines are not printed, but the text following the word NOTE is printed.

Size of Input and Output

IW = K spaces

IW specifies the input width. The first K spaces on each line are read by Minitab. The rest of the line is ignored. The value of K can be from 10 to 160. The default value depends on the brand of computer, but is at least 80 on most computers.

OW = K spaces
OH = K lines

OW specifies the output width from the commands PRINT, TABLE, CORRELATION, AOVONEWAY, and ONEWAY. Output from all other commands is always 65 or fewer spaces. The value of K may be from 65 to 132 spaces. The default value of K is 79 (120 in batch mode).

OH specifies the output height, that is the number of lines on a page or a screen. When you use Minitab interactively, Minitab prints a page of output, then asks CONTINUE? If you type YES or Y or push RETURN, the next page is printed. If you type NO or N, no more output from this command is printed. All storage, however, is done.

When you use Minitab in batch mode or put output into a file with the OUTFILE command, the output is divided into pages of K lines each. If you do not want Minitab to divide your output into pages, use OH = 0.

The default value of K is 24 on most computers using terminals with screens. (It is 60 in batch mode or when output is to PAPER or a file.)

Putting Output in a Computer File

The OUTFILE command puts all Minitab output that follows into a computer file. You may then print the file or use your editor to edit it. (This command is similar to the command PAPER, described on p. 6.)

```
READ 'STUDY1' into C1-C4
LET C5 = C1/C2
PLOT C5 versus C3
OUTFILE 'REPORT'
PLOT C2 versus C1
DESCRIBE C1-C3
```

```
HISTOGRAM C1-C3
NOOUTFILE
ONEWAY C2 C4
OUTFILE 'REPORT'
ONEWAY C2 C4
```

Output from the first three commands is printed on your terminal. OUTFILE says to print the output from the next three commands on your terminal and also put it into the file REPORT. The next command, NOOUTFILE, tells Minitab to stop putting output into the file REPORT. Therefore the output from ONEWAY is printed only on the screen. Now suppose you decide you want the ONEWAY output added to REPORT. The last two commands do this.

Notice, unlike the commands SAVE and WRITE, OUTFILE does not erase the previous contents of a file; it appends the new output to the old output.

B.10 Matrix Calculations

In addition to columns and stored constants, Minitab has matrices, denoted M1, M2, M3, and so forth. The total number of matrices available to you depends on your computer. In addition, columns can be used as column matrices.

There are two ways to enter data with a matrix. You can use READ, as in the following example:

```
READ 3 BY 4 MATRIX M3
  1  4  6  3
  2  8  1  8
 16 12 14 10
END
```

and you can COPY columns into a matrix as in the following example:

```
COPY C2-C4 into M2
```

C2	*C3*	*C4*
16	2	40
18	4	35
20	3	42

→

M2

16	2	40
18	4	35
20	3	42

Commands to do matrix calculations are described in boxes. Here is a simple example using them.

```
COPY        C1-C3 into M1
MULTIPLY    M1 by C5 put in M2
TRANSPOSE   C4 into M3
MULTIPLY    M3 by M1 put in M4
MULTIPLY    M3 by M2 put in K1
ADD         2 to M1 put in M5
```

C1	*C2*	*C3*	*C4*	*C5*	*K1*
1	4	2	10	5	334
1	0	3	12	0	
				4	

M1 $\begin{bmatrix} 1 & 4 & 2 \\ 1 & 0 & 3 \end{bmatrix}$ M2 $\begin{bmatrix} 13 \\ 17 \end{bmatrix}$ M3 $[10 \quad 12]$ M4 $[22 \quad 40 \quad 56]$ M5 $\begin{bmatrix} 3 & 6 & 4 \\ 3 & 2 & 5 \end{bmatrix}$

READ into K by K matrix M

Data follow, one line for each row of the matrix.

PRINT M, ..., M

Prints out the matrices.

COPY C, ..., C into M
COPY M into C, ..., C

The first version of COPY forms a matrix M out of columns. The second version puts a matrix into columns.

ADD M to M put into M
SUBTRACT M from M put into M
MULTIPLY M by M put into M

TRANSPOSE M put into M
INVERT M put into M

These five commands do the usual matrix arithmetic calculations. You can also do scalar addition, subtraction and multiplication with ADD, SUBTRACT and MULTIPLY. For example:

```
ADD       K  to   M  put in M
SUBTRACT M from K  put in M
MULTIPLY M  by   K  put in M
```

You can use a column as a column matrix and a constant as a 1 by 1 matrix when doing matrix arithmetic. For example, suppose M1 is a 4 by 3 matrix, and C2 contains three numbers.

```
MULTIPLY    M1   C2    put in M2
TRANSPOSE   C2   into  M3
MULTIPLY    M3    by   C2, put in K1
TRANSPOSE   M1   into  M4
MULTIPLY    M4    by   M2, put in C1
```

DIAGONAL is in C put into M
DIAGONAL of M put into C

In the first form of DIAGONAL, the matrix M will have the entries in C on its diagonal and zeros elsewhere. The second form puts the entries on the diagonal of M into the columnC.

EIGEN for M, put values in C [vectors in M]

EIGEN calculates eigenvalues (also called characteristic values and latent roots) and eigenvectors for a symmetric matrix. The eigenvalues are stored in decreasing order of magnitude down the column. The eigenvectors are stored as columns of the matrix; the first column corresponds to the first eigenvalue (largest magnitude), the second column to the second eigenvalue, etc.

B.11 Stored Command Files and Simple Loops

You can store a block of Minitab commands in a computer file. Anytime you want to use this block, you type one command, EXECUTE. An instructor may store commands to do a complicated exercise; his students then would not have to type them. A researcher may have an analysis that she does each week, when she gets a new data set. If she stores the commands to do the analysis, she will not have to retype them each week.

As an example, we will use the commands in Exercise 7–7, which simulate fifty 90% confidence intervals.

```
STORE 'INTERVAL'
RANDOM 50 C1-C5;
  NORMAL 30 2.
RMEAN C1-C5  C10
LET K1 = 1.645*2/SQRT(5)
LET C21 = C10-K1
LET C22 = C10+K1
SET C23
  1:50
  END
MPLOT C21 C23, C22 C23
END
```

STORE says to put all the Minitab commands that follow into a computer file called INTERVAL. When you have typed all commands to be stored, then type END to tell Minitab you are finished typing stored commands. Now, if sometime in the same session or in another session, you type

```
EXECUTE 'INTERVAL'
```

Minitab will execute the commands in the file. (If a command file is very long, this may take a few moments.) When Minitab is done, C1–C5, C21–C23 and K1 will contain the calculated results and an MPLOT will be printed.

You can EXECUTE a file several times just by specifying the number of times you wish to do so. For example,

```
EXECUTE 'INTERVAL'  3
```

would do the simulation three times, with different random data each time, and print three MPLOTS. C1–C5, C21–C23 and K1 would contain the calculations from the last simulation.

As another example of using EXECUTE to loop through a block of commands several times, we will calculate a moving average of length 2 for data in C1. Here are the commands.

```
STORE 'MOVEAVG'
NOECHO
LET C2(K1) = (C1(K1)+C1(K1+1))/2
LET K1=K1+1
```

```
ECHO
END
LET K1 = 1
LET K2 = COUNT(C1)-1
EXECUTE 'MOVEAVG' K2
```

There are two new commands in the stored file: NOECHO says do not (echo) print the commands that follow, just do the calculations and print output. ECHO returns to the usual state, where commands in a STORE file are printed when the file is executed. Without NOECHO the two LET commands would be printed every time through the file. But we do not need to see them.

After EXECUTE, the worksheet is

C1	*C2*	*K1*	*K2*
4	3	6	5
2	2.5		
3	4.5		
6	5.5		
5	4.5		
4			

The calculations we did are very simple, although the Minitab commands may at first look somewhat complicated. C1 contains the original data. C2 contains the moving average of C1. The first entry in C2 is the average of the first and second entries in C1, the second entry in C2 is the average of the second and third entries in C1, and so on. Notice that since C1 contains six numbers, C2 contains 6 − 1 = 5 numbers. Before we used EXECUTE we set K1 = 1 and K2 = 5. We EXECUTED the file K2 = 5 times, once to calculate each value in C2. The first time K1 = 1 and we calculated

```
C2(1) = (C1(1)+C1(2))/2
```

and increased K1 to 2. The second time, we calculated

```
C2(2) = (C1(2)+C1(3))/2
```

and increased K1 to 3. This continued until we looped five times.

If you use STORE to create a command file and make a typing error, there is, unfortunately, no way to correct the file in Minitab. You must start over. Thus STORE is most useful for short command files, when you are not likely to make an error. If you want to store a long file, then it is best to create the file outside of Minitab, using your computer's editor. You also can create the file in Minitab using STORE, then leave Minitab and use your editor to make corrections.

A Minitab file of stored commands must have an appropriate extension to its file name. Exactly how this is done varies with computer brand. Usually the extension is MTB. If you use STORE, Minitab automatically adds the appropriate extension. If you use your computer's editor (or operating system) to list the names of your files, you will see what extension Minitab uses on your computer. If you create a file of Minitab commands with your editor, make sure the name of the file contains the appropriate extension.

Looping Through Columns

There is a little device, called the CK capability, that allows you to loop through columns. As a simple example, suppose we want to plot each of C1 through C25 against C30. This would require 25 separate PLOT commands. With the CK capability, we do the following:

```
STORE 'PLOTS'
PLOT   CK1  C50
LET K1 = K1+1
END
LET K1 = 1
EXECUTE 'PLOTS' 25
```

The file PLOTS is EXECUTED 25 times, once to produce each plot. The first time, K1 = 1, so Minitab substitutes a 1 for the K1 in the PLOT command and does PLOT C1 C50. Then K1 is increased to 2. The second time through the file, Minitab substitutes a 2 for K1 and does PLOT C2 C50, and so on.

Anytime a column appears in a command, you may use CK1 (or CK2 or CK3, ...) for the column. When Minitab executes the command, it substitutes the correct value for K1.

NOECHO
ECHO

These commands are used mostly in a STORED file; NOECHO is the first command in the file, right after STORE, and ECHO is the last, just before END. When the file is EXECUTED, no commands are printed out, only output is printed. This is what you usually want.

There is one time when you may not want to use NOECHO. When you first create a file, especially one that is a little complicated, you should EXECUTE it to check for errors. Having the commands printed out will help you interpret any error messages. Once you are sure the commands in your file are correct, you can add NOECHO and ECHO to it.

B.12 Formatted Input and Output of Data

Minitab has five commands that allow FORTRAN formats: READ, SET, INSERT, PRINT, and WRITE. In this section we will give just a brief introduction to the use of formats. You may need to consult a FORTRAN manual for more details.

Minitab can use all format specifications except I (for integers). These include F (for real numbers), E (for numbers using exponential notation), A (for alphabetic data), X (to skip over numbers), and / (to go to the next line).

Formatted Input with READ, SET, and INSERT

Suppose you have data that were typed without any blanks or commas between the numbers. READ by itself cannot enter these, but it can if you use a FORMAT subcommand. For example, suppose we have the data

```
231615.411821143  9
55431.8911941138128
```

and we want these numbers read into C1–C6 as follows:

C1	*C2*	*C3*	*C4*	*C5*	*C6*
231	61	5.4	1182	1143	9
554	31	.89	1194	1138	128

First we determine how wide each number is. In this case, the first number on each line uses three spaces, the second number uses two spaces, the third uses three spaces, the fourth and fifth both use four spaces, and the last uses three spaces. We use a FORMAT subcommand with READ as follows:

```
READ C1-C6;
  FORMAT (F3.0,F2.0,F3.0,2F4.0,F3.0).
```

This FORMAT describes one row of data. In F3.0, the 3 says the number to be read uses three spaces; F2.0 says the second number uses two spaces; and F3.0 says the third number uses three spaces. Notice that in this instance, one of the spaces holds the decimal point. The next two numbers both use four spaces. To save typing we can write 2F4.0 instead of F4.0, F4.0.

The last F3.0 says the last number uses three spaces. Notice that the number on the first line, 9, is just one space wide. The remaining two spaces are blank. When the field a number uses is wider than the number, the number must be right justified; that is, all blanks must appear before the number.

You can skip over numbers with formats. For example, suppose we wanted the data read into the worksheet as follows:

C1	*C2*	*C3*	*C4*	*C5*	*C6*
231			1182		
554			1194		

We would use the command

```
READ C1 C4;
  FORMAT (F3.0,5X,F4.0).
```

The 5X says to skip over five spaces. Thus we read a number from the first three spaces, skip five spaces, then read a number from the next four spaces. The rest of the line is ignored since there are no more format specifications in the FORMAT subcommand.

Formats must be typed according to strict rules. You must use the parentheses and you must put a comma between format specifications.

You can enter several data lines into one row of the worksheet by using a slash. For example, suppose a file continues data for 200 people, three lines for each person. Here are the data for the first two people:

```
2431442012
1421431422
02011
0531340013
0531021421
11120
```

Suppose we want to enter these data into C1–C25, one digit per column. The following format does this:

```
READ C1-C25;
  FORMAT(10F1.0/10F1.0/5F1.0).
```

The first 10F1.0 reads ten numbers from the first line. The slash says go to the next line. The second 10F1.0 says to read ten numbers from this line. The second slash says to go to another new line. The 5F1.0 says to read five numbers from this line. This procedure is repeated for each person, always using three lines of data. Notice that FORMAT describes one row of the worksheet.

You also can use slashes to skip over data lines. For example, suppose we use the following format on the data for 200 people:

```
READ C1-C3;
  FORMAT (3F1.0//).
```

This would take three numbers from the first line for each person and skip over the other two lines.

Formatted Output with PRINT and WRITE

Formats also can be used to output data. This is done if you want to control the spacing of the numbers or if you want to print the numbers to a different (usually larger) number of decimal places than Minitab would choose.

If you use FORMAT with PRINT you must use a carriage control (see a FORTRAN manual for a complete discussion of carriage controls) at the beginning of the format statement. The most common is 1X, which gives single-spaced output.

As an example, consider the data that we read into C1–C6 in the first example of this section. The command

```
PRINT C1-C6;
  FORMAT (1X,F5.0,F4.0,F5.2,2F6.0,F5.0).
```

would print out the data as follows:

```
 231 61 5.40 1182 1143    9
 554 31 0.89 1194 1138  128
```

The first specification, F5.0, describes the first five spaces: the 5 says to use five spaces and the 0 says to print no digits after the decimal point. The decimal is printed, however. The number, 231., is then right justified in the field of five spaces. Similarly the second number, 61., is right justified in a field of four spaces. The next specification, F5.2, says to use five spaces and put two digits after the decimal point.

You may use the X format to skip spaces on the printed page. For example,

```
PRINT C1-C6;
  FORMAT (1X,F5.0,F4.0,12X,F5.2,2F6.0,F5.0).
```

would print two numbers, then skip 12 spaces, then print the last four numbers.

You may use slashes to print one row of the worksheet on several lines of output. You must put a carriage control at the beginning of each line. For example,

```
PRINT C1-C6;
  FORMAT (1X,F5.0,F4.0,F5.2/1X,2F6.0,F5.0).
```

puts three numbers on the first line and three on the second.

APPENDIX C
A List of Minitab Commands in Releases 5–7

This appendix summarizes all Minitab commands in Releases 5 through 7. Features that are in Release 6 but not in Release 5 are marked with an *. Release 7 features are marked with **. In Minitab, type HELP COMMANDS for full documentation. Release 8 commands are listed in Appendix E.

Notation:

K denotes a constant such as 8.3 or K14

C denotes a column, such as C12 or 'Height'

E denotes either a constant or column

M denotes a matrix, such as M5

[] encloses an optional argument

Subcommands are shown indented under the main command.

1. General Information

```
HELP   explains Minitab commands, can be a
       command or a subcommand
INFO   [C...C] gives the status of worksheet
STOP   ends the current session
```

2. Input and Output of Data

```
READ      the following data      into C...C
READ      data [from 'filename'] into C...C
SET       the following data      into C
SET       data [from 'filename'] into C
INSERT    data [from 'filename'] between rows
          K and K of C...C
INSERT    data [from 'filename'] at the end
          of C...C
  READ, SET and INSERT have the
             subcommands:
    FORMAT (Fortran format)
    NOBS = K
END       of data (optional)
NAME      for C is 'name', for C is 'name'...
          for C is 'name'
PRINT     the data in E...E
```

```
 WRITE    [to 'filename'] the data in C...C
   PRINT and WRITE have the subcommand:
     FORMAT (Fortran format)
 SAVE     [in 'filename'] a copy of the
          worksheet
*     PORTABLE
 RETRIEVE the Minitab saved worksheet [in
          'filename']
*     PORTABLE
```

3. Editing and Manipulating Data

```
 LET      C(K) = K # changes the number in
          row K of C
 DELETE   rows K...K of C...C
 ERASE    E...E
 INSERT   (see Section 2)
 COPY     C...C into C...C
 COPY     C into K...K
      USE  rows K...K
      USE  rows where C = K...K
      OMIT rows K...K
      OMIT rows where C = K...K
 COPY     K...K into C
 CODE     (K...K) to K ... (K...K) to K for
          C...C, store in C...C
 STACK    (E...E) ... on (E...E), store in
          (C...C)
       SUBSCRIPTS into C
 UNSTACK (C...C) into (E...E) ... (E...E)
       SUBSCRIPTS are in C
 CONVERT using table in C C, the data in C,
          and store in C
*CONCATENATE C...C put in C
*ALPHA C...C
```

4. Arithmetic

```
 LET = expression
```

Expressions may use arithmetic operators + − * / ** (exponentiation),
**Comparison operators = ~= < > <= >=
**Logical operators & | ~
and any of the following: ABSOLUTE, SQRT, LOGTEN, LOGE, EXPO,

ANTILOG, ROUND, SIN, COS, TAN, ASIN, ACOS, ATAN, SIGNS, NSCORE, PARSUMS, PARPRODUCTS, COUNT, N, NMISS, SUM, MEAN, STDEV, MEDIAN, MIN, MAX, SSQ, SORT, RANK, LAG, **EQ, NE, LT, GT, LE, GE, AND, OR, NOT.

Examples:
```
LET C2 = SQRT(C1 - MIN(C1))
LET C3(5) = 4.5
```

Simple Arithmetic Operations

```
ADD       E to E ...    to E, put into E
SUBTRACT E from            E, put into E
MULTIPLY E by E ...    by E, put into E
DIVIDE    E by              E, put into E
RAISE     E to the power  E, put into E
```

Columnwise Functions

```
ABSOLUTE value         of E, put into E
SQRT                   of E, put into E
LOGE                   of E, put into E
LOGTEN                 of E, put into E
EXPONENTIATE              E, put into E
ANTILOG                of E, put into E
ROUND to integer          E, put into E
SIN                    of E, put into E
COS                    of E, put into E
TAN                    of E, put into E
ASIN                   of E, put into E
ACOS                   of E, put into E
ATAN                   of E, put into E
SIGNS                  of E, put into E
PARSUMS                of C, put into C
PARPRODUCTS            of C, put into C
```

Normal Scores

```
NSCORES                of C, put into C
```

Columnwise Statistics

```
COUNT the number of values in        C [put into K]
N (number of nonmissing values in)  C [put into K]
NMISS (number of missing values in) C [put into K]
SUM      of the values in            C [put into K]
MEAN     of the values in            C [put into K]
STDEV    of the values in            C [put into K]
```

```
MEDIAN   of the values in              C [put into K]
MINIMUM  of the values in              C [put into K]
MAXIMUM  of the values in              C [put into K]
SSQ (uncorrected sum of sq.) for       C [put into K]
```

Rowwise Statistics

```
RCOUNT              of E...E put into C
RN                  of E...E put into C
RNMISS              of E...E put into C
RSUM                of E...E put into C
RMEAN               of E...E put into C
RSTDEV              of E...E put into C
RMEDIAN             of E...E put into C
RMINIMUM            of E...E put into C
RMAXIMUM            of E...E put into C
RSSQ                of E...E put into C
```

Indicator Variables

```
INDICATOR variables for subscripts in C, put
          into C...C
```

5. Plotting Data

```
   HISTOGRAM  C...C
   DOTPLOT    C...C
     HISTOGRAM and DOTPLOT have the subcommands:
       INCREMENT = K
       START at K [end at K]
       BY C
       SAME scales for all columns
   PLOT    C vs C
       SYMBOL = 'symbol'
   MPLOT   C vs C, and C vs C, and ... C vs C
   LPLOT   C vs C using tags in C
   TPLOT   C vs C vs C
     PLOT, MPLOT, LPLOT and TPLOT have the
                             subcommands:
**     TITLE       = 'text'
**     FOOTNOTE    = 'text'
**     YLABEL      = 'text'
**     XLABEL      = 'text'
       YINCREMENT = K
```

```
        YSTART      at K [end at K]
        XINCREMENT = K
        XSTART      at K [end at K]
    TSPLOT  [period K] of C
        ORIGIN = K
    MTSPLOT [period K] of C...C
        ORIGIN = K for C...C [... origin K for
                       C...C]
      TSPLOT and MTSPLOT have the subcommands:
        INCREMENT = K
        START at K [end at K]
        TSTART at K [end at K]
    GRID    C [K to K] C [K to K]
    CONTOUR C vs C and C
        BLANK bands between letters
        YSTART = K [up to K]
        YINCREMENT = K
    WIDTH  of all plots that follow is K spaces
    HEIGHT of all plots that follow is K lines
```

High Resolution Graphics

```
    GOPTIONS
       +DEVICE = 'device'
        HEIGHT = K inches
        WIDTH  = K inches
    GHISTOGRAM C...C
        INCREMENT = K
        START at K [end at K]
        BY C
        SAME scales for all columns
    GPLOT    C vs C
        SYMBOL = 'symbol'
    GMPLOT   C vs C, and C vs C, and ... C vs C
    GLPLOT   C vs C using tags in C
    GTPLOT   C vs C vs C
      GPLOT, GMPLOT, GLPLOT and GTPLOT have the
                                  subcommands:
**      TITLE       = 'text'
**      FOOTNOTE    = 'text'
**      YLABEL      = 'text'
**      XLABEL      = 'text'
```

+ Not implemented in the microcomputer version

```
    YINCREMENT = K
    YSTART     at K [end at K]
    XINCREMENT = K
    XSTART     at K [end at K]
    LINES [style K [color K]] connecting
          points in C C
    COLOR C
```

All high resolution graphics commands have the subcommand:

```
    FILE 'filename' to store graphics output
```

6. Basic Statistics

```
DESCRIBE C...C
    BY C
ZINTERVAL [K% confidence] assuming sigma = K
              for C...C
ZTEST [of mu = K] assuming sigma = K for
      C...C
    ALTERNATIVE = K
TINTERVAL [K% confidence] for data in C...C
TTEST [of mu = K] on data in C...C
    ALTERNATIVE = K
TWOSAMPLE test and c.i. [K% confidence]
          samples in C C
    ALTERNATIVE = K
    POOLED procedure
TWOT test and c.i. [K% confidence] data in
      C, groups in C
    ALTERNATIVE = K
    POOLED procedure
CORRELATION between C...C [put into M]
COVARIANCE  between C...C [put into M]
CENTER the data in C...C put into C...C
    LOCATION [subtracting K...K]
    SCALE    [dividing by K...K]
    MINMAX   [with K as min and K as max]
```

7. Regression

```
REGRESS C on K predictors C...C [store st.
        resids in C [fits in C]]
    NOCONSTANT    in equation
```

```
      WEIGHTS          are in C
      MSE              put into K
      COEFFICIENTS     put into C
      XPXINV           put into M
      RMATRIX          put into M
      HI               put into C (leverage)
      RESIDUALS        put into C (observed -
                       fit)
      TRESIDUALS       put into C (deleted
                       studentized)
      COOKD            put into C (Cook's
                       distance)
      DFITS            put into C
      PREDICT          for E...E
      VIF              (variance inflation
                       factors)
      DW               (Durbin-Watson statistic)
      PURE             (pure error lack-of-fit
                       test)
      XLOF             (experimental lack-of-fit
                       test)
      TOLERANCE        K [K]
   STEPWISE regression of C on the predictors
            C...C
      FENTER           = K (default is four)
      FREMOVE          = K (default is four)
      FORCE            C...C
      ENTER            C...C
      REMOVE           C...C
      BEST             K alternative predictors
                         (default is zero)
      STEPS            = K (default depends on
                            output width)
  *BREG C on predictors C...C
      INCLUDE predictors C...C
      BEST K models
      NVARS K [K]
      NOCONSTANT in equation
   NOCONSTANT in REGRESS, STEPWISE and BREG
              commands that follow
   CONSTANT   fit a constant in REGRESS,
              STEPWISE and BREG
   BRIEF      K
```

8. Analysis of Variance

```
 AOVONEWAY aov for samples in C...C
 ONEWAY    aov, data in C, subscripts in C
           [store resids in C [fits in C]]
 TWOWAY    aov, data in C, subscripts in C C
           [store resids in C [fits in C]]
     ADDITIVE model
*    MEANS for the factors C [C]
*ANOVA model
     RANDOM    factorlist
     EMS
     FITS      put into C...C
     RESIDUALS put into C...C
     MEANS     for termlist
     TEST      for termlist/errorterm
     RESTRICT
*ANCOVA model
     COVARIATES are in C...C
     FITS       put into C...C
     RESIDUALS  put into C...C
     MEANS      for termlist
     TEST       for termlist/errorterm
**GLM model
     COVARIATES   are in C...C
     WEIGHTS      are in C
     FITS         put into C...C
     RESIDUALS    put into C...C
     SRESIDS      put into C...C
     TRESIDS      put into C...C
     HI           put into C
     COOKD        put into C...C
     DFITS        put into C...C
     XMATRIX      put into M
     COEFFICIENTS put into C...C
     MEANS        for termlist
     TEST         for termlist/errorterm
     BRIEF        K
     TOLERANCE    K [K]
```

9. Multivariate Analysis

```
*PCA principal component analysis of C...C
     COVARIANCE matrix
     NCOMP       = K (number of components)
     COEF        put into C...C
     SCORES      put into C...C
*DISCRIMINANT groups in C, predictors in
              C...C
     QUADRATIC discrimination
     PRIORS    are in K...K
     LDF       coef put in C...C
     FITS      put in C [C]
     XVAL      cross-validation
     PREDICT   for E...E
     BRIEF     K
```

10. Nonparametrics

```
  RUNS    test [above and below K] for C
  STEST   sign test [median = K] for C...C
      ALTERNATIVE = K
  SINTERVAL sign c.i. [K% confidence] for
         C...C
  WTEST   Wilcoxon one-sample rank test
         [median = K] for C...C
      ALTERNATIVE = K
  WINTERVAL Wilcoxon c.i. [K% confidence] for
         C...C
  MANN-WHITNEY test and c.i. [K% confidence]
         on C C
**    ALTERNATIVE = K
  KRUSKAL-WALLIS test for data in C,
         subscripts in C
**MOOD median test, data in C, subscripts in C
         [put res. in C [fits in C]]
**FRIEDMAN data in C, treatment in C, blocks
         in C [put res. in C [fits in C]]
  WALSH averages for C, put into C [indices
         into C C]
 *WDIFF for C and C, put into C [indices into
         C C]
 *WSLOPE y in C, x in C, put into C [indices
         into C C]
```

11. Tables

```
TALLY the data in C...C
    COUNTS
    PERCENTS
    CUMCOUNTS   cumulative counts
    CUMPERCENTS cumulative percents
    ALL         four statistics above
CHISQUARE test on table stored in C...C
TABLE the data classified by C...C
    MEANS       for C...C
    MEDIANS     for C...C
    SUMS        for C...C
    MINIMUMS    for C...C
    MAXIMUMS    for C...C
    STDEV       for C...C
    STATS       for C...C
    DATA        for C...C
    N           for C...C
    NMISS       for C...C
    PROPORTION of cases = K [through K] in
                C...C
    COUNTS
    ROWPERCENTS
    COLPERCENTS
    TOTPERCENTS
    CHISQUARE analysis [output code = K]
    MISSING level for classification
            variable C...C
    NOALL in margins
    ALL for C...C
    FREQUENCIES are in C
    LAYOUT K rows by K columns
```

12. Time Series

```
ACF   [with up to K lags] for series in C
      [put into C]
PACF  [with up to K lags] for series in C
      [put into C]
CCF   [with up to K lags] between series in
      C and C
```

```
DIFFERENCES [of lag K] for data in C, put
     into C
LAG [by K] for data in C, put into C
ARIMA p = K, d = K, q = K, data in C [put
     resids in C [preds in C [coefs in C]]]
ARIMA p = K, d = K, q = K, P = K, D = K,
     Q = K, S = K, data in C [put resids
     in C [preds in C [coefs in C]]]
   CONSTANT    term in model
   NOCONSTANT  term in model
   STARTING    values are in C
   FORECAST    [origin = K] up to K leads
               [put in C [limits in C C]]
```

**13. Statistical Process Control

**High resolution commands begin with the letter G.

```
**XBARCHART  for C...C, subgroups are in E
**GXBARCHART for C...C, subgroups are in E
     MU = K
     SIGMA = K
     RSPAN = K
     TEST K...K
     SUBGROUP size is E
     See additional subcommands listed below
**RCHART  for C...C, subgroups are in E
**GRCHART for C...C, subgroups are in E
     SIGMA = K
     SUBGROUP size is E
     See additional subcommands listed below
**SCHART  for C...C, subgroups are in E
**GSCHART for C...C, subgroups are in E
     SIGMA = K
     SUBGROUP size is E
     See additional subcommands listed below
**ICHART  for C...C
**GICHART for C...C
     MU = K
     SIGMA = K
     RSPAN = K
     TEST K...K
     See additional subcommands listed below
```

```
**MACHART  for C...C, subgroups are in E
**GMACHART for C...C, subgroups are in E
      MU = K
      SIGMA = K
      SPAN = K
      RSPAN = K
      SUBGROUP size is E
      See additional subcommands listed below
**EWMACHART  for C...C, subgroups are in E
**GEWMACHART for C...C, subgroups are in E
      MU = K
      SIGMA = K
      WEIGHT = K
      RSPAN = K
      SUBGROUP size is E
      See additional subcommands listed below
**MRCHART  for C...C
**GMRCHART for C...C
      SIGMA = K
      RSPAN = K
      See additional subcommands listed below
**PCHART  number of nonconformities are in
         C...C, sample size = E
**GPCHART number of nonconformities are in
         C...C, sample size = E
      P = K
      TEST K...K
      SUBGROUP size is E
      See additional subcommands listed below
**NPCHART  number of nonconformities are in
          C...C, sample size = E
**GNPCHART number of nonconformities are in
          C...C, sample size = E
      P = K
      TEST K...K
      SUBGROUP size is E
      See additional subcommands listed below
**CCHART  number of nonconformities are in
         C...C
**GCCHART number of nonconformities are in
         C...C
      MU = K
      TEST K...K
      See additional subcommands listed below
```

```
**UCHART  number of nonconformities are in
          C...C, sample size = E
**GUCHART number of nonconformities are in
          C...C, sample size = E
      MU = K
      TEST K...K
      SUBGROUP size is E
      See additional subcommands listed below
```

**All statistical process control charts have the subcommands:

```
      SLIMITS are K...K
      HLINES at E...E
      ESTIMATE using just samples K...K
      TITLE = 'text'
      FOOTNOTE = 'text'
      YLABEL = 'text'
      XLABEL = 'text'
      YINCREMENT = K
      YSTART at K [end at K]
      XSTART at K [end at K]
```

**All high resolution charts have the subcommand:

```
      FILE 'filename'
```

14. Exploratory Data Analysis

```
STEM-AND-LEAF display of C...C
    TRIM outliers
    INCREMENT = K
    BY C
BOXPLOT  for C
GBOXPLOT for C (high resolution version)
  BOXPLOT and GBOXPLOT have the subcommands:
    INCREMENT = K
    START at K [end at K]
    BY C
    LINES = K
    NOTCH [K% confidence] sign c.i.
    LEVELS K...K
    FILE 'filename' to store GBOXPLOT output
LVALS of C [put lvals in C
      [mids in C [spreads in C]]]
MPOLISH C, levels in C C [put residuals in C
        [fits in C]]
    COLUMNS     (start iteration with column
                medians)
```

```
    ITERATIONS = K
    EFFECTS  put common into K, rows into C,
             cols into C
    COMPARISON values, put into C
RLINE y in C, x in C [put resids in C [fits
      in C [coefs in C]]]
    MAXITER = K (max. number of iterations)
RSMOOTH C, put rough into C, smooth into C
    SMOOTH by 3RSSH, twice
CPLOT (condensed plot) C vs C
    LINES = K
    CHARACTERS = K
    XBOUNDS = from K to K
    YBOUNDS = from K to K
CTABLE (coded table) data in C, row C,
      column C
    MAXIMUM value in each cell should be
            coded
    MINIMUM value in each cell should be
            coded
ROOTOGRAM data in C [use bin boundaries in C]
   BOUNDARIES store them in C
   DRRS store them in C
   FITTED values store them in C
   COUNTS store them in C
   FREQUENCIES are in C [bin boundaries are
               in C]
   MEAN = K
   STDEV = K
```

15. Distributions and Random Data

```
RANDOM K observations into C...C
    BERNOULLI trials p = K
PDF    for values in E [store results in E]
CDF    for values in E [store results in E]
INVCDF for values in E [store results in E]
  RANDOM, PDF, CDF, INVCDF have the
  subcommands:
    BINOMIAL  n = K, p = K
    POISSON   mu = K
    INTEGER   discrete uniform on integers K
              to K
```

```
    DISCRETE  dist. with values in C and
              probabilities in C
    NORMAL    [mu = K [sigma = K]]
    UNIFORM   [continuous on the interval K
              to K]
    T         degrees of freedom = K
    F         df numerator = K,
              df denominator = K
    Additional subcommands are BETA, CAUCHY,
    LAPLACE, LOGISTIC, LOGNORMAL, CHISQUARE,
    EXPONENTIAL, GAMMA, WEIBULL
  SAMPLE  K rows from C...C put into C...C
**    REPLACE (sample with replacement)
  BASE    for random number generator = K
```

16. Sorting

```
  SORT  C [carry along C...C] put into C [and
        C...C]
**    BY C...C
**    DESCENDING C...C
  RANK  the values in C, put ranks into C
```

17. Matrices

```
READ    [from 'filename'] into a K by K
        matrix M
PRINT     M...M
TRANSPOSE M into M
INVERT    M into M
DEFINE    K into K by K matrix M
DIAGONAL  is C, form into M
DIAGONAL  of M, put into C
COPY      C...C into M
COPY      M into C...C
COPY      M into M
    USE   rows K...K
    OMIT  rows K...K
EIGEN     for M put values into C [vectors
          into M]
```

In the following commands E can be either C, K or M

```
ADD      E to E,   put into E
SUBTRACT E from E, put into E
MULTIPLY E by E,   put into E
```

18. Miscellaneous

```
ERASE       E...E
OUTFILE     'filename' put all output in file
    OW       = K output width of file
    OH       = K output height of file
    NOTERM no output to terminal
NOOUTFILE output to terminal only
PAPER       output to printer
    OW       = K output width of printer
    OH       = K output height of printer
    NOTERM no output to terminal
NOPAPER     output to terminal only
JOURNAL     ['filename'] record Minitab
            commands in this file
NOJOURNAL cancels JOURNAL
NOTE        comments may be put here
NEWPAGE     start next output on a new page
UC          use only upper case letters on
            output
LC          use mixed case letters on output
OW          = K number of spaces for width of
            output
OH          = K number of lines for height of
            one page (or screen) of output
IW          = K number of spaces for width of
            input
BRIEF       = K controls amount of output from
            REGRESS, GLM, DISCRIM, ARIMA,
            RLINE
RESTART     begin fresh Minitab session
SYSTEM      provides access to operating
            system commands
TSHARE      interactive or timesharing mode
BATCH       batch mode
```

19. Stored Commands and Loops

The commands STORE and EXECUTE provide the capability for simple macros (stored command files) and loops.

```
EXECUTE 'filename' [K times]
STORE   [in 'filename'] the following
        commands (Minitab commands go here)
END     of storing commands
```

```
  NOECHO   the commands that follow
  ECHO     the commands that follow
**YESNO    K
```

The CK capability. The integer part of a column number may be replaced by a stored constant.

Example:
```
LET K1 = 5
PRINT C1 - CK1
```

Since K1 = 5, this PRINTS C1 through C5.

20. Symbols

* Missing Value Symbol. An * can be used as data in READ, SET and INSERT and in data files. Enclose the * in single quotes in commands and subcommands.

Example: `CODE (-99) to '*' in C1, put into C3`

Example:
```
COPY C6 INTO C7;
  OMIT C6 = '*'.
```

\# Comment Symbol. The symbol # anywhere on a line tells Minitab to ignore the rest of the line.

& Continuation Symbol. To continue a command onto another line, end the first line with the symbol &. You can use + + as a synonym for &.

21. Worksheet and Commands

Minitab consists of a worksheet for data and over 200 commands. The worksheet contains columns of data denoted by C1, C2, C3 . . . , stored constants denoted by K1, K2, K3 . . . , and matrices denoted by M1, M2, M3, ...

A column may be given a name with the command NAME (Section 2). A name may be up to eight characters long, with any characters except apostrophes and #. Names may be used in place of column numbers. When a name is used, it must be enclosed in apostrophes (single quotes).

Example: `PLOT 'INCOME' vs 'AGE'`

Each command starts with a command word and is usually followed by a list of arguments. An argument is a number, a column, a stored constant, a matrix or a file name. Only the command word and arguments are needed. All other text is for the readers' information.

22. Subcommands

Some Minitab commands have subcommands. To use a subcommand, put a semicolon at the end of the main command line. Then type the subcommands. Start each on a new line, then end it with a semicolon. When you are done, end the last subcommand with a period.

The subcommand ABORT cancels the whole command.

APPENDIX D
Changes in Release 6 and Release 7

Most of the major changes in Release 6 and Release 7 are additions of capabilities beyond the scope of this *Handbook*. For a more comprehensive description of new Minitab features, mini/mainframe users can type NEWS at the MTB > prompt. PC users can PRINT or TYPE the README file which is included with the Minitab package.

The new commands (all listed in Appendix C) for Release 6 are ANOVA/ANCOVA (general balanced analysis of variance and analysis of covariance), DISCRIM (linear and quadratic discriminant analysis), PCA (principal component analysis), BREG (best subset of variables for regression), WDIFF and WSLOPES (for computing certain robust estimates) and several commands for dealing with alphanumeric data. High resolution graphics were in some versions of Release 5 and in more versions of Release 6 (including the standard PC version); these commands have been added to Appendix C. Subcommands have been added to TWOWAY, SAVE, RETRIEVE, and the high resolution graphics commands. All commands listed in Appendix C of previous printings continue to work.

Many Release 6 commands have changes in output format (such as number of decimal digits printed for numbers), changes in spacing, and changes from all upper case to mixed upper and lower case for text. These should not cause any difficulties for users. Other changes that may affect users are listed below by chapter.

Additional capabilities in Release 7 (also listed in Appendix C) include the GLM command to fit the general linear model (analysis of variance and of covariance for unbalanced designs), eleven commands to produce statistical process control charts, and two new nonparametric commands, MOOD (performs the mood median test) and FRIEDMAN. Release 7 also allows the use of comparison operators (EQ, NE, LT, GT, LE, GE) and logical operators (AND, OR, NOT) in the LET command. The new subcommands TITLE, FOOTNOTE, YLABEL, and XLABEL are used with graphics commands to label plots. The SORT command has two new subcommands BY (allows sorting by multiple columns) and

DESCENDING (allows sorting columns in descending order). In addition, the YESNO command allows interruption of a macro to input a YES or NO value. Release 7 also improves Minitab's alpha data handling capabilities. Other changes that may affect users are listed below by chapter.

Chapter 1.

In Release 7, the INFO command identifies alpha data columns.

Chapter 6.

In Release 7, the SAMPLE command has a subcommand REPLACE which allows sampling with replacement.

Chapter 9.

In Release 6, the AOVONEWAY and ONEWAY commands print the p-value (significance level) for the F-test.

TWOWAY has a new subcommand, MEANS, which prints marginal means and individual 95% confidence intervals for the factors.

A three factor model (as in Exercise 9-11, p. 214) can be fit with the new ANOVA command. See HELP ANOVA for details.

Chapter 10.

In Release 6, the output from the REGRESS command now contains, for each coefficient, the p-value for the test whether the true coefficient is zero. This is computed from the t for each coefficient, as explained on pages 232 and 239. In addition, the F-statistic for overall significance of regression is printed in the analysis of variance table, along with the associated p-value.

In Release 7, BRIEF 3 controls the amount of output for the GLM command and also prints the p-value for unequal spreads in the MOOD command output.

Chapter 12.

In Release 6, the SINTERVAL output now contains an additional line. In addition to the intervals with confidence levels below and above, there is a "middle" line with (approximately) the desired confidence. This is found using a "nonlinear interpolation" procedure, abbreviated NLI, developed by Thomas P. Hettmansperger and Simon J. Sheather (Confidence Intervals Based on Interpolated Order Statistics, Statistics and

Probability Letters, Volume 4, Number 2, 1986, pp. 75–79). The NLI procedure is complicated, and explaining how to calculate it is beyond the scope of this *Handbook,* but it can be used easily. More assumptions are needed than for the other two lines, it is nearly exact if the distribution being sampled from is symmetric.

The example shown in Exhibit 12.4 was redone using Release 6, with the results shown below.

```
MTB > SINT 'SALES'
SIGN CONFIDENCE INTERVAL FOR MEDIAN
                              ACHIEVED
        N   N*   MEDIAN   CONFIDENCE   CONFIDENCE   INTERVAL   POSITION
SALES  29    1    144.0     0.9386     (   110.0;     210.0)         10
                            0.9500     (   108.5;     211.7)        NLI
                            0.9759     (   101.0;     220.0)          9
```

In Release 7, the MANN-WHITNEY command has a subcommand ALTERNATIVE (similar to STEST and WTEST) and the test statistic adjusted for ties is also output.

APPENDIX E
Using this Book with Minitab Release 8 on a PC or a Macintosh®

MINITAB Statistical Software runs on PCs, Macintosh computers, and most of the leading workstations, minicomputers, and mainframe computers. In addition, there are different releases of Minitab available for different computers. While Minitab differs across releases and computer platforms, most notably in ways that take advantage of platform-specific capabilities, the core of Minitab, the worksheet and commands, is the same. Thus, if you know how to use Minitab using one release on one platform, you can easily switch to a different release or a different platform.

This *Handbook*, based on Release 5 of Minitab, contains appendixes explaining how to use this book with different releases of the software. This appendix explains how to use the *Handbook* with Release 8 of Minitab, which is available only on a PC or a Macintosh.

E.1 New Release 8 Commands

For details on using these commands, see the Release 8 *MINITAB Reference Manual* or use on-line help. There are several ways to access help information: Type HELP at the command prompt in the Session window; choose the <?> button in the lower left corner of each dialog box; or choose Help from the menu.

Multiple Comparisons for One-Way Analysis of Variance

When you do one-way analysis of variance with ONEWAY, you can perform any of four multiple comparison procedures: TUKEY, FISHER, DUNNET, and MCB.

```
ONEWAY aov, data in C, levels in C, [put
resids in C [fits in C]
  TUKEY       [family error rate K]
  FISHER      [individual error rate K]
  DUNNETT     [family error rate K] for
              control group K
  MCB         [family error rate K] best is K
```

Design of Experiments

Two new commands, FFDESIGN and PBDESIGN, generate optimal fractional factorial and Plackett-Burman designs, based on your specifications. You can use another new command, FFACTORIAL, to analyze experiments based on these designs.

```
FFDESIGN K factors [K runs]
  BLOCKS        K or termlist
  REPLICATES    K (replicate points)
  CENTER        K (add center points)
  FRACTION      K (use this fraction)
  FOLD          [on factors fact...fact]
  ADD           fact=term...fact=term
  ALIAS         info up to order K
  RANDOMIZE     [using base K]
  XMATRIX       put in C...C
  LEVELS        are in K...K or C
  TERMS         in design, put in C
  BRIEF         K
FFACTORIAL model
  BLOCKS        are in C
  COVARIATES    are in C...C
  EFFECTS       put in C...C
  COEFFICIENTS put in C...C
  FITS          put in C...C
  RESIDUALS     put in C...C
  SRESIDS       put in C...C
  TRESIDS       put in C...C
  HI            put in C
  COOKD         put in C...C
  DFITS         put in C...C
  XMATRIX       put in M
  MEANS         for termlist
  CUBE          for termlist
  EPLOT
  ALIAS         info [up to order K]
  TOLERANCE     K [K]
PBDESIGN K factors [K runs]
  REPLICATES    K (replicate points)
  CENTER        K (add center points)
  RANDOMIZE     [using base K]
  XMATRIX       put in C...C
  LEVELS        are in K...K or C
```

Quality Control and Improvement Macros

Release 8 includes the following macros: analysis of means (ANOM); process capability analysis (CAPA); cumulative sum control charts (CUSUM); Pareto charts (PARETO); a macro which generates Central Composite and Box-Behnken designs for 2 to 6 factors (RSDESIGN); and a macro which fits a quadratic model to response surface designs and draws contour plots (RSMODEL).

To use these macros, type the EXECUTE command followed by the macro name in single quotation marks. For example, to execute the CAPA macro, type: EXECUTE 'CAPA'.

Your copy of Minitab includes notes on where the macro files are stored and additional details on using them.

Statistical Process Control Charts

By default, Minitab uses the pooled standard deviation to estimate sigma. RBAR is a new subcommand, which you may use with XBARCHART, RCHART, SCHART, MACHART, and EWMACHART, which tells Minitab to estimate sigma using a function of the sample means or ranges instead.

Miscellaneous

The command BRIEF 0 suppresses normal output to your screen for all commands. This is useful primarily when writing macros.

E.2 Differences Between Minitab on a Mainframe and on a Macintosh or PC

The following general descriptions of menus, dialog boxes, and the Data screen apply to both PCs and Macintoshes. For specific instructions on actually working with these features, see the two sections after this one: Using Minitab on a PC and Using Minitab on a Macintosh.

Menus

A *menu* is simply a list of available command categories. Almost all of Minitab's commands are logically grouped by topic into menus. For example, all the commands having to do with files (like SAVE, RETRIEVE, WRITE, READ, and so on) are grouped together on the File menu. Similarly, you will find all commands that produce graphs (like HISTOGRAM, BOXPLOT, SCATTER PLOT, and so on) on the Graph menu.

The advantage of issuing commands through a menu interface is that you don't have to memorize Minitab's command language, and, if you have a mouse, you can execute some commands with just a few clicks. Often menus use language that is more explanatory than the corresponding Minitab command word. For example, the RETRIEVE command appears on the File menu as "Open Worksheet."

Dialog Boxes

When you choose a menu command, Minitab either executes it immediately or responds by presenting you with a *dialog box*. A dialog box is just what it says: a box that appears on the screen that you hold a dialogue with. For example, if you choose the Histogram command from the Graph menu, Minitab presents you with a dialog box that asks which variable or variables you want to plot, whether you want high-resolution graphics, and so on.

In the dialog boxes Minitab lists many of the available options that accompany a given command (often these would be subcommands in Minitab's command language). You can fill in most information very simply, often with just a click of the mouse button or a two-stroke keyboard shortcut. You tell Minitab you are done by choosing "OK." Minitab then executes your commands and displays the results on the screen.

Data Screen

The Data screen is a special area providing immediate access to your worksheet. All the commands described in this manual for working with data do function in the Mac or PC computer environment, but often it is easiest to enter or manipulate data using the Data screen.

On a mainframe, although you can use the PRINT command to view the worksheet by column and row, you cannot alter or add to the values you see before you without using commands like INSERT or DELETE. The Data screen, on the other hand, presents all the data for easy editing on your screen. It displays your data, organized by column and row, in a spreadsheet or grid format. The intersection of every row and column is called a *cell;* each cell contains one piece of data. You can reach any cell using the mouse (simply click the cell and you are there) or the arrow keys (press the directional arrows until you've arrived). Once in a cell, you can change it any way you want. To enter data, you can just type it in, cell by cell, without having to use the READ, SET or INSERT commands.

Getting Help

The next sections give you tips for learning to use these handy features, but they do not cover all the available functions that make using Minitab on a Mac or PC so convenient. You should become familiar very early on with Minitab's Help function; it offers full documentation on Minitab's features. On a PC, pressing the [F1] key at any point opens the Help menu, or type HELP to get help on commands. On a Mac, you may choose Help from the apple menu on the far left of the menu bar. All dialog boxes on both PC's and Mac's have a button <?> in the lower left corner that you can select to get help with using that particular dialog box.

E.3 Using Minitab on a PC

You will do most of your work in two places on the PC, the Session window and the Data screen. The Data screen allows you to view and edit your worksheet. The Session window contains Minitab's main menus and displays the output from most commands.

The Session Window

The first thing that appears when you start Minitab is the Session window, with the menu bar on top and the MTB > prompt below it.

Exhibit E.1 Session Window

```
                                    Minitab
  File  Edit  Calc  Stat  Graph                                       F1=Help
Worksheet size: 16174 cells

Press ALT + a highlighted letter to open a menu

MTB >
```

Most commands are executed through the Session window. You can use either the menus, the command language, or a combination of both, as you prefer. To use Minitab's command language, type the commands after the MTB > prompt as you would in the mainframe environment. To issue commands through the menus, first select the menu that lists the command you want.

Menus. To open a menu, click it with your mouse, or press the [Alt] key plus the highlighted letter of the menu name. To open the File menu, for example, press the [Alt] key in combination with the highlighted letter "F."

Most main menus contain submenus from which you select the actual commands. The symbol >> to the right of a menu choice indicates that a submenu will appear. The ellipsis (. . .) following a menu choice indicates that, when you choose it, a dialog box will appear.

The Other Files command on the File menu is followed by a >> on the right; it has been selected and its submenu opened. A number of commands on the File menu have dialog boxes (indicated by . . .). A few commands have keyboard shortcuts assigned, shown on the menu itself. Notice that the first command on the File menu, Open Worksheet, is followed by Alt+O on the right. This means that pressing [Alt] in combination with "O" executes the Open Worksheet command, without having to work through the menus.

Say you wanted to use Minitab's correlation function. The Stat menu lists all of Minitab's statistical commands, so press [Alt]+S. The Basic Statistics submenu contains the Correlation command. Once a menu is open, you need only type the highlighted letter to access the submenu. In this case, press "B" for Basic Statistics.

Exhibit E.2 File Menu

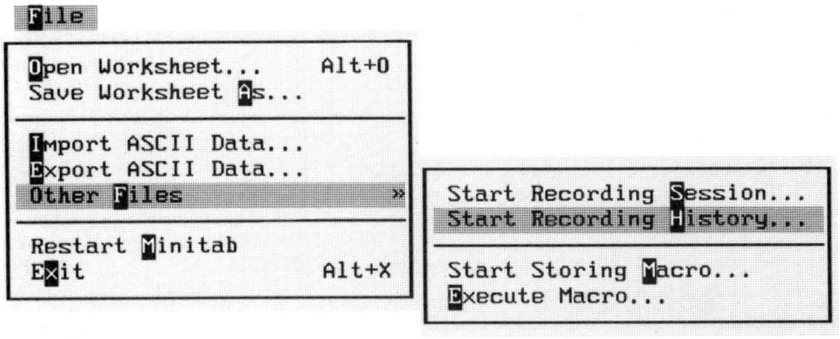

```
File
Open Worksheet...      Alt+O
Save Worksheet As...

Import ASCII Data...
Export ASCII Data...
Other Files               »   Start Recording Session...
                              Start Recording History...
Restart Minitab
Exit                   Alt+X  Start Storing Macro...
                              Execute Macro...
```

Exhibit E.3 Correlation Menu Command

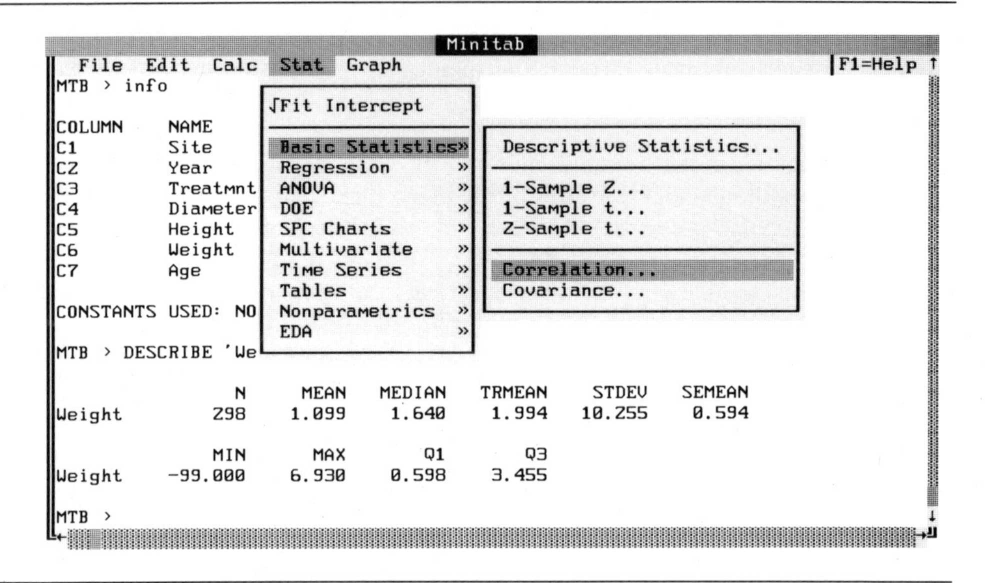

To execute the Correlation command, press "C," and Minitab then displays the Correlation dialog box.

Exhibit E.4 Correlation Dialog Box

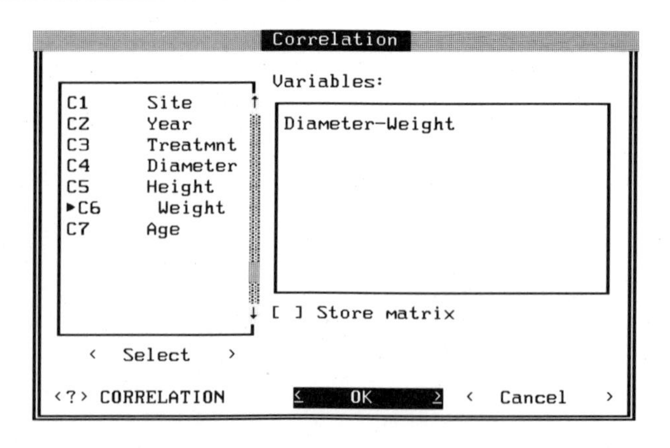

After you have completed the dialog box, Minitab executes the command and displays the correlation output in the Session window.

Dialog Boxes. Most menu commands display a dialog box allowing you to choose variables and options. The section Menu Quick Reference for the PC illustrates the key points of working with dialog boxes: moving around; selecting variables; getting help; choosing buttons; choosing files.

Example: The Normal Distribution Dialog Box. Choose **Calc ▸ Random Data ▸ Normal. . .** (short for choose the Normal command from the Random Data submenu in the Calc menu).

Type the number of rows of data you want in the first text box.

Move the insertion point, indicated by the blinking vertical bar, to the second text box by clicking or by pressing [Tab].

Type column names or numbers in the second text box, to tell Minitab where to store the random data. Separate them with blanks, not commas.

Exhibit E.5 Normal Distribution Dialog Box

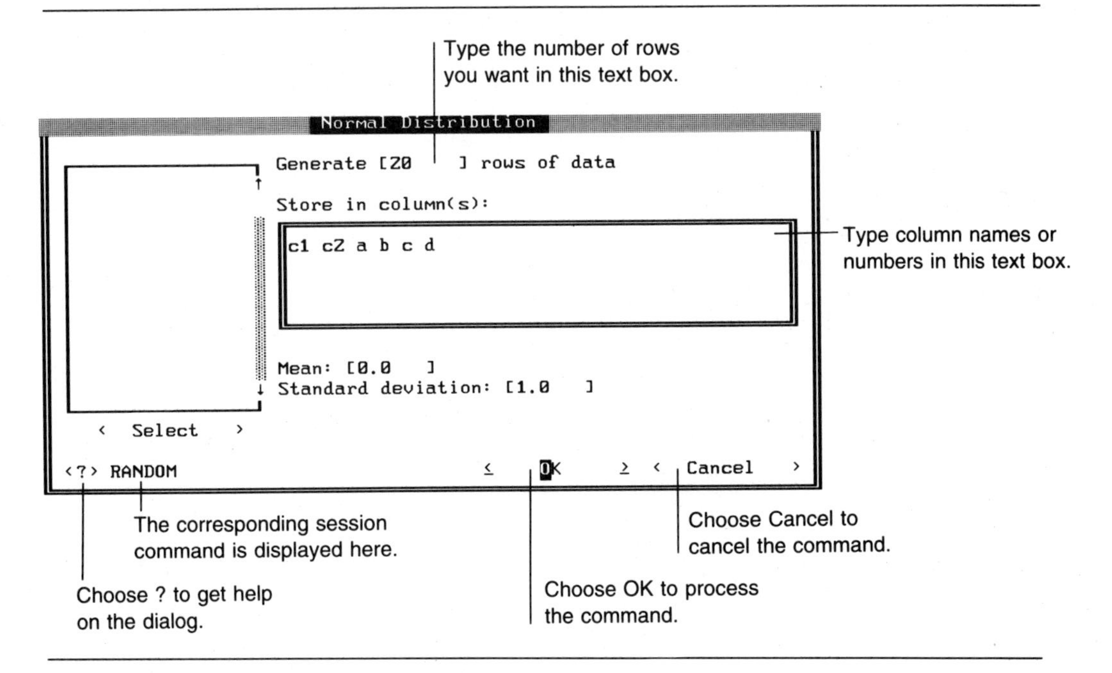

If you type a name which has not already been assigned to a column, Minitab automatically chooses an unused column and assigns it the name.

The text boxes labeled "Mean" and "Standard deviation" contain default values. If you want some other value, just type it in the appropriate box. To get help, press [F1] or click the ? button. To cancel the command, press [Esc] or click Cancel. To begin processing the command, press [Alt]+[O] (the letter O, not the digit zero) or click OK.

The Data Screen. The Data screen is most useful for entering data from the keyboard, editing individual values, or viewing your data. To enter patterned data, erase columns, or manipulate large blocks of data, use the Edit and Calc menus through the Session window. It's easy to switch back and forth: to reach the Data screen from the Session window, press the key combination [Alt]+D. To return to the Session window, press [Alt]+M.

Exhibit E.6 Data Screen

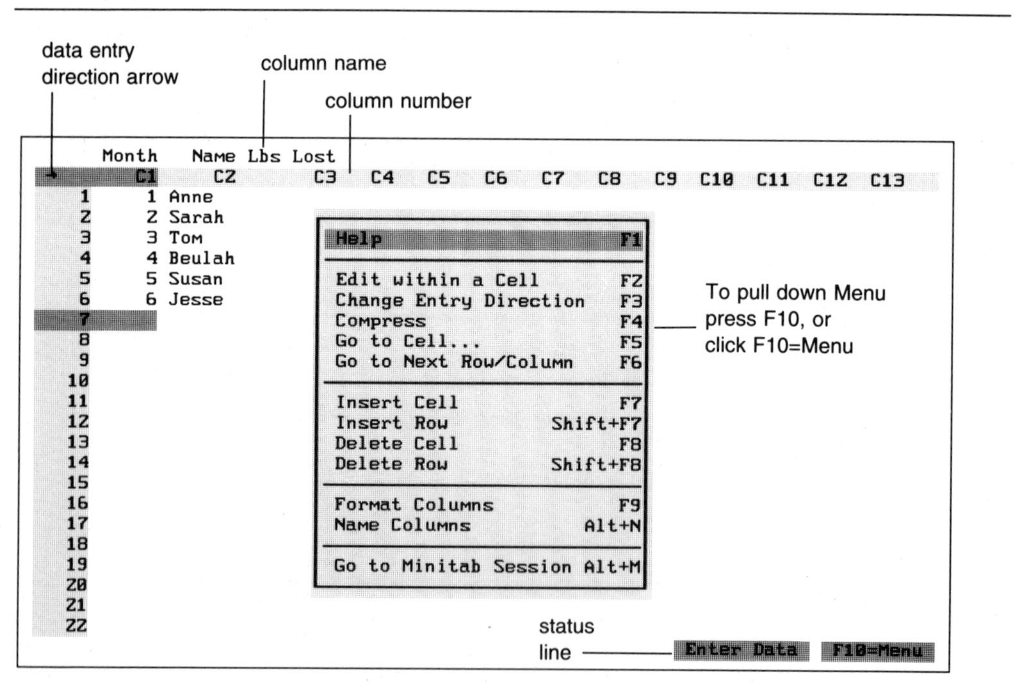

To enter the Data screen, choose Edit ► Data Screen or press Alt+D

The Data screen has its own menu (opened by pressing the [F10] key) that includes commands that operate in the Data screen. All of these commands have keyboard shortcuts (function keys like [F1], [F2], or [F3]) and once you've gained experience with the Data screen, you will probably use the menu only to remind yourself of the keyboard shortcuts.

When you enter the Data screen, one cell will be highlighted (usually the upper left cell); this is the *active cell*. To highlight a different cell, press the arrow keys until you reach the cell you want. To enter a new value into the active cell, just type the value. It overwrites the previous contents of the cell.

To edit the active cell (add to it without changing it), press [F2] for the command, Menu ▸ Edit within a Cell.

Menu Quick Reference for the PC. Exhibit E.7 summarizes menu operations for the PC keyboard. If you have a mouse, click to open a menu or choose a menu command; double click to select variables, directories or files. (See pages 384–389.)

Exhibit E.7 PC Menu Quick Reference

This section summarizes menu operations for the keyboard. If you have a mouse, click to open a menu or choose a menu command; double click to select variables, directories or files.

▶ Starting MINITAB

To start MINITAB, at the DOS prompt, enter:

MINITAB

After you clear the opening screen, you'll see the Session window.

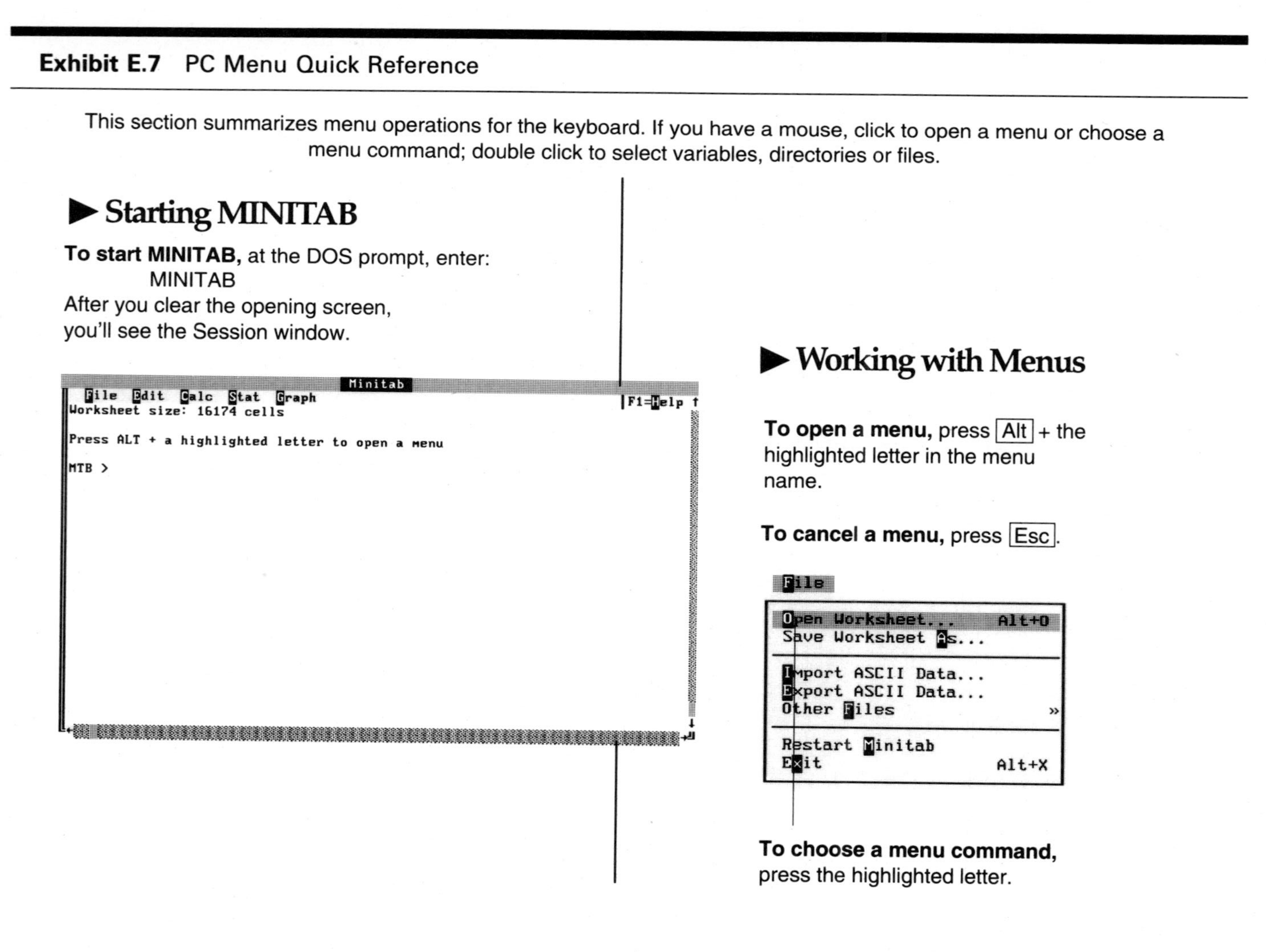

▶ Working with Menus

To open a menu, press Alt + the highlighted letter in the menu name.

To cancel a menu, press Esc.

To choose a menu command, press the highlighted letter.

▶ Stopping MINITAB

To stop MINITAB, choose File ▶ Exit:

- ❑ Press [Alt]+[F] to open the File menu
- ❑ Press [x] to choose Exit

```
File
Open Worksheet...        Alt+O
Save Worksheet As...
-------------------------------
Import ASCII Data...
Export ASCII Data...
Other Files                  »
-------------------------------
Restart Minitab
Exit                     Alt+X
```

Exhibit E.7 *(continued)*

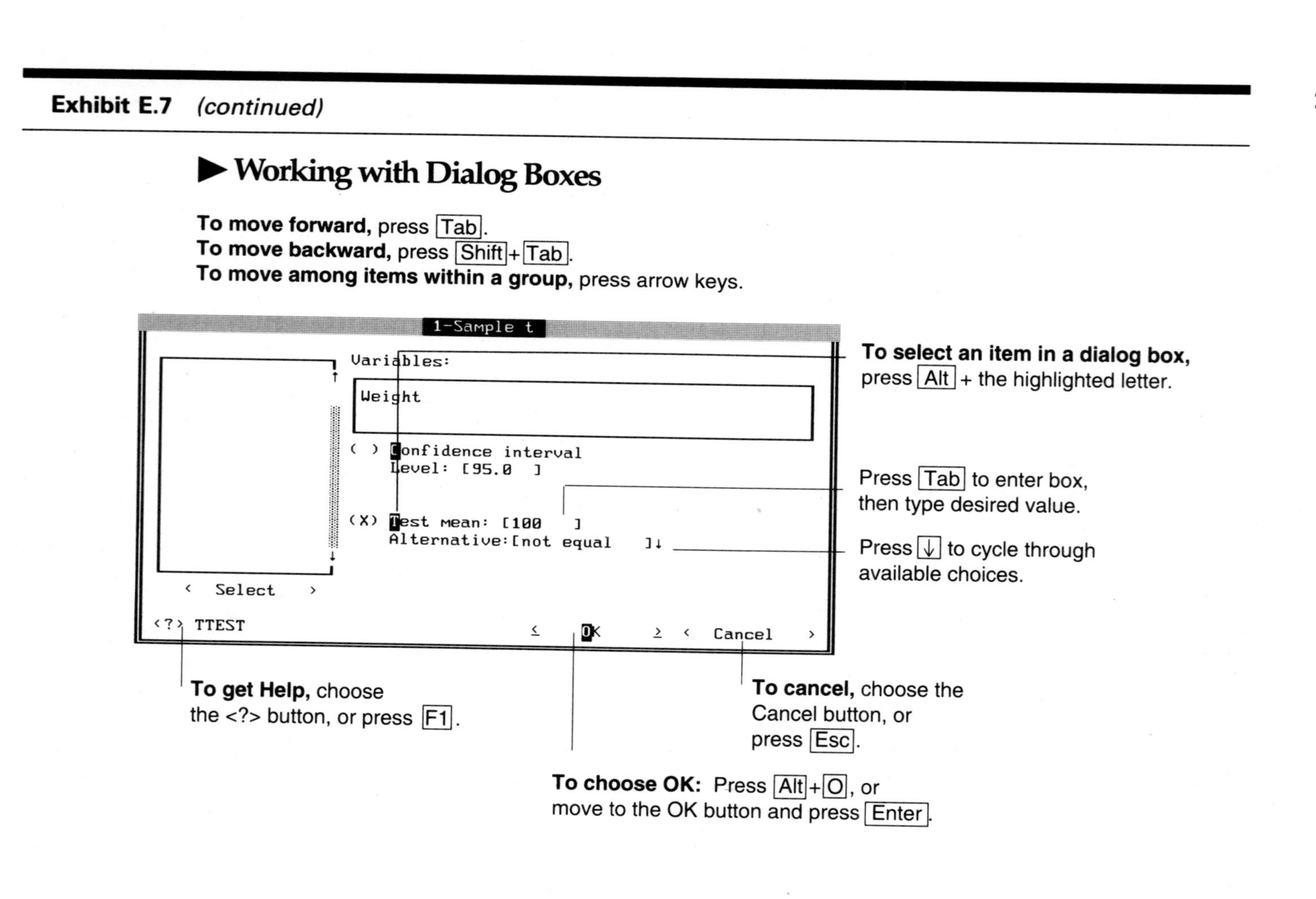

variable list box

To select a variable when you can select only one:

- Make sure your cursor is in appropriate text box
- Press F2 to make variable list box active
- Use arrow keys to highlight variable of interest
- Press F2 to select

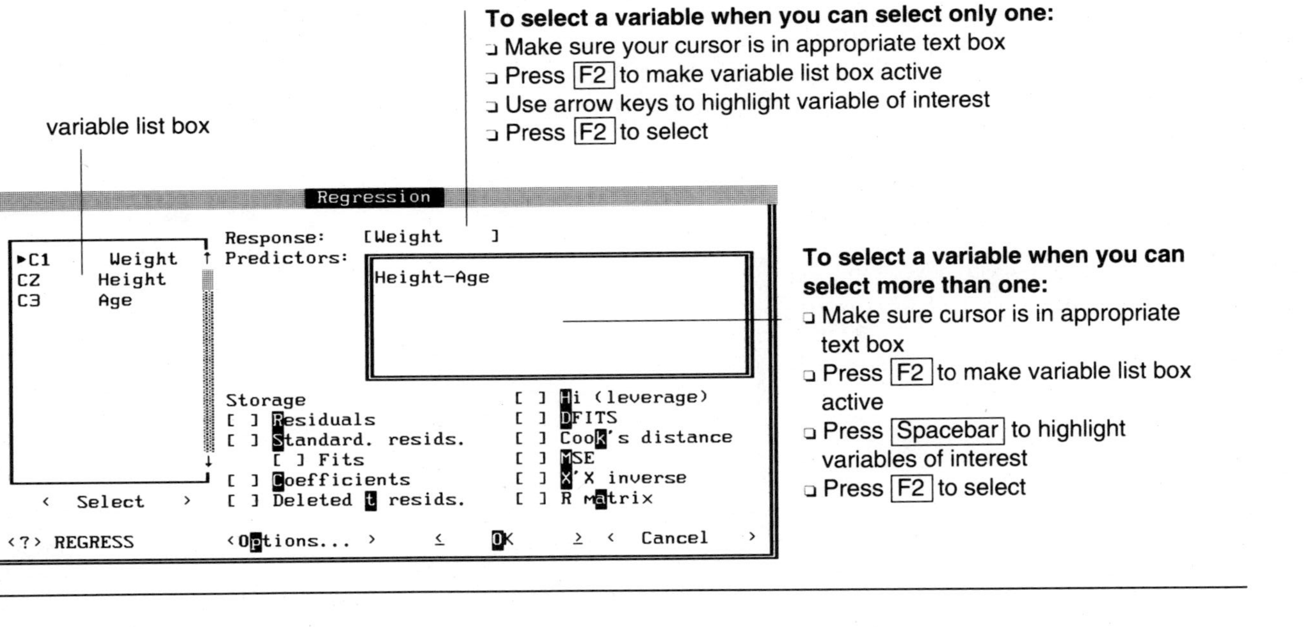

To select a variable when you can select more than one:

- Make sure cursor is in appropriate text box
- Press F2 to make variable list box active
- Press Spacebar to highlight variables of interest
- Press F2 to select

Exhibit E.7 *(continued)*

▶ Working with the File Dialog Box

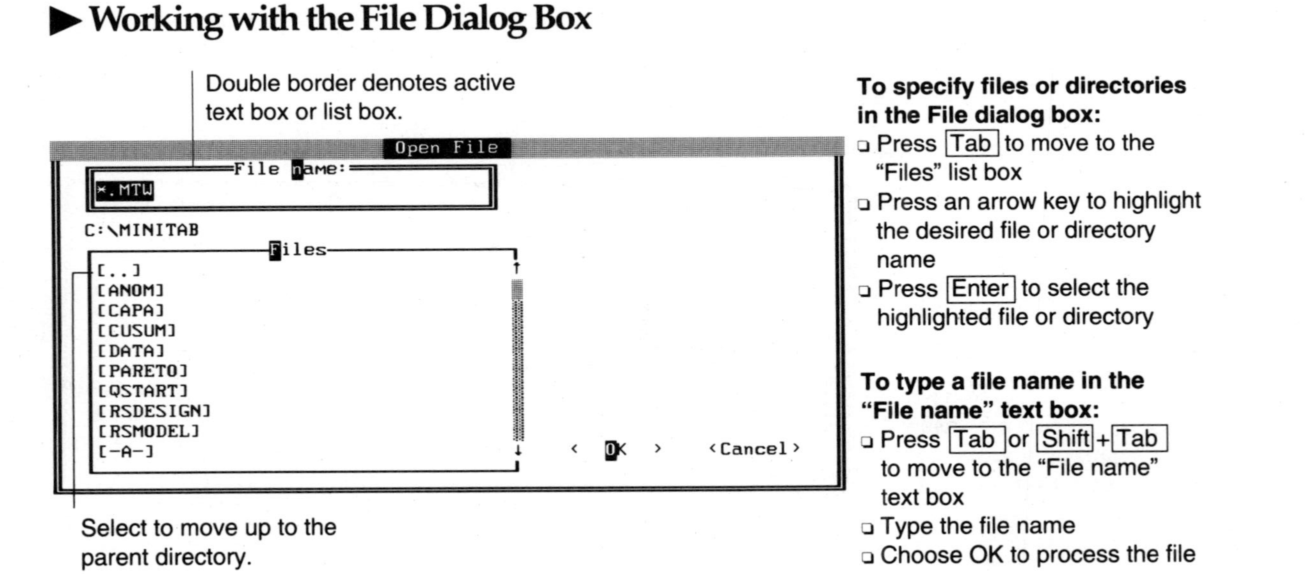

To specify files or directories in the File dialog box:

- Press Tab to move to the "Files" list box
- Press an arrow key to highlight the desired file or directory name
- Press Enter to select the highlighted file or directory

To type a file name in the "File name" text box:

- Press Tab or Shift+Tab to move to the "File name" text box
- Type the file name
- Choose OK to process the file

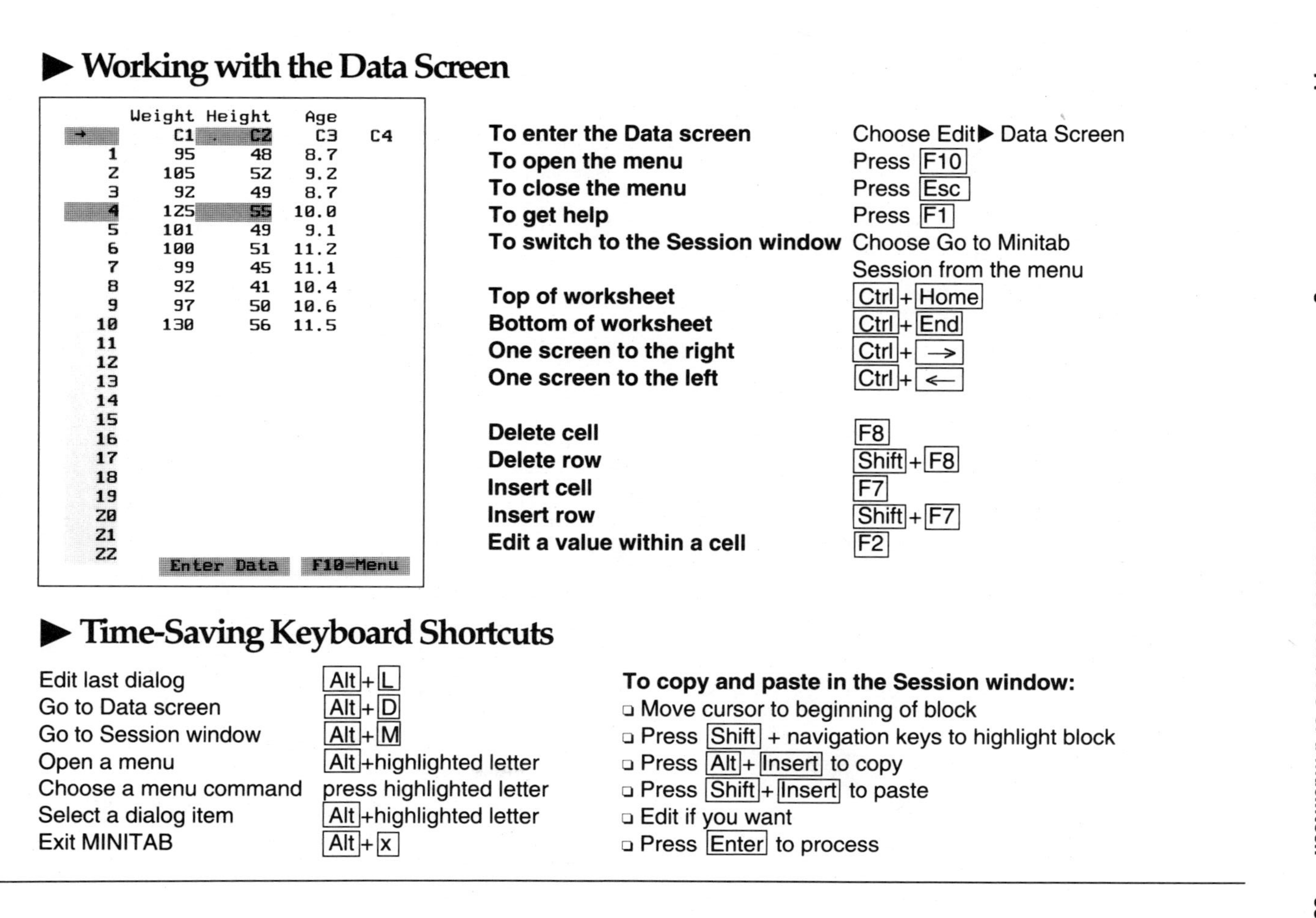

▶ Working with the Data Screen

→	C1	C2	C3	C4
	Weight	Height	Age	
1	95	48	8.7	
2	105	52	9.2	
3	92	49	8.7	
4	125	55	10.0	
5	101	49	9.1	
6	100	51	11.2	
7	99	45	11.1	
8	92	41	10.4	
9	97	50	10.6	
10	130	56	11.5	
11				
12				
13				
14				
15				
16				
17				
18				
19				
20				
21				
22				

Enter Data F10=Menu

To enter the Data screen	Choose Edit ▶ Data Screen
To open the menu	Press F10
To close the menu	Press Esc
To get help	Press F1
To switch to the Session window	Choose Go to Minitab Session from the menu
Top of worksheet	Ctrl + Home
Bottom of worksheet	Ctrl + End
One screen to the right	Ctrl + →
One screen to the left	Ctrl + ←
Delete cell	F8
Delete row	Shift + F8
Insert cell	F7
Insert row	Shift + F7
Edit a value within a cell	F2

▶ Time-Saving Keyboard Shortcuts

Edit last dialog	Alt + L
Go to Data screen	Alt + D
Go to Session window	Alt + M
Open a menu	Alt +highlighted letter
Choose a menu command	press highlighted letter
Select a dialog item	Alt +highlighted letter
Exit MINITAB	Alt + x

To copy and paste in the Session window:

- Move cursor to beginning of block
- Press Shift + navigation keys to highlight block
- Press Alt + Insert to copy
- Press Shift + Insert to paste
- Edit if you want
- Press Enter to process

Exhibit E.8 Overview of Minitab Windows on the Macintosh

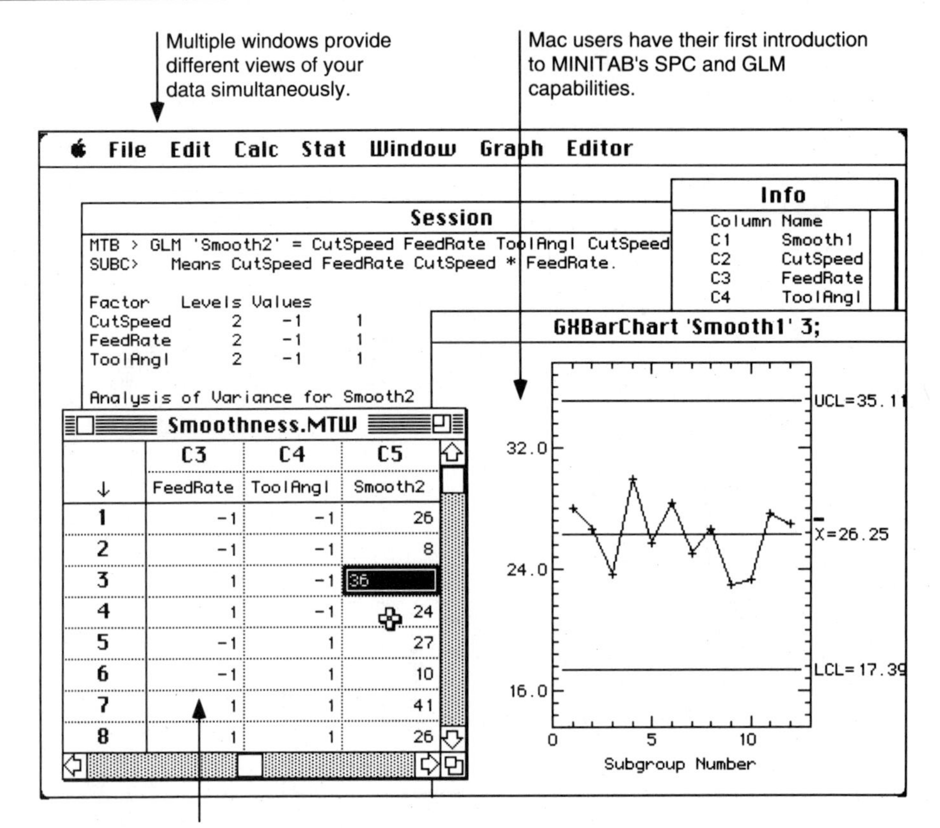

The Data window provides a convenient way to enter, edit and browse through your data. It is automatically updated as you execute commands, so you can see worksheet changes as they happen.

E.4 Using Minitab on a Macintosh

On the Macintosh, you will do your work primarily in the Data window, in which you can view and edit your worksheet, and the Session window, in which Minitab displays most output.

Using Minitab's Windows

Minitab for the Macintosh has six types of windows, and they all can be open at the same time: the Data window gives you a view of your worksheet, the Session window displays output, Graph windows contain high-

Exhibit E.9 Data Window

Untitled Worksheet

↓	C2	C3	C4	C5	C6	C7	C8
		a	b	c	d	RESI1	
1	0.89539	0.87818	0.30354	-0.50902	-1.07590	0.67404	
2	-0.93631	0.15642	1.01751	0.50118	-0.97415	1.65992	
3	-1.69225	-2.57048	0.34429	-0.70370	0.33429	0.69497	
4	0.44016	-0.14621	-0.74197	0.57051	2.21395	1.59487	
5	1.37542	-1.15981	-0.62462	0.06475	-0.76071	0.30163	
6	-1.00456	-0.53509	-1.65741	-1.44865	-1.81732	-0.65140	
7	-0.12331	1.29035	-1.13038	-1.91886	0.56293	-0.63604	
8	0.29836	1.58498	-0.29951	1.16288	0.66422	-0.21350	
9	-0.56115	1.40530	-0.55208	0.90228	1.86588	-0.04733	
10	0.69558	0.61382	-0.34700	0.57014	-0.58985	1.42284	

resolution graphs, the History window contains a record of previously-executed commands, the Info window summarizes the data in the current worksheet, and the Help window gives you instant access to Minitab's Help function.

In this figure, the Data window is currently the active window; you can tell because it has the dark scroll bars around it. You can activate any window by simply clicking it, as in any Macintosh program.

The Data window always opens in front when you start Minitab. You can access any other window through the Window menu (see the next section). The Data window is a convenient way to view, enter and edit data.

The Session window displays the results of all commands that produce numerical summaries, character-style graphs, and statistical analyses. When the output from a command exceeds the size of the Session window, you can press the mouse button to pause the display. Release the button to continue it. (See Exhibit E.10 on page 392.)

You can also use the Session window to type commands or data after the MTB > prompt as you would on the mainframe if you prefer to execute commands using Minitab's command language. You can also use a combination of menu commands and typed commands (called *session* commands).

The Graph windows contain high-resolution graphs, one graph per window. Each time you create a high-resolution graph, Minitab stores it

Exhibit E.10 Session Window

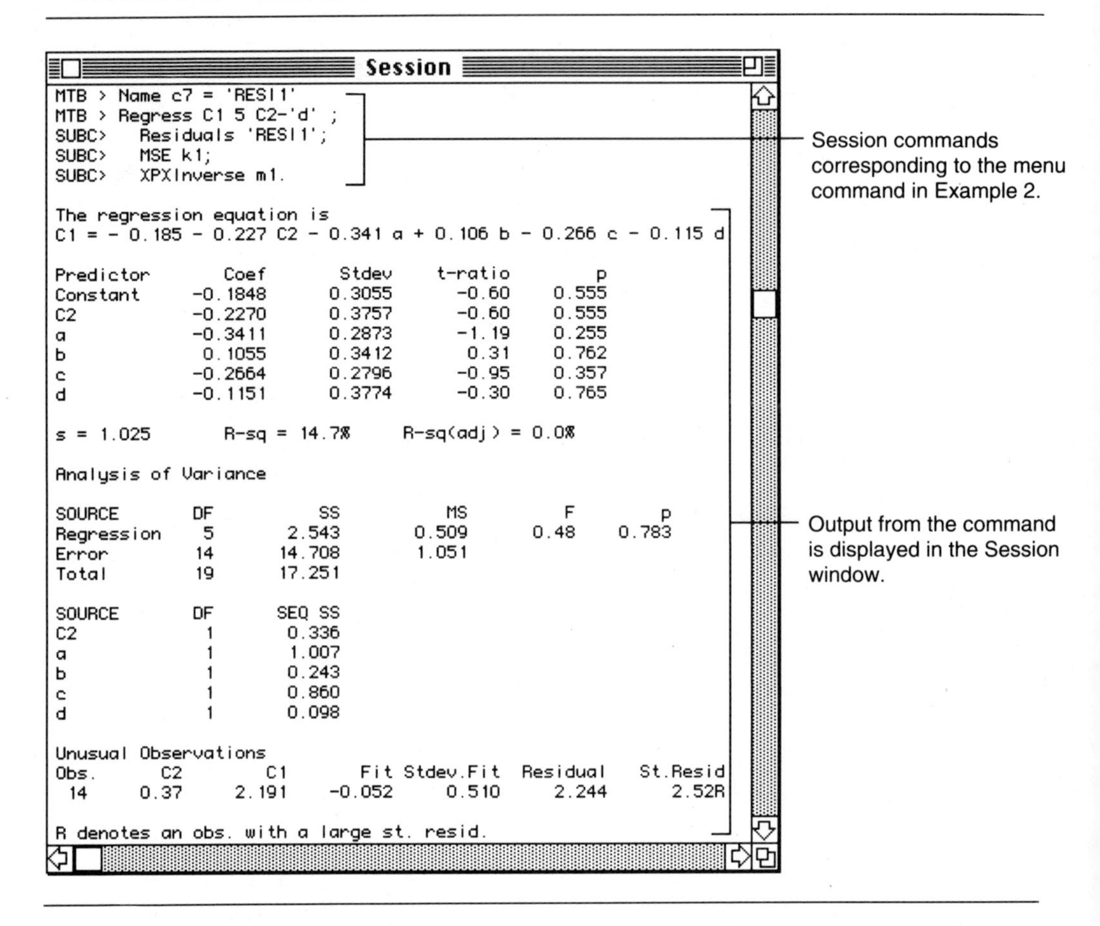

```
Session
MTB > Name c7 = 'RESI1'
MTB > Regress C1 5 C2-'d' ;
SUBC>   Residuals 'RESI1';
SUBC>   MSE k1;
SUBC>   XPXInverse m1.

The regression equation is
C1 = - 0.185 - 0.227 C2 - 0.341 a + 0.106 b - 0.266 c - 0.115 d

Predictor       Coef       Stdev     t-ratio        p
Constant     -0.1848      0.3055       -0.60    0.555
C2           -0.2270      0.3757       -0.60    0.555
a            -0.3411      0.2873       -1.19    0.255
b             0.1055      0.3412        0.31    0.762
c            -0.2664      0.2796       -0.95    0.357
d            -0.1151      0.3774       -0.30    0.765

s = 1.025       R-sq = 14.7%     R-sq(adj) = 0.0%

Analysis of Variance

SOURCE       DF          SS          MS         F        p
Regression    5       2.543       0.509      0.48    0.783
Error        14      14.708       1.051
Total        19      17.251

SOURCE       DF      SEQ SS
C2            1       0.336
a             1       1.007
b             1       0.243
c             1       0.860
d             1       0.098

Unusual Observations
Obs.       C2         C1        Fit  Stdev.Fit   Residual    St.Resid
 14      0.37      2.191     -0.052      0.510      2.244       2.52R

R denotes an obs. with a large st. resid.
```


in its own window that you can access through the Graph menu or by simply clicking the window. Minitab can retain up to 15 Graph windows at a time.

The History window displays all the session commands and data from the Session window, without the output. This provides a convenient overview of your session. If you choose File ▸ Other Files ▸ Start Storing History, Minitab sends this information to a file as well as to the History window; this is equivalent to the JOURNAL session command. You can cut and paste commands from the History window to the Session window and then execute them there by pressing [Return]. You cannot edit the contents of this window.

Exhibit E.11 Graph Window

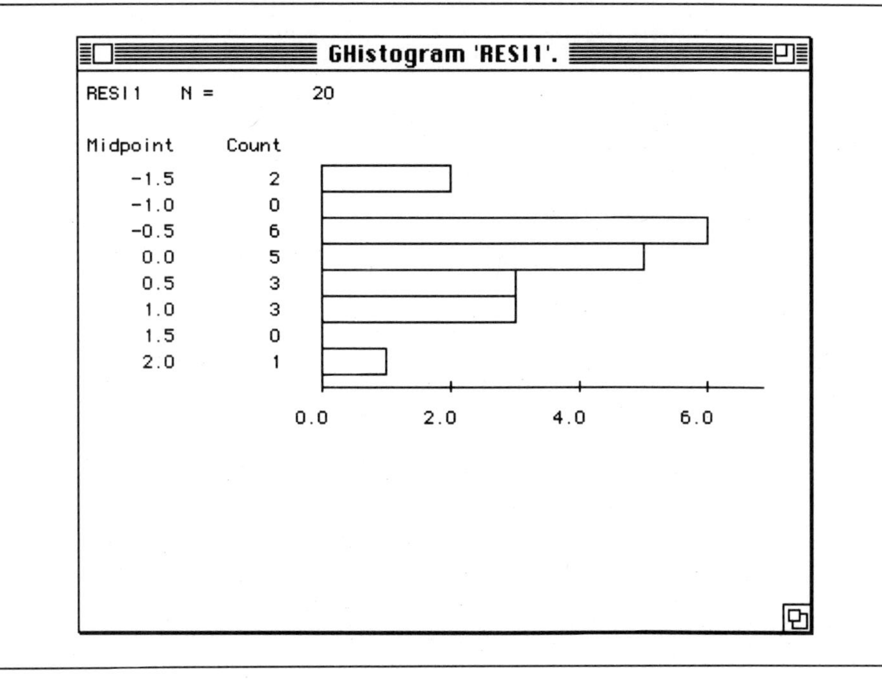

Exhibit E.12 History Window

```
History
Name c3 = 'a' c4 = 'b' c5 = 'c' c6 = 'd'
Random 20 c1 c2 'a' 'b' 'c' 'd' ;
  Normal 0.0 1.0.
Name c7 = 'RESI1'
Regress C1 5 C2-'d' ;
  Residuals 'RESI1';
  MSE k1;
  XPXInverse m1.
TTest 0.0 'a' 'b' ;
  Alternative 1.
```

Exhibit E.13 Info Window

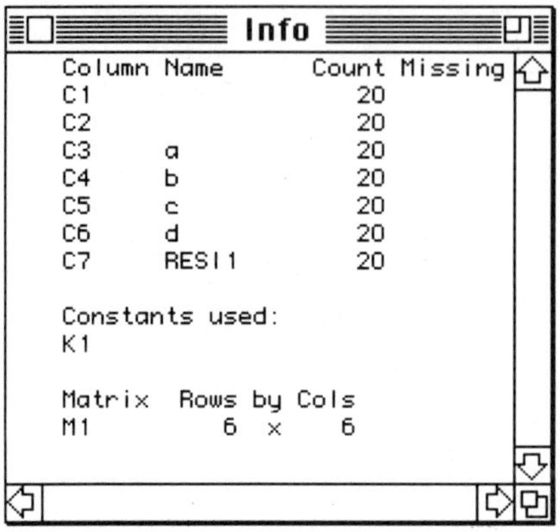

```
Info
Column Name        Count Missing
C1                    20
C2                    20
C3     a              20
C4     b              20
C5     c              20
C6     d              20
C7     RESI1          20

Constants used:
K1

Matrix  Rows by Cols
M1         6  x   6
```

The Info window displays essentially the same information provided by the INFO session command: a summary of the columns, stored constants, and matrices you are using in the current worksheet.

The Help window display depends on how you open the window. You can open it from a dialog box by clicking the ? button in the lower left corner; this automatically opens the Help window to the Help information available for the topic of the dialog box. When you are not working in a dialog box, choose Help from the [apple] menu or type HELP after a prompt in the Session window. For example, if you click the ? button in the Regression dialog box (Stat ▸ Regression ▸ Regression), the Help window looks like Exhibit E.14 on page 395.

There are four categories of Help information: Overview (for general Minitab topics), Menus (for menu commands), COMMANDS by topic (for session commands and subcommands organized by topic), COMMANDS by name (for session commands and subcommands organized alphabetically).

Menus

The menu bar across the top lists Minitab's seven main menus (File, Edit, Calc, Stat, Window, Graph, and Editor) and the Apple menu that contains special Macintosh functions and Minitab's Help. You can execute nearly all of Minitab's commands through these menus by clicking the menu open and dragging diagonally to the desired submenu command. (See Exhibit E.15 on page 396.)

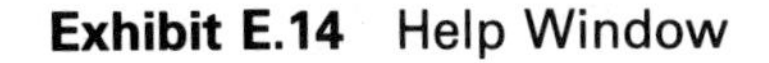

Exhibit E.14 Help Window

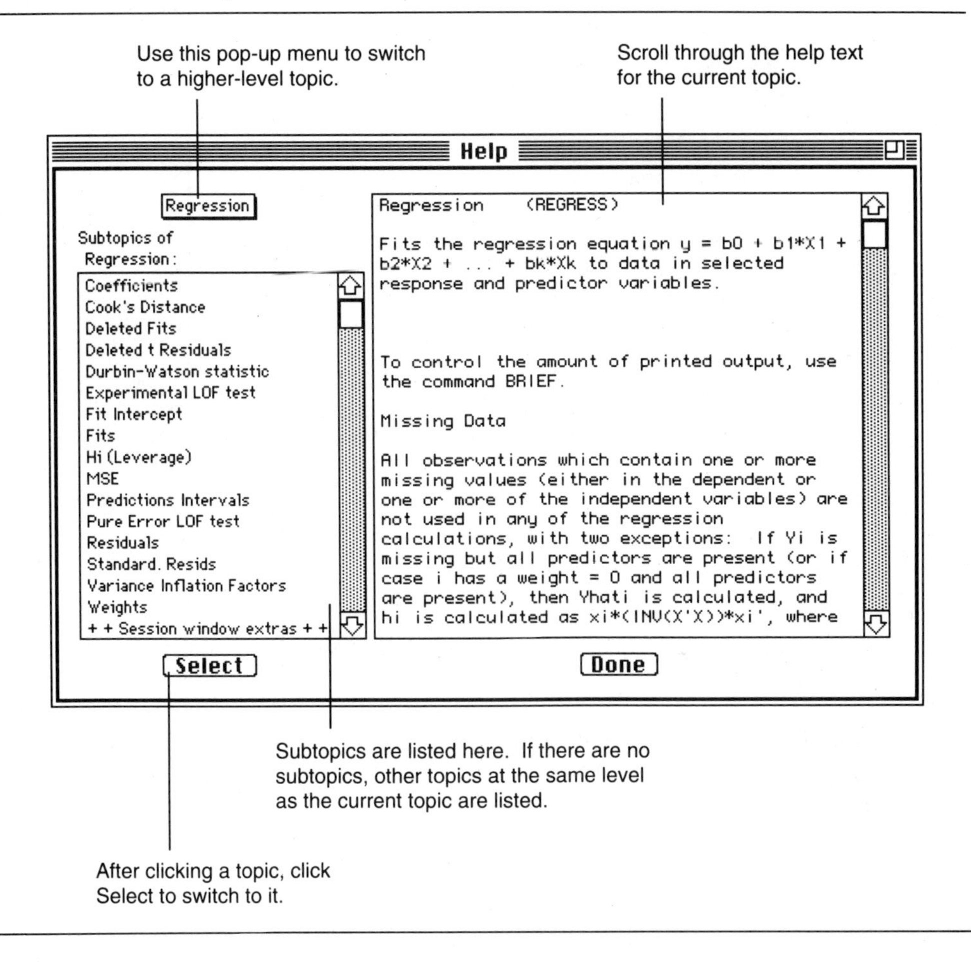

In the Stat menu, for example, you can drag diagonally into the Basic Statistics submenu to choose the Correlation command. The ▸ to the right of the main menu option indicates that a submenu will open if you choose this option. The ellipses (. . .) to the right of the commands on the submenu indicates that those commands use dialog boxes to get further information from you.

Dialog Boxes. Most menu commands display a dialog box allowing you to choose variables and options. Many dialog boxes have subdialog boxes

Exhibit E.15 Working with Menus

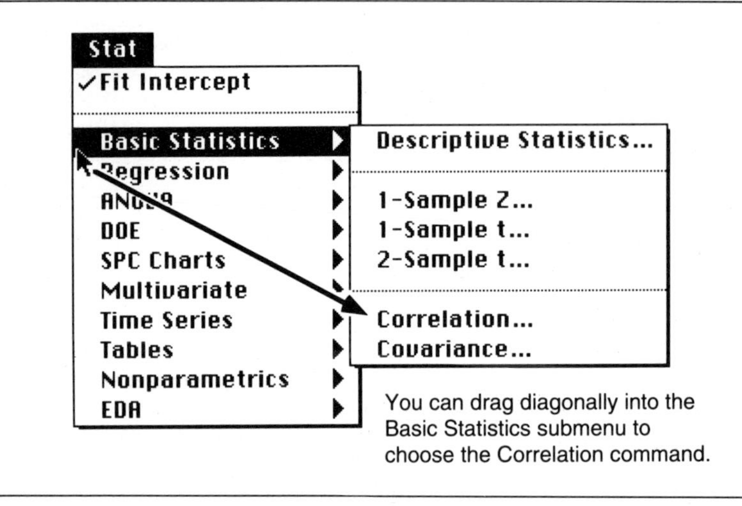

containing less frequently used options. For the most part, you use the components in Minitab's dialog boxes the same as you do in any other Macintosh applications.

Example: The Normal Distribution Dialog Box. Choose **Calc ▸ Random Data ▸ Normal. . .** (short for choose the Normal command from the Random Data submenu in the Calc menu).

Type the number of rows of data you want in the first text box.

Move the insertion point, indicated by the blinking vertical bar, to the second text box by clicking or by pressing [Tab].

Type column names or numbers in the second text box, to tell Minitab where to store the random data. Separate them with blanks, not commas. If you type a name which has not already been assigned to a column, Minitab automatically chooses an unused column and assigns it the name.

The text boxes labeled "Mean" and "Standard deviation" contain default values. If you want some other value, just type it in the appropriate box. To get help, click the ? button. To cancel the command, click Cancel. To begin processing the command, click OK.

Exhibit E.16 Normal Distribution Dialog Box

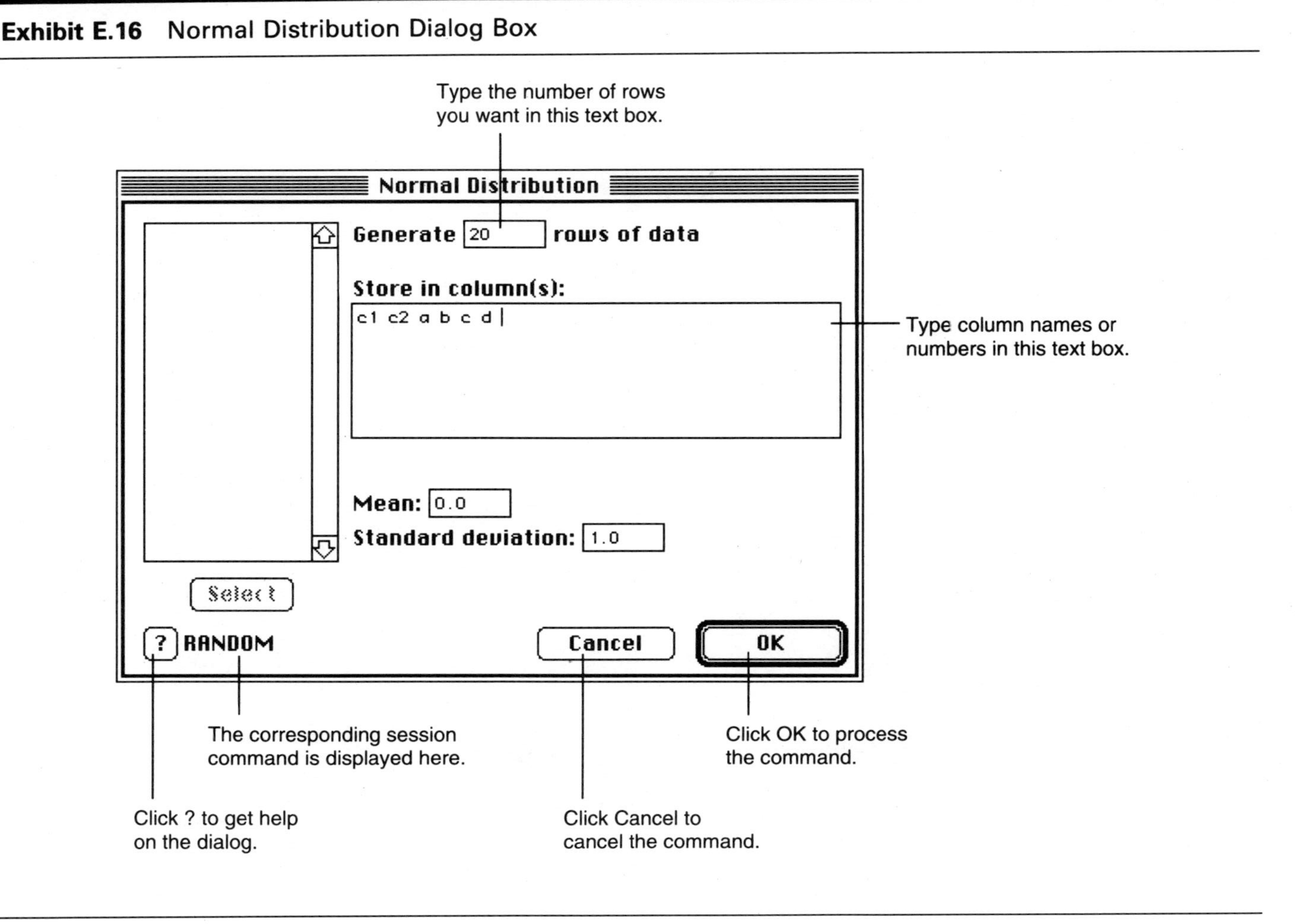

References

Chapter 1

The cholesterol data in Table 1.1 are from a study conducted by a major northeastern medical center.

The men's track records in Table 1.2 and the women's track records in Exercise 1–13 were obtained from

The World Almanac & Book of Facts, 1984 (New York: Newspaper Enterprise Association, Inc., 1984). Copyright © Newspaper Enterprise Association, Inc., 1984, New York, N.Y. 10166.

Chapter 2

The pulse data in Table 2.1 are from a classroom experiment by Brian L. Joiner.

The lake data set was collected by the Wisconsin Department of Natural Resources and provided to us by Alison K. Pollack and Joseph M. Eilers.

The furnace data are from a study conducted by Tim LaHann and Dick Mabbott, the Wisconsin Power and Light Company.

Chapter 3

The cartoon data set was obtained from

Stephan Kauffman, "An Experimental Evaluation of the Relative Effectiveness of Cartoons and Realistic Photographs in Both Color and Black and White Visuals in In-Service Training Programs," Master's thesis, The Pennsylvania State University, 1973.

The plywood data in Table 3.3 and Exercise 9–1 came from a study directed by Frank J. Fronczak of the U.S. Forest Products Research Laboratory.

The employment data in Table 3.4 are from

R. B. Miller and Dean W. Wichern, *Intermediate Business Statistics* (New York: Holt, Rinehart & Winston, 1977). Copyright © 1977 by Holt, Rinehart & Winston. Reprinted by permission of CBS College Publishing.

The radiation data used in Exhibit 3.8 came from the Las Vegas Radiation Facility, Las Vegas, Nevada.

The distance versus speed data used in Exhibit 3.10 were obtained from the National Bureau of Standards.

The trees data were obtained from

H. Arthur Meyer, *Forest Mensuration* (State College, Pennsylvania: Penns Valley Publishers, Inc., 1953).

The temperature data used in Exercise 3–5 were obtained from

The World Almanac & Book of Facts, 1984 (New York: Newspaper Enterprise Association, Inc., 1984). Copyright © Newspaper Enterprise Association, Inc., 1984, New York, N.Y. 10166.

The wind chill data used in Exercise 3–11 were obtained from the U.S. Army.

The automobile accident data used in Exercise 3–14 were obtained from

Paul M. Hurst, "Blood Test Legislation in New Zealand," *Accident Analysis and Prevention,* 10, pp. 287–296. Copyright © 1978, Pergamon Press, Ltd. Reprinted with permission.

The leaf-surface resistance data in Exercise 3–18 were provided by Thomas Starkey of The Pennsylvania State University.

The magnifying glass data in Exercise 3–19 were collected by a student in an introductory physics class.

The cadmium data used in Exercise 3–20 were obtained from

Robert C. Paul and John Mandel, National Bureau of Standards, Special Publication 260, 1971.

Chapter 4

The Wisconsin Restaurant Survey was done by William A. Strang of the University of Wisconsin, Madison, Small Business Development Center, 1980. We also thank Robert Miller for his help with the analysis of these data and with the writing of Chapter 4.

The psychiatric commitment data in Exercise 4–4 were provided by a Health Care Center in Wisconsin.

Chapter 5

The cholesterol data for heart-attack victims in Table 5.1 were collected by a Pennsylvania medical center in 1971.

The shoe-sole materials data in Table 5.2 were obtained from

G. E. P. Box, W. G. Hunter, and J. S. Hunter, *Statistics for Experimentation* (New York: Wiley & Sons, 1978). Copyright © 1978 by John Wiley & Sons, Inc. Reprinted by permission.

The radiation data in Table 5.3 came from the Las Vegas Radiation Facility, Las Vegas, Nevada.

The tool wear data in Exercise 5–6 were provided by Michael R. Delozier of Kennametal Inc.

The Peru data were obtained from

P. T. Baker and C. M. Beall, "The Biology and Health of Andean Migrants: A Case Study in South Coastal Peru," *Mountain Research and Development,* 2(1), 1982. Reprinted by permission of the author.

The teacher data set in Exercise 5–14 was obtained from

The Capital Times, Madison, Wisconsin, December 12, 1975. Reprinted by permission.

The Alfalfa data are from an experiment conducted at the University of Wisconsin.

The billiard ball data are from Albert Romano, *Applied Statistics for Science and Industry* (Boston: Allyn and Bacon, Inc., 1977, p. 300).

Chapter 6

The chest-girth data in Table 6.1 were obtained from

Roger Carlson, "Normal Probability Distributions," in *Statistics by Example: Detecting Patterns,* by F. Mosteller, W. H. Kruskal, R. F. Link, R. S. Pieters, and G. R. Rising, editors (Reading, Mass.: Addison-Wesley, 1973). Copyright © 1973, Addison-Wesley Publishing Company, Inc.

The Grades data were obtained from a university that wishes to remain anonymous.

The Penn State nuclear reactor data in Exercise 6–25 were obtained from testimony presented to the Atomic Energy Commission in 1971.

Chapter 9

The fabric flammability test (Table 9.1) was conducted by the American Society for Testing Materials. Related analyses are

"ASTM Studies DOC Standard FF-30-71 on Flammability of Children's Sleepwear,"*Materials Research Standards* (May, 1972) pp. 38–39.

John Mandel, Mary N. Steel, and L. James Sharman, "National Bureau of Standards Analysis of the ASTM Interlaboratory Study of DOC/FF

3-71 Flammability of Children's Sleepwear," *ASTM Standardization News* (May, 1973) pp. 9–12.

The pendulum data in Exercise 9–2 were obtained from John Mandel. These data also appear in his article in

Frederick Mosteller, William H. Kruskal, Richard S. Pieters, Gerald R. Rising, and Richard F. Link, editors, "The Acceleration of Gravity," *Statistics by Example: Detecting Patterns* (Reading, Mass.: Addison-Wesley, 1973). Copyright © 1973, Addison-Wesley Publishing Company, Inc.

The meat loaf data in Table 9.2 were obtained in a study at the University of Wisconsin conducted by Barbara J. Bobeng and Beatrice David.

The pancake data in Table 9.3 were given to us by Gregory Mack and John Skilling.

The potato rot data in Table 9.4 and Exercise 9–11 were collected by the University of Wisconsin.

The manganese in cow liver data in Exercise 9–4 were obtained from the National Bureau of Standards.

Chapter 10

The test score data set in Table 10.1 was collected in a statistics class at The Pennsylvania State University.

The stream data in Exhibit 10.8 and Exercises 10–22 and 10–23 were gathered as part of an environmental impact study.

The SAT scores data set was obtained from "The Digest of Educational Statistics: 1983–1984 Volume," U.S. Government Publication.

The samara velocity data in Exercise 10–17 were provided by Erik V. Nordheim of the University of Wisconsin.

The regression data in Exercise 10–28 were obtained from

F. J. Anscombe, "Graphs in Statistical Analysis," *The American Statistician,* 27(1) (February 1973), pp. 17–21. Reprinted by permission of the American Statistical Association.

Chapter 11

The data on migratory geese in Table 11.1 are discussed in

R. K. Tsutakawa, "Chi-Square Distribution by Computer Simulation," in *Statistics by Example: Detecting Patterns* by F. Mosteller, W. H. Kruskal, R. F. Link, R. S. Pieters, and G. R. Rising, editors (Reading,

Mass.: Addison-Wesley, 1973). Copyright © 1973, Addison-Wesley Publishing Company, Inc.

Further discussions of the ESP data in Table 11.2 are given in

Marvin Lee Moon, "Extrasensory Perception and Art Experience," Ph.D. thesis, The Pennsylvania State University, 1973. An additional article by Moon appeared in *American Journal of Psychical Research* (April, 1975).

The accidental deaths data set in Exercise 11–3 was obtained from

The World Almanac & Book of Facts, 1984 (New York: Newspaper Enterprise Association, Inc., 1984). Copyright © Newspaper Enterprise Association, Inc., 1984, New York, N.Y. 10166.

The penny tossing data in Exercise 11–4 are described in

W. J. Youden, *Risk, Choice and Prediction* (North Scituate, Mass.: Duxbury Press, 1974).

The precipitation data in Exercise 11–5 were collected in State College, Pennsylvania.

The Harrisburg, Pennsylvania, infant mortality data in Exercises 11–7 and 11–8 are given in

Vilma Hunt and William Cross, "Infant Mortality and the Environment of a Lesser Metropolitan County: A Study Based on Births in One Calendar Year," *Environmental Research* 9 (1975) pp. 135–151.

The car defects data in Exercises 11–9 and 11–10 were obtained from

L. A. Klimko and Camil Fuchs, "An Analysis of the Data from Wisconsin's Dealer Based Demonstration Motor Vehicle Inspection Program," Statistical Laboratory Report 77/7, Department of Statistics, University of Wisconsin-Madison, 1977.

The Mark Twain data in Exercise 11–11 are given in

Claude Brinegar, "Mark Twain and the Quintus Curtius Snodgrass Letters: A Statistical Test of Authorship," *Journal of the American Statistical Association* 58 (1963) pp. 85–96.

Chapter 12

The acid data set in Exercises 12–3 and 12–5 was collected in a chemistry class.

The Parkinson's disease data in Table 12.1 were obtained by Gastone Celesia in a study at the University of Wisconsin.

The presidents data set in Example 12–10 was obtained from

The World Almanac & Book of Facts, 1984 (New York: Newspaper Enterprise Association, Inc., 1984). Copyright © Newspaper Enterprise Association, Inc., 1984, New York, N.Y. 10166.

The bird fat data in Exercise 12–18 were obtained from

Jack Hailman, "Notes on Quantitative Treatments of Subcutaneous Lipid Data," *Bird Banding* 36 (1965) pp. 14–20.

The azalea data in Exercise 12–19 are from a study done at The Pennsylvania State University by Professor Stanley Pennypacker, Department of Plant Pathology.

Index